Formaldehyde and Cognition

Rongqiao He

Formaldehyde and Cognition

Springer

Rongqiao He
State Key Laboratory of Brain
and Cognitive Science
Institute of Biophysics
Chinese Academy of Sciences
Beijing, China

University of Chinese Academy of Sciences
Beijing, China

中文地址：
北京市朝阳区大屯路15号中国科学院生物物理研究所，
脑与认知科学国家重点实验室 邮编：100101
赫荣乔
手机号：13552534019

ISBN 978-94-024-1503-2 ISBN 978-94-024-1177-5 (eBook)
DOI 10.1007/978-94-024-1177-5

Softcover re-print of the Hardcover 1st edition 2017

Printed on acid-free paper

This Springer imprint is published by Springer Nature
The registered company is Springer Science+Business Media B.V.
The registered company address is: Van Godewijckstraat 30, 3311 GX Dordrecht, The Netherlands

Preface

Formaldehyde, the simplest carbonyl compound, is colorless, strong-smelling, and toxic. Human beings never want to be exposed to formaldehyde, but have to use it widely in daily life and industries. We live in an environment where formaldehyde exists more or less. Aware or not, our body is producing, processing, and degrading formaldehyde internally at each moment. As the saying puts, everything has two sides, and formaldehyde is not exceptive. On one hand, formaldehyde as a devil is detrimental to our brain, liver, kidney, and other critical organs and even induces and promotes carcinogenesis (Kuchenbaecker et al. 2017; Tan et al. 2017). On the other hand, under some conditions, formaldehyde may also be a good factor (Chap. 11), which could prolong the lifespan of *Drosophila* (Li and He 2016). Reduction of endogenous formaldehyde concentration by knocking in formaldehyde dehydrogenase (FAD) impaired the learning and memory in *Drosophila* (Hou et al. 2011). During the in vitro cell culture, lower concentration of formaldehyde could enhance cell viability, although higher concentration suppressed it (Chap. 5). Furthermore, in the experiment with rat brain sections, the viability of astrocytes was enhanced and then weaken with the increase of formaldehyde concentration, exhibiting an optimal range of the formaldehyde concentration for the glial activity (Tong et al. 2013). Among many roles formaldehyde played, in this book, we would like to focus on the association between formaldehyde and cognition because this toxic baddy plays a role in human cognitive ability and impairment.

When discussing cognitive ability and impairment, Alzheimer's disease (AD) is the most common form. Except familial dementia (FD), AD is a syndrome often present with other diseases and sporadically appears in the elderly people. AD features a progressive and irreversible neurodegeneration that gradually impairs memory and thinking capabilities and eventually destroys the ability to perform the simplest of daily tasks. Amyloid-β (Aβ) deposition (senile plaque formation), one of the typical lesions, occurs at the early onset of AD. Another typical lesion of AD is neurofibrillary tangles (NFTs) which are composed of hyperphosphorylated Tau protein, which affects the connections between neurons in our brain. The phosphorylated Tau is also used as a biomarker to detect and assess preclinical and clinical AD. Amyloid-β deposition and Tau hyperphosphorylation deteriorate neuronal transmission and worsen cognitive impairment. We will discuss the relation of formaldehyde with Aβ deposition, Tau hyperphosphorylation, and cognitive impairment in this book.

Since aging is a natural progress followed by a progressive decrease of water intake or disturbance in the water metabolic balance (Stout et al. 1999; Luckey and Parsa 2003), we reviewed dehydration and endogenous formaldehyde. AD brain is dry due to patients suffering from decreased sensitivity to thirst and an afterward chronic dehydration (Chap. 4). Dehydration induces the production of endogenous formaldehyde (Li et al. 2016). For an analogy, the aged human brain is just like the lake in which water has gradually evaporated, followed by fishes dying. Therefore, in Chap. 4, we address how water intake is beneficial as it could dilute and reduce formaldehyde and other metabolic products.

In fact, senile plaques and neurofibrillary tangles in the brains of Alzheimer's patients are involved in protein misfolding, aggregation, and deposition. In order to study protein misfolding and aggregation, as a Ph.D. student, I participated in the group supervised by Prof. Chen-Lu Tsou in the State Key Laboratory of Biomacromolecules, Institute of Biophysics, Chinese Academy of Sciences (CAS), in Beijing, after graduating (master of science) from the Institute of Microbiology, CAS, in 1986. We studied the relationship between denaturation (aggregation) and inactivation (dysfunction) of proteins (enzymes) by using chemical modifications and denaturants disturbance (He and Tsou 1991). In the modification of peptide with an aldehyde derivative (glyoxylic acid), we observed the fluorescence of peptide N-terminal 2-oxoacyl derivative (He and Tsou 1992). Later, we found an asynchronous unfolding procedure among the different regions of D-glyceraldehyde-3-phosphate dehydrogenase (GAPDH) and maltotetraose-forming amylase (MTFA) from an *Alcaligenes* species (He et al. 1993). After isolating glycated GAPDH and non-glycated GAPDH from rabbit muscle (He et al. 1995a), we observed that glycation affects the stabilization of the enzyme (He et al. 1995b). In 1994, I organized my own scientific group and studied protein modification with aldehyde, especially neuronal proteins. It was supposed that a long-term exposure to low concentrations of formaldehyde, namely, chronic formaldehyde exposure to proteins of the central nervous system (CNS), may be related to neurodegeneration.

In 1996, I paid a visit to Hemant K. Paudel's laboratory in the University of McGill in Quebec (Canada) for 3 months. Paudel and his coworkers studied the relationship between Tau protein phosphorylation and AD at that time (Paudel 1997a, b). On the visit, I was gifted *tau40* plasmid (which expresses the longest Tau isoform) back to Beijing. First, according to Cullen and Holiday (1995), neurofibrillary tangles existed in the brain of patients with chronic alcoholics. Jianying Luo (a Ph.D. student in my group, Institute of Biophysics, CAS) incubated Tau in solution and observed the protein aggregation in the presence of acetaldehyde rather than ethanol solution (Luo and He 1999). Acetaldehyde affected phosphorylation of human neuronal Tau in the presence of neuronal cdc2-like kinase (NCLK) (Chen et al. 1999). Later, we focused on the interaction between formaldehyde and Tau protein, because formaldehyde showed much more toxicity in cell experiments and clinical toxicosis. The patho-mechanism of alcoholics is different from AD even though neurodegeneration is involved in alcoholic brain, especially in white matter (Martin et al. 1986). Tau aggregation with acetaldehyde needed higher protein concentrations than with formaldehyde. In general, overexpression of protein in cells may result in protein aggregation, especially the formation of inclusion body (Fink 1998), and formaldehyde could induce Tau aggregation in vivo and in vitro even though the protein concentration was low (Hua and He 2002; Nie et al. 2005, 2007).

In this book, at the beginning, Meihua Qu and Jing Lu discuss formaldehyde from the environment, indoor and outdoor, and show formaldehyde exposure ubiquitously, especially indoor exposure, in Chap. 1. Rong Xiao shows that formaldehyde comes from various pathways and leaves in limited ways, in which lysosome acts as a central cellular organelle in the metabolism of formaldehyde in vivo, in Chap. 2. Tao Su reviews the function of endogenous formaldehyde in methylation and demethylation of DNA, RNA, and histone, showing how formaldehyde affects memory in Chap. 3. Ting Li and Tong Ge emphasize that there is a vicious circle in the interaction of age-related cognitive impairment with chronic dehydration and endogenous formaldehyde in Chap. 4.

Weichuan Mo reviews the role of formaldehyde in cell proliferation and death. He shows that the effects of formaldehyde on cells depend on the concentration of the aldehyde and reactive time in Chap. 5. Kaili Liu discusses gut microbiota, formaldehyde dysmetabolism, and cognitive impairment, exhibiting a significant elevation of formaldehyde concentration in the intestinal contents of APP/PS1 transgenic mouse compared with that of C57BL/6j mouse in Chap. 6. I discuss the effects of formaldehyde on protein (Tau) aggregation and cytotoxicity, showing a possibility of a direct interaction between formaldehyde and Tau protein in Chap. 7. Afterward, I review cognitive ability and impairment related to formaldehyde. Endogenous formaldehyde is closely related with the educational levels of elderly individuals and age-related cognitive impairment of patients with either AD or vascular dementia (VD). Furthermore, I try to discuss the consciousness impairment of a patient with age-related cognitive impairment in Chap. 8. Symptoms that cannot be explained by memory loss or other patho-mechanisms alone, except for consciousness impairment. Jing Lu studied Tau phosphorylation in this laboratory, and she reviews Tau hyperphosphorylation and Aβ deposition in the presence of formaldehyde. She emphasizes that Tau isomers exist in neuronal nuclei in Chap. 9. Xiumei Wang discusses formaldehyde exposure and neuropsychiatric disorders except for cognitive impairment in Chap. 10. Formaldehyde affects not only cognitive ability but also other behaviors in humans. Yining Li proposes an interesting topic that formaldehyde affects lifespan and stress resistance in *Drosophila*. A proper concentration of formaldehyde can prolong the lifespan of *Drosophila*, but higher concentration shortens the lifespan in Chap. 11. Zijian Zhang explores some potential interventions or drugs on neuronal toxicity induced by formaldehyde in Chap. 12. He places his emphasis on the traditional Chinese medicine (TCM) used to treat the patients with age-related cognitive impairment. Min Qiang, Tao Su, and Beibei Wu review molecular, cellular, and animal experiments in formaldehyde study in Chap. 13. They discuss different cells and animal models in the metabolism of formaldehyde, as well as age-related cognitive impairment. Endogenous formaldehyde can be used as a criterion for age-related cognitive impairment, but not used to distinguish AD from VD. It can be used to screen senile dementia in a large population. Eventually, AD can be distinguished from VD by the patient's history and further examination. Finally in Chap. 14, Tao Su, who is interested in analytic methods for formaldehyde and ribose in this laboratory, discusses the methods for the determination of formaldehyde.

Dysmetabolism of endogenous formaldehyde is supposed to be related to sporadic AD because formaldehyde metabolism is complicated with different metabolic pathways. Sporadic AD patients have got various etiological factors and symptoms besides memory loss. As shown in Chap. 8, approximately 40% of AD patients had high levels of endogenous formaldehyde. That is to say, dysmetabolism of formaldehyde is not for all patients with age-related cognitive impairment. Each etiological risk factor or pathogenic factor is only associated with a part or a small proportion of AD cases. FD links with the gene mutation, showing typical pathological features in Aβ deposition, Tau hyperphosphorylation, and cognitive impairment. FD is "pure" in pathological characteristics, but sporadic AD is "complex," which should not be caused by one gene, one protein, or one compound dysfunction. In other words, sporadic AD is more pathologically and clinically inclined. Under this complicated situation, let us see how to use TCM to understand, diagnose and treat the sporadic AD (Chap. 12).

As described in the classical TCM work *Huangdi Neijing* (99–26, B.C.), humans contain "five internal organs" and "six viscera." In terms of the five internal organs, they are *Heart* (brain and heart) controlling mental activities and circulation, *Liver* (nerve and nutrition) regulating conveyance and dispersion, *Spleen* (digestion and immune) governing transportation and transformation, *Lung* (respiration, blood, and fluid) providing energy and connecting all vessels, and *Kidney* (reproduction, endocrine, and body fluid) storing and providing the essence for life (Zhu 2016). All these five internal organs are more or less involved in human emotion and intelligence. For example, *Heart* houses spirit, *Lung* maintains corporeal soul, *Liver* stores mood, *Spleen* processes idea, and *Kidney* accommodates aspiration, as described in *Plain Question-Declared Five Gas* in *Huangdi Neijing* (Zhao and Tian 2016). This is to say, age-related cognitive impairment may be resulted from dysfunctions of different internal organs. Particularly for sporadic AD, it could be resulted from different failures of metabolisms of sugar, lipid, and protein or other substances. Compared with FD, sporadic AD is definitely combined with other diseases such as hypertension, diabetes, vasculopathy, hyperlipidemia, and endocrine and nutrient disorders. To solve these complicated disorders, TCM does not identify AD, VD, or mixed dementia, but recognizes the personalized symptoms of each patient, to diagnose and treat him/her in different ways. A patient with sporadic dementia can be diagnosed with one of the following types of age-related cognitive impairment: the kidney-marrow deficiency, the Qi deficiency and blood stasis, the phlegm obstructing intelligence, the stagnated fire agitating the heart, and the spleen-yang impaired by dampness. Then, the medical doctors prescribe a special recipe (a group of drugs) according to the personalized clinical features (see Chap.

2). A patient with cognitive impairment suffering from a high level of endogenous formaldehyde belonging to whichever type of dementia needs further investigation.

According to TCM, a comprehensive and systematic treatment should be performed to manage the age-related cognitive impairment at the early stage. To treat a sporadic AD patient, we should identify whether the patient with mild cognitive impairment (MCI) suffers from hypertension, diabetes, heart disease, kidney disorder, and/or dysmetabolism of protein, lipid, formaldehyde, or ribose. If we detected a patient suffers from memory loss and dysmetabolism of formaldehyde, we should not only intervene with age-related cognitive impairment, but also regulate the imbalance of formaldehyde. When an MCI patient is complicated with hypertension, we should help him/her to regulate the blood pressure besides trying to improve cognition. Of course, when we detected a patient is suffering from high levels of D-glucose or D-ribose (Wei et al. 2015; Wu et al. 2016) and cognitive impairment, we should intervene with his/her cognition as well as dysmetabolism of sugar. This is the so-called personalized treatment, which may be an effective invention for senile dementia at the early stage.

Another purpose of this book is to remind all people to pay attention to formaldehyde exposure, especially from sources such as indoor pollution as described in Chap 1, and to prevent formaldehyde pollution and its exposure. Never forget those tragedies that happened in the past, for example, the calamity that happened in Wenshui of Shanxi province, China, in 1998. Hundreds of people drank the adulterated alcohol containing a high concentration of methanol by mistake, resulting in 27 people dead and 222 wounded (Chi and Wu 1998). The important organs of the victims such as the brain, eye, liver, and kidney were seriously damaged by the acute toxicosis. Accordingly, their cognitive ability was also severely impaired. Another grave incidence happened from 11 May to 28 May 2004 in Guangzhou, China. 115 people drank the adulterated alcohol containing a high concentration of methanol. The victims suffered from methanol (formaldehyde) toxicosis and formic acid accumulation, resulting in the death of 14 victims and serious wounds for 21 persons (Wen et al. 2006). On 23 March 2010 in Wufeng county in Hubei province, four people died of adulterated alcohol drinking (Tian et al. 2011). From 01 Jan to 23 Jan in 2017, 148 cases (1 person died) suffering from methanol acute toxicosis by wrong intake of adulterated alcohol were reported in Luzhou city of Sichuan province, China. The victims complained of vision impairment and other disorders involved in the central nervous system such as headache and dizziness (Luzhou News: http://news.lzep.cn/2017/0125/221391.shtml). So far, the incidence of the toxicosis still persists. Furthermore, many cases of formaldehyde exposure happened in the working place or at home (Xu et al. 2017), especially when the apartment or house is newly decorated (Amiri et al. 2015, Ye et al. 2017). We still have a long way to go, to completely avoid and eliminate formaldehyde exposure and thoroughly protect our health in daily work and life.

Here, I acknowledge all the authors who participated in writing this book! They are Meihua Qu, Jing Lu, Rong Xiao, Tao Su, Ting Li, Tong Ge, Weichuan Mo, Kaili Liu, Xiumei Wang, Yining Li, Zijian Zhang, Min Qiang, and Beibei Wu. I am also grateful to my collaborators who have studied aldehyde in this laboratory: Jianying Luo who studied effects of acetaldehyde and glutaraldehyde on Tau protein; Qian Hua and Chunlai Nie who performed and induced Tau aggregation and cytotoxicity with formaldehyde; Yonghui Chen, Wei Zhang, and Yang Liu who studied Tau aggregation in the presence of acetaldehyde; Zhiqian Tong and Jinling Zhang who investigated the correlation between endogenous formaldehyde and MMSE scores for AD patients; Zhiqian Tong who also clarified the relation among formaldehyde, DNA methylation, and memory; Fangxu Li who observed the production of formaldehyde from oxidation of myelin and malondialdehyde decomposition; Chanshuai Han who worked and contributed to the formaldehyde project in this laboratory; Jing Lu, Junye Miao, and Yang Lu who demonstrated that formaldehyde triggers Tau hyperphosphorylation and dysfunction; Ping Sun and Jinyan Chen who found the protective effect of geniposide on human neuroblastoma cells in the presence of formaldehyde; Jiawan Wang and Yue Yun who collaborated with us to reveal the role of formaldehyde in postoperative cognitive dysfunction (POCD); XiXi Chen and Tao Su who studied accommodation and transporation of formaldehyde by lysosome; and Yingge He who analyzed and determined thousands of formaldehyde samples. I would like to thank Meifeng Yang and Xintian Hu who are excellent collaborators and have confirmed the formaldehyde-induced senile plaques, Tau hyperphosphorylation, and memory loss in monkeys; Xiaping He who studied the effect of resveratrol on formaldehyde-treated cells; Jing Yu and Juan Li who collaborated to investigate the relationship between endogenous formaldehyde and cognitive ability in the elderly population of Beijing communities; and Jun Zhou and Jing Wang who studied neurites and processes of neural cells under low concentrations of formaldehyde, which was simulated to pathological values for AD patients. I thank Jihui Lu, Wenhong Luo, Jiangning Zhou, Dai Zhang, Jinshun Qi, Jianping Jia, Wenshan Wang, Shouzhi Zhang, Xiaohui Wang, Hui Li, Hongjun Luo, Hui Li, Hongjun Luo, Dehua Cui, Yanye Ma, and Xiaomin Wang. Many sincere thanks to Xinyong Chen, Martyn C. Davies, and Saul J. B. Tendler who supported me to observe formaldehyde-induced Tau protein aggregation with atomic force microscopy in the Laboratory of Biophysics and Surface Analysis, School of Pharmacy, University of Nottingham, UK. I deeply thank Paudel Heman for supporting me to study Tau phosphorylation and also Iqbal for his collaboration to study the interaction between Tau protein and DNA. I am also grateful to Zhigang Xue who works at Biologie Moleculaire de la Differenciation, Universite Paris. In 1994, he came to my institute and established the group of molecular neurodevelopment where I collaborated with him. Ying Liu is highly appreciated for her honest collaboration and indispensable work, and also Yan Wei and Xiumei Wang for their excellent work in this group.

I deeply thank Dr. Peng Zhang from Springer Beijing Office for his language editing for each chapter in this book. Many thanks to Miss Yiman Li who is in the Department of Biomedical Sciences in Imperial College London, UK, and Jing Lu who works at the Monash Institute of Cognitive and Clinical Neurosciences, Monash University, Melbourne, Australia, for their language editing of Chaps. 7 and 8 as well as the Preface.

I would like to show my gratitude to the National Natural Science Foundation of China (NSFC) for the Director Fund in 1996 and many grants that support my work (NSFC 30170297, 30370453, 90206041, 39970236, 39770254, 39770254, 30621004, 30870544, 30970695, 31270868, and 31470036), to the Ministry of Science and Technology (MOST) for supporting my group (2006CB500703, 2010CB912303, and 2012CB911004 as well as 2016YFC1306300), and to the Beijing Municipal Science and Technology Project (Z161100000217141; Z161100000216137). I show my gratitude to the Institute of Biophysics, Chinese Academy of Sciences, and the University of CAS for supporting me from the beginning to my retirement!

I have to say that there may be problems and mistakes in this book because formaldehyde is involved in a wide aspect from gene to molecule to cell to organ to human and disease as well as our environment. Our international and domestic colleagues have done many excellent work on formaldehyde at different aspects. We may have unintentionally missed some important information and contributions of our colleagues in the related chapter because our knowledge and abilities are limited. Here, I apologize for all the shortages in this book from the bottom of my heart.

Beijing, China Rongqiao He

References

Amiri A, Pryor E, Rice M, Downs CA, Turner-Henson A, Fanucchi MV (2015) Formaldehyde exposure during pregnancy. MCN Am J Matern Child Nurs 40(3):180–185

Chen YH, Luo JY, Li W, He RQ (1999) Effect of acetaldehyde on phosphorylation of human neuronal tau. J Biochem Mol Biol Biophys 3:197–202

Chi RH, Wu L (1998) The case of catastrophic producing counterfeit spirits in Shanxi province. Zhi Liang Tian Di 2:13–13

Cullen KM, Halliday GM (1995) Neurofibrillary tangles in chronic alcoholics. Neuropathol Appl Neurobiol 21(4):312–318

Fink AL (1998) Protein aggregation: folding aggregates, inclusion bodies and amyloid. 3(1):R9–R23

He RQ, Tsou CL (1991) Comparison of inactivation and conformational changes of the yeast D-glyeraldehyde-3-phosphate dehydrogenase during denaturation by Guanidine hydrochloride. Acta Biophysica Sinica 7(2):126–131

He RQ, Tsou CL (1992) Fluorescence of peptide N-terminal 2-oxoacyl and quinoxaline derivatives. Biochem J 287:1001–1008

He RQ, Zhao KY, Yan ZZ, Li M (1993) An asynchronous unfolding procedure among the different regions of GAPDH and maltotetraose-forming amylase from an *Alcaligenes* species, Biochim. Biophysi. Acta 1163:315–320

He RQ, Yang MD, Zheng X, Zhou JX (1995a) Some properties and isolation of glycated D-glyceraldehyde-3-phosphate dehydrogenase (GAPDH) of rabbit muscle and human erythrocyte. Biochem J 309:133–139

He RQ, Li YG, Wu XQ, Li L (1995b) Inactivation and conformation changes of the glycated and non-glycated GAPDH during guanidine-HCl denaturation. Biochim Biophys Acta 1253:47–56

Hou Q, Jiang H, Zhang X, Guo C, Huang B, Wang P, Wang T, Wu K, Li J, Gong Z, Du L, Liu Y, Liu L, Chen C (2011) Nitric oxide metabolism controlled by formaldehyde dehydrogenase (fdh, homolog of mammalian GSNOR) plays a crucial role in visual pattern memory in Drosophila. Nitric Oxide 24(1):17–24

Hua Q, He RQ (2002) Effect of phosphorylation and aggregation on tau binding to DNA. Prot Pept Lett 9:349–357

Kuchenbaecker KB et al (2017) Risks of breast, ovarian, and contralateral breast cancer for BRCA1 and BRCA2 mutation carriers. JAMA 317(23):2402 doi:10.1001/jama.2017.7112

Li T, Su T, He YG, Lv JH, Mo WC, Wei Y, He RQ (2016) Brain formaldehyde is related to water intake behavior. Aging Dis 7(5):561–584

Li Y, He R (2016) The effects of formaldehyde on life span and stress resistance in *Drosophila melanogaster*. Prog Biochem Biophys 43(4):420–428

Luckey AE, Parsa CJ (2003) Fluid and electrolytes in the aged. Arch Surg 138(10):1055–1060

Luo JY, He RQ (1999) Effect of acetaldehyde on aggregation of neuronal tau. Prot Pept Lett 6(2):105–110

Martin PR, Adinoff B, Weingartner H, Mukherjee AB, Eckardt MJ (1986) Alcoholic organic brain disease: nosology and pathophysiologic mechanisms. Prog Neuropsychopharmacol Biol Psychiatry 10(2):147–164

Nie CL, Zhang W, Zhang D, He RQ (2005) Changes in conformation of human neuronal Tau during denaturation in formaldehyde solution. Prot Pept Lett 12(1):75–78

Nie CL, Wei Y, Chen XY, Liu YY, Dui W, Liu Y, Davies MC, Tendler SJ, He RQ. (2007) Formaldehyde at low concentration induces protein tau into globular amyloid-like aggregates in vitro and in vivo. PLoS ONE 2(7):e629

Paudel HK (1997a) Phosphorylation by neuronal cdc2-like protein kinase promotes dimerization of Tau protein in vitro. J Biol Chem 272(45):28328–28334

Paudel HK (1997b) The regulatory Ser262 of microtubule-associated protein tau is phosphorylated by phosphorylase kinase. J Biol Chem 272(3):1777–1785

Stout NR, Kenny RA, Baylis PH (1999) A review of water balance in ageing in health and disease. Gerontology 45(2):61–66

Tan SLW, Chadha S, Liu Y et al (2017) Class of environmental and endogenous toxins induces BRCA2 Haplo insufficiency and genome instability. Cell 169(6):1105–1118. e15

Tian GB, Yang H, Chen YF, Zhao JL, Yang ZD, Xiong B, Feng SG (2011) Journal of Public Health and Preventive Medicine 6:112–113

Tong ZQ, Han CS, Luo WH, Wang XH, Li H, Luo HG, Zhou JN, Qi JS, He RQ. (2013) Accumulated hippocampal formaldehyde induces age-dependent memory decline. AGE 35(3):583–596

Wen YZ, Wang SX, Liao HL, Cao JR, Zhong GH, He XD, He ZH, Liang WH, Zhen ZM, Huang JZ (2006) Analysis of 115 cases suffering methanol toxicosis by wrongly drinking alcohol containing methanol. Guangdong Trace Elements Sci 13(6):13–22

Wei Y, Han CS, Wang YJ, Wu BB, Su T, Liu Y, He RQ. (2015). Ribosylation triggering Alzheimer's disease-like Tau hyperphosphorylation via activation of CaMKII. Aging Cell 14(5):754–763

Wu BB, Wei Y, Wang YJ, Zhou L, Su T, Liu Y, He RQ (2015) Gavage of D-Ribose induces Aβ-like deposits, Tau hyperphosphorylation as well as memory loss and anxiety-like behavior in mice. Oncotarget 6(33):34131–342142

Xu W, Zhang W, Zhang X, Dong T, Zeng H, Fan Q (2017) Association between formaldehyde exposure and miscarriage in Chinese women. Medicine (Baltimore) 96(26):e7146

Ye W, Zhang X, Gao J, Cao G, Zhou X, Su X (2017) Indoor air pollutants, ventilation rate determinants and potential control strategies in Chinese dwellings: a literature review. Sci Total Environ 586:696–729

Zhao JX, Tian YX (2016) Huangdi Neijing 1st edn. Beijing Joint Publishing Company, Beijing, pp 28–33

Zhu SP (2016) Explore, 1st edn. Huangdi Neijing Beijing Science Publisher, Beijing, pp 6–13

Contents

1 **Formaldehyde from Environment** 1
Meihua Qu, Jing Lu, and Rongqiao He

2 **Metabolism of Formaldehyde In Vivo** 21
Rong Xiao and Rongqiao He

3 **Formaldehyde Playing a Role in (De)methylation for Memory** 47
Tao Su and Rongqiao He

4 **Dehydration, Formaldehyde, and Age-Related Cognitive Impairment** 63
Ting Li, Tong Ge, and Rongqiao He

5 **The Role of Formaldehyde in Cell Proliferation and Death** 79
Weichuan Mo and Rongqiao He

6 **Gut Microbiota, Formaldehyde Dysmetabolism, and Cognitive Impairment** 99
Kaili Liu and Rongqiao He

7 **Effects of Formaldehyde on Protein (Tau) Aggregation and Cytotoxicity** 121
Rongqiao He

8 **Cognitive Ability and Impairment Related to Formaldehyde** 143
Rongqiao He

9 **Tau Phosphorylation and Amyloid-β Deposition in the Presence of Formaldehyde** 167
Jing Lu and Rongqiao He

10 **Formaldehyde Exposure and Neuropsychiatric Disorders** 191
Xiumei Wang and Rongqiao He

11 Formaldehyde Affecting Lifespan and Stress Resistance in *Drosophila* 209
Yining Li and Rongqiao He

12 Potential Interventions on Formaldehyde-Induced Neuronal Toxicity 221
Zijian Zhang and Rongqiao He

13 Molecular, Cellular, and Animal Experiments in Formaldehyde Study 245
Min Qiang, Tao Su, Beibei Wu, and Rongqiao He

14 Methods in Determination of Formaldehyde 271
Tao Su and Rongqiao He

Glossary 297

Index 307

Chapter 1
Formaldehyde from Environment

Meihua Qu, Jing Lu, and Rongqiao He

Abstract Formaldehyde is widely present from the universe to a single cell. It also comes from our daily life, including food, water, clothing and building materials. As a main material or by-product from industrial and commercial production, formaldehyde is the most common aldehyde in the environment. The most common formaldehyde exposure is from contaminated air. In general, human beings spend around 90% of their time indoors, where there are complex mixtures of pollutants, including formaldehyde. Indoor formaldehyde exposures can be extremely dangerous, such as residential exposing to renovated homes, offices and public shopping centres. Another main method of formaldehyde immersion is through occupational exposure, for instance, in some factories, hospitals and laboratories, where formaldehyde concentrations are much higher and the people working there are highly vulnerable. Although the toxicity of formaldehyde has been extensively studied and its detections have been well developed, the exposure of formaldehyde is still serious all over the world, especially the countries that are producing and consuming huge amounts of construction materials which contain and emit aldehydes. People are exposed to formaldehyde through breathing and gastric intestinal digestion or by skin contact. The exposure to formaldehyde can cause respiratory irritation, asthma, tumours and multiple neuropsychological abnormalities or even cause death. Therefore, it is of great importance to control possible sources of formaldehyde and set up international standards to avoid formaldehyde insult.

Keywords Formaldehyde • Occurrence • Chemical characteristics • Exposure • Indoor exposure • Outdoor exposure • Occupational exposure • Daily life exposure

---

M. Qu
Key Laboratory of Applied Pharmacology in Universities of Shandong Province, Department of Pharmacology, Weifang Medical University, Weifang 261053, China

J. Lu
Monash Institute of Cognitive and Clinical Neurosciences, Monash University, Melbourne, Australia

R. He (✉)
State Key Laboratory of Brain and Cognitive Science, Institute of Biophysics, Chinese Academy of Sciences, Beijing 100101, China

University of Chinese Academy of Sciences, Beijing 100049, China
e-mail: rongqiaohe@163.com

R. He, *Formaldehyde and Cognition*, DOI 10.1007/978-94-024-1177-5_1

1 Introduction

Formaldehyde is the smallest aldehyde with two hydrogen atoms and one oxygen atom bonded to a carbon atom in the centre (CH_2O). As described by the US Department of Health and Human Services (ATSDR and Registry 1999b), formaldehyde (CAS#50-00-0) is one of the intermediate products in the oxidative reaction from methanol to carbon dioxide and water (Base et al. 1999). Formaldehyde was first described in 1855 by Alexander Michailowitsch Butlerow (Brooks 1998). German chemist August Wilhelm von Hofmann technically synthesise formaldehyde in 1867 (Loeb 2010). The properties of formaldehyde were extensively investigated in the following decades (Tulpule and Dringen 2013).

Formaldehyde ubiquitously exists in the environment, for it is primarily formed by abundant natural sources and anthropogenic activities (World Health Organization 2010). Natural formaldehyde is usually synthesised by photochemical reactions in the Earth's primitive atmosphere or forest and bush fires that produced biomass combustion or emissions of volcanoes and molecular processes as biological decomposition (National Research Council Committee to Review 2011). Anthropogenic sources of formaldehyde include direct emissions by fuel combustion of all kinds of vehicles, plants and combustion processes (such as power plants, incineration, etc.). Moreover, formaldehyde is produced worldwide as fixatives and disinfectants as well as preservatives in consumable products (Tang et al. 2009). Thirty senven percent of formaldehyde solution is widely used as a preservative to maintain biological samples due to its ability in protein cross-linking (Fox et al. 1985; Hoffman et al. 2015; Leon and Pletcher 1995; Sekine et al. 2016; Toth and Biggin 2000).

Formaldehyde, with a characteristic pungent, irritating odour, is highly toxic to all animals, no matter what methods of exposure. Most formaldehyde exposures occur by inhalational, oral and dermic route. We live in an environment with formaldehyde which is present both indoor and outdoor, thereofore, we are exposed to formaldehyde when breathe. We are also exposed to formaldehyde when we eat and drink. In some specific conditions, formaldehyde could enter the human body by direct skin contact. Studies have shown that formaldehyde exposure can cause multiple problems in the nervous, immune and reproductive system, with consequence of systemic, developmental, genotoxic and carcinogenic effects and even death (Acheson et al. 1984; Johannsen et al. 1986; Kerns et al. 1983; Til et al. 1989; Vaughan et al. 1986a, b). Furthermore, the toxicity of formaldehyde was also extensively investigated and evaluated by scientific researchers and government agencies (Agency for Toxic Substances and Disease Registry 1999a, b; Binazzi et al. 2015; Bosetti et al. 2008; Madureira et al. 2016; McLaughlin 1994; Nielsen et al. 2017). In this chapter, we would like to discuss the formaldehyde sources and exposures in the environment where we live.

2 Formaldehyde Sources

2.1 Natural Sources of Formaldehyde

The route of formaldehyde exposure to human mainly including inhalation exposure and dietary exposure from indoor and outdoor air, or food and drinking. The main source of formaldehyde for exposure is the indoor air, which includes residential, office, public and occupational place. The formaldehyde indoor exposure is resulted from the regulated and non-regulatedsources (Stranger et al. 2007). Sanriannis and collaborators emphasized the regulation of the occurrence of formaldehyde as the prior pollutants. They studied on the correlation between formaldehyde and tumorigenesis when exposure to formaldehyde in dwellings and public buildings in European Union (EU) countries (Sarigiannis et al. 2011). Some sources of the chemical formaldehyde in the indoor air is not yet regulated within the European Union yet. As described by Kotzias and collaborators, identification formaldehyde as prior pollution compounds developed a strategy in further formaldehyde regulation by the European Commission (Kotzias 2005; Kotzias et al. 2005).

Formaldehyde can be produced naturally. Up to 90% of naturally produced formaldehyde is synthesised in the Earth's primitive atmosphere by photochemical reactions (Pinto et al. 1980). Pinto and colleagues presumed that the precipitation of formaldehyde from the atmosphere provides 10^{11} moles organic carbon per year to the oceans. Formaldehyde is also a part of smog produced by the reaction of oxygen and atmospheric methane and other hydrocarbons in the air under sunlight (Cheremisinoff 2016). Snyder and colleagues first discovered formaldehyde as the polyatomic organic molecule in the universe (interstellar space) using the National Radio Astronomy Observatory (Weaver 1970; Zuckerman et al. 1970). Since then, formaldehyde has been detected in outer space and found in many areas of the galaxy (Blair et al. 2008). Due to extensive interest in interstellar formaldehyde, the study of extragalactic sources of formaldehyde has been yielded (Mangum et al. 2007). One of the proposed mechanisms for the formation of formaldehyde in the interstellar dimension is the hydrogenation of iced CO, shown as:

$$H + CO \rightarrow HCO + H \rightarrow H_2CO\,(\text{rate constant} = 9.2 * 10^{-3}\,s^{-1})$$

Formaldehyde is an abiogenic compound, and it is also part of plant/atmosphere exchanges and plant physiological processes (Seco and Filella 2008).

Formaldehyde is prone to break down in the sunlight and in soil or water where bacteria are present (Zuckerman et al. 1970). Formaldehyde can be metabolised to formate by FDH/class III alcohol dehydrogenase consisted in almost all tissues and removed quickly in vivo (Casanova-Schmitz et al. 1984a, b; Heck et al. 1982). A normal person metabolises formaldehyde quickly and does not accumulate it in the body, except for dysmetabolism of formaldehyde.

2.2 *Anthropogenic Sources of Formaldehyde*

Fuel combustion is a major source of formaldehyde from vehicle exhaust (Hoekman 1992), refineries and factories of stone, clay and glass production (ARB 1997). Road transport is responsible for over 70% of transport greenhouse gas emissions and much of the air pollution in the European Union (EU) (Suarez-Bertoa et al. 2017).

As a feedstock for numerous industrial products, formaldehyde is produced over 5 million tons each year (Viegas et al. 2013). Formaldehyde is widely used to synthesise formaldehyde-based resins including urea-formaldehyde resin, phenol formaldehyde resin, melamine resin and other complex compounds. Formaldehyde-based resins are utilised in the decorative materials, household products, textiles and building materials (Reuss et al. 2000). Urea-formaldehyde adhesives are the most widely used products to produce wood-based furniture and other materials, for their rapid curing, compatibility with additives and low cost (Pizzi 2003). Furniture with artificial board, carpets and clothing and even the wall with paintings are releasing formaldehyde continually, making formaldehyde one of the major indoor pollutants (Salthammer et al. 2010). Formaldehyde is also an ingredient of numerous consumable household products, including daily usage of cosmetics and dishwashing liquids, carpet cleaners, shoe-care agents, fabric softeners, glues and adhesives, lacquers, antiseptics and even medicines. It would be necessary for consumers to know the content of formaldehyde in the materials used daily and to avoid using them. Furthermore, the application of formaldehyde is extended to food preservative processing, and it can be detected in Italian cheeses, dried foods, fish, etc. (Restani and Galli 1991).

3 Formaldehyde Inhalation Exposure

The route of formaldehyde exposure to humans mainly includes inhalation exposure and dietary exposure. The formaldehyde inhalation exposure resulted from the regulated and non-regulated sources showed in Fig. 1.1 (Stranger et al. 2007). Sanriannis and collaborators emphasised the regulation of the occurrence of major organic compounds classified by the European Commission's INDEX strategy. Formaldehyde was regarded as the prior pollutant to be regulated due to its cancer risks when exposed in home or indoor public buildings from European Union countries (Sarigiannis et al. 2011). Some sources of formaldehyde emission in indoor air have not been regulated within the European Union yet. As described by Kotzias and collaborators, the development of a strategy leading to identification of prior compounds including formaldehyde has been highlighted in their further regulation by the European Commission (Kotzias 2005; Kotzias et al. 2005).

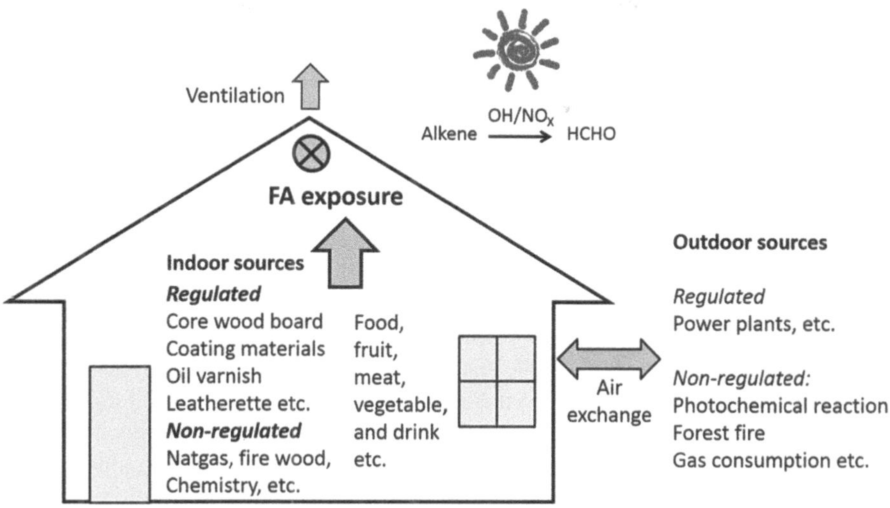

Fig. 1.1 Formaldehyde sources for indoor exposure. Indoor formaldehyde comes from different sources such as outdoor air, food, drink as well as indoor itself. Indoor air is the most common for formaldehyde exposure. Indoor formaldehyde exposure resulted from regulated and non-regulated sources in human daily life (Salthammer 2015). Effective ventilation is necessary for special working places, for instance, anatomy laboratories. The Environmental Protection Agency (EPA) of United States indicates 0.2 mg/kg body weight as the acceptable daily intake (ADI) of formaldehyde and it is potentially a harm to health for formaldehyde exposure over the ADI (EPA 1999). The data are referred to the Agency for Toxic Substances and Disease Registry (1999b) and Institute for Health and Consumer Protection (2005) (Kotzias et al. 2009; Lai et al. 2006; Salthammer et al. 2010; Saowakon et al. 2015)

3.1 Indoor Formaldehyde Inhalation Exposure

Indoor formaldehyde inhalation exposure includes residential, office, public and occupational exposure.

3.1.1 Indoor Residential Formaldehyde Exposure

The highest levels of airborne formaldehyde are detected in indoor air. As an important chemical widely used, formaldehyde is a very general indoor pollutant (Sekine 2005; World-Health-Organization 2014). The major emission sources of formaldehyde indoor are the interior building materials, including adhesives for walls, furniture and textiles finished with crease-resistant fabrics (Tohmura 2001). Electric appliances such as photocopiers and laser printers are also reported to release a considerable amount of formaldehyde (Matsuo et al. 2008). Therefore, the amount of formaldehyde emission from interior materials has become essential for indoor air pollution control (Dunky 2004).

The concentration of indoor formaldehyde is correlated with ventilation and the emission rate of formaldehyde from construction materials. Bruinen showed that building age and urea-formaldehyde foam insulation (UFFI) usage were important factors regarding to sources of formaldehyde (Institute for Health and Consumer Protection 2005). The average concentration of formaldehyde inside the building without any specification was 79.3 μg/m^3 and as high as 94 μg/m^3 with UFFI compared to 32 μg/m^3 without UFFI. The average concentration of formaldehyde in buildings for less than 5 years was as high as 196 μg/m^3 compared to 46.8 μg/m^3 for more than 5 years (Institute for Health and Consumer Protection, USA 2005). It is recommended in Sweden that the median concentration of formaldehyde for personal exposure in the general population was 23 μg/m^3 in the value range of 12–60 μg/m^3 (Gustafson et al. 2004; Victorin 1998). A study carried out in Germany from 2001 to 2004 showed that the median indoor formaldehyde concentration was 28 μg/m^3. Kirchner and colleagues monitored 567 randomly selected dwellings from 2003 to 2005 and showed that the median concentration of formaldehyde in bedrooms was 19.6 μg/m^3 (Kirchner et al. 2008). Statistical concentration of formaldehyde (mean ± SD) and the daily exposure rate were referred to IHCP (Institute for Health and Consumer Protection 2005).

China is a developing country with a rapid economic growth. The construction booms accompanying huge consumption of construction materials including decoration materials containing formaldehyde (Salthammer et al. 2010). Furthermore, the residential renovation is frequent and common in both urban and rural regions. In the renovated buildings, the indoor formaldehyde problem is more serious than the older buildings because of higher emission rates from recently manufactured materials. Certain furniture and decoration and building materials are sources of indoor formaldehyde (Du et al. 2014; Du et al. 2013). Engineered wood products using urea-formaldehyde resin adhesive, such as particle board and chipboard, are important formaldehyde sources (He et al. 2012; Xing et al. 2005). Indoor pollution with formaldehyde becomes serious in China (Ye et al. 2017).

Meng et al. measured 6000 recently refurbished dwellings selected from 1999 to 2006 in urban areas in China, and the results showed that the average concentration of indoor formaldehyde was 238 μg/m^3, while the average outdoor formaldehyde level was around 12 μg/m^3 (Meng and Wu 2015). Huang and colleagues measured the formaldehyde concentrations in indoor air of 410 dwellings and 451 offices in Beijing area during 2013; the results showed that 85% of dwellings and 67% of offices have indoor formaldehyde concentrations higher than the OEHHA recommended acute Reference Exposure Level (REL) (Huang et al. 2013).

The emission rates of formaldehyde from building materials are influenced by surrounding air temperature and humidity (Huang et al. 2017). As described by many studies (Andersen et al. 1975; Crawford and Lungu 2011; Lin et al. 2009; Myers 1985; Parthasarathy et al. 2011), the emission rate of formaldehyde is positively related to both temperature and humidity (Huang et al. 2015; Xiong et al. 2013; Xiong and Zhang 2010) and negatively correlated with ventilation rate (Hult et al. 2015; Liu et al. 2015; Salthammer et al. 1995). Huang and colleagues investigated and analysed factors on the concentration of indoor formaldehyde in urban areas in 39 cities of southern and northern China. Their results showed that the concentration of winter indoor formaldehyde in the northern cities is about 4.0-folds higher than that in the southern ones. The reason mainly is the different living habits between the northern and southern people adapting to the climate. To keep warm, usually a central heating system is used and the doors and windows are closed in northern China in winter. Thus, the ventilation is worse and the temperature is higher in rooms in the north than those in the south. According to Salthammer (2013), the future ventilation strategy may be regulated because levels of formaldehyde in ambient (outdoor) air have been increasing with time.

3.1.2 Indoor Office Exposure

According to the EXPOLIS study in Helsinki, the average level of formaldehyde at the workplace was 15 μg/m^3 (Oglesby 2000). Formaldehyde concentrations in offices or in public buildings varied from 3 to 33 μg/m^3, detected from 2001 to 2006 at offices from southern Finland (Salonen et al. 2009). But the office formaldehyde exposure seems more serious in developing countries. A study showed that the average formaldehyde concentration in 351 offices in China (data from 1996–2005) was 256 ± 2.7 μg/m^3 which was similar as in recently refurbished dwellings (Mui et al. 2009). In the developed countries, environmental friendly materials and better Occupational Health and Safety (OHS) system are used to protect their employees, while in the developing countries, cost remains the first priority and the office air pollution remains a serious problem.

3.1.3 Indoor Public Exposure

Over the period of 2003–2008, Italian scientists monitored formaldehyde in European public buildings for children such as schools and kindergartens. The results showed that in schools/kindergartens, the concentration of formaldehyde was 17.4 μg/m^3, with a range from 1.5 to 49.7 μg/m^3 (Kotzias et al. 2009). The indoor air quality was evaluated in classrooms in Germany during 2004–2005, which showed that indoor formaldehyde concentrations varied from 3.1 to 46.1 μg/m^3 (Fromme et al. 2008). The data from Japan showed that formaldehyde concentrations were around 14 μg/m^3 in winter and 30 μg/m^3 in summer as measured in 50 Japanese schools in 2000 (Uchiyama 2008). Bradman measured the air samples in 2010–2011 from 40 early childhood education (ECE) facilities serving children <6 years old in California; the data showed that the median indoor formaldehyde levels (μg/m^3) were 17.8 while the maximum were 48.8 (Bradman et al. 2016). All the results showed that the public indoor formaldehyde exposure is also a serious problem all over the world.

3.1.4 Indoor Occupational Exposure

In some workplaces where excessive formaldehyde is applied or produced, people are exposed in higher concentration of formaldehyde (Dreyfuss 2010; Mirabelli et al. 2011). For example, factories produce chemicals which will be degraded to formaldehyde including production of formaldehyde itself. The occupational exposure to formaldehyde involves not only individuals employed in the direct manufacture of products containing certain levels of formaldehyde but also those actively utilising such products (e.g. construction and decoration). Formaldehyde exposure occurs primarily by breathing air polluted by formaldehyde or by absorbing liquids containing formaldehyde through the skin (d'Ettorre et al. 2017; Lee et al. 2017; Peters et al. 2017). Those working in certain job sectors (e.g. manufacturers of resins, plywood and particle board, laboratory faculties, certain health-care professionals, fire fighters and mortuary employees) are exposed to higher doses of formaldehyde.

A study carried out in Portugal showed the positive correlation between occupational formaldehyde exposure time (years of exposure) and micronucleus frequency in peripheral blood lymphocytes ($r = 0.401$; $p < 0.001$) and in epithelial cells ($r = 0.209$; $p < 0.01$) (Viegas et al. 2010).

In medical area, thirty-seven percent formaldehyde solution is a preservative for dead bodies to prevent them from decaying. Cadavers are usually preserved in embalming solution composed of formaldehyde. Medical students and instructors face a higher risk of inhalation exposure to formaldehyde from cadavers during dissection. The indoor air and personal air samples in breathing zone were collected in anatomy dissection classes in 2014 with sorbent tubes and analysed by high-performance liquid chromatography (HPLC). The results showed that the concentration of formaldehyde was 140–498 μg/m^3 in the indoor air and was 151.2–1411.2 μg/m^3 in the breathing zone of students and instructors. All the personal exposure data ran over the threshold limit of the National Institute for Occupational Safety and Health (NIOSH) and WHO agencies (Kimbell et al. 2001; Saowakon et al. 2015). Domingo et al. investigated formaldehyde concentration in 15 Chinese medicine clinics (CMCs) with frequent use of moxibustion. The results showed that the mean values of formaldehyde in the CMCs' indoor air were 654 μg/m^3 and 155 μg/m^3 in the therapy rooms and in the waiting rooms. The high concentration of formaldehyde is a threat to the health of medical staffs significantly and to the patients' health slightly, when applying moxibustion in the CMCs (Domingo et al. 2014; Ladeira et al. 2010, 2011; Mirabelli et al. 2011; Ochs et al. 2012; Viegas et al. 2010). The elevated risk values for occupational formaldehyde exposure suggested that the negative health impact for a certain group of people represented a valid concern, and precautions should be taken to protect people from that risks (Ho et al. 2013).

3.2 *Outdoor Air Exposure*

Outdoor formaldehyde pollution is less serious than that of indoor pollution, since ambient levels are generally low, typically around 1–4 μg/m^3. The human exposure characterisation of chemical substances (HEXPOC) report, which data were collected from Germany, Italy, the Netherlands, Brazil, Canada, Mexico and the United States, showed that the ambient concentrations of formaldehyde are 1.5–16.4 μg/m^3 with a medium value of 7.2 μg/m^3 (SD = 5.1 μg/m^3). The indoor-outdoor ratio is always far above 1 (Bruinen de Bruin et al. 2008; Institute for Health and Consumer Protection 2005). That is, normal air is safe unless the air is polluted. Marcon and colleagues showed that children living near the largest chipboard manufacturing district had the highest average exposure to formaldehyde. In addition, the exposure to pollutants was associated with genotoxic markers in exfoliated buccal cells (Marcon et al. 2014).

4 Dietary Exposure of Formaldehyde

Formaldehyde is also a natural ingredient in a variety of fruits, vegetables, meat, milk products and fish (Jurvelin et al. 2001). The study by Bruinen and collaborators provides data of formaldehyde content in fruits and vegetables (Bruinen de Bruin et al. 2008). The concentration of formaldehyde in solid food and liquid drink listed was reviewed by IHCP (Institute for Health and Consumer Protection 2005). The food and drink exposure varies, and the high concentration exposure was as high as 41,900 μg per day or low to 4350 μg per day depending on the dietary habit of people. However, whether formaldehyde is a natural presence or adulterant is controversial. Human exposure to formaldehyde was calculated with the most represented food in Bangladeshi diet.

The average formaldehyde uptake per person in solid food is 14.3–50.5 μg/kg, 24–436 μg/kg and 0.52–50.5 μg/kg in meat and poultry, fruit and vegetables and fish and shellfish, respectively (Institute for Health and Consumer Protection 2005). Formaldehyde uptake for people through liquid diet as drinking water is less than 0.1 μg/kg, alcoholic beverages 0.08–15 μg/kg, coffee 14–65 μg/kg and carbonated soft drinks (cola) 15–18 μg/kg (Institute for Health and Consumer Protection 2005). For an adult with average weight 64 kg, the average formaldehyde intake was estimated to be 7.01 mg/kg food or 0.12 mg/kg body weight per day (Safety 2014). This level of exposure to formaldehyde from the average diet in adults is lower than the maximum limits suggested by the European Food Safety Authority (EFSA) (<100 mg/kg food per day) and 11 mg/kg food per person per day (Wahed et al. 2016). There seems to be no health-associated risk when consuming food with traces of formaldehyde.

5 Formaldehyde Exposure and Health Problems

5.1 Acute Toxicosis

There was a case in the Wenshui County of Shanxi Province in China on 26 February 1998. Hundreds of people accidently drank adulterated alcohol containing a high concentration (36.1%) of methanol (Chi and Wu 1998). Ambulances were roaring and quickly passing through the street to rescue the victims whose organs such as the eye, brain, liver and kidney were damaged by drinking the methanol. In the accident, 27 people died and 222 were injured. To be more precise, the first symptom of their CNS was eye impairment which was strikingly visible because of the acute toxicosis of the metabolic intermediate products from methanol. Methanol is converted into formaldehyde, formic acid and finally carbon dioxide and water. The toxic formaldehyde is quickly metabolised into formic acid that is slowly degraded to CO_2 and H_2O. In acute toxicosis, formic acid is accumulated and leads to acidosis in brain tissues and neural cells (Bruckner and Warren 2001). The local acidosis that resulted from formic acid accumulation is one of the key causes to impair CNS cells, especially in acute toxicosis with high concentrations of methanol and formaldehyde although the patho-mechanism for slow toxicosis with low concentrations of formaldehyde is complex and still needed to by investigated.

5.2 Impairment and Symptoms of Intoxication

Azari and collaborators showed a relation between the concentration of indoor formaldehyde and symptoms of the participants (Azari et al. 2012). In their study, the general indoor concentration of formaldehyde was measured as 62.5 μg/m^3 in a school library. They found that prevalence of symptoms due to formaldehyde exposure was greater in study groups than in controls. Under the conditions, the most commonly reported complaints were unpleasant odour (68%), cough (64%), sore throat and runny nose (56%), nasal irritation and itching (52%) and eye irritation (48%). Acute exposure to formaldehyde can cause various health-related issues such as irritation on various body parts (eyes, nose, throat and skin) (ATSDR and Registry 1999b). Moreover, sustained exposure can lead to certain types of cancers (e.g. nasopharyngeal) and asthma (Kim et al. 2011). Formaldehyde is also classified as a human carcinogen that possibly leads to nasopharyngeal cancer and probably leukaemia and especially linked to sick building syndrome and multiple chemical sensitivities (Coggon et al. 2014).

For the group with occupational contact of formaldehyde, protection is compulsory. The level of formaldehyde in an anatomy lab was much higher than many of other environments. Considering the existence of clinical symptoms when the participants were exposed to 62.5 μg/m^3 formaldehyde, better control of the ventilation system should be facilitated and furnished to protect those staffs and instructors in anatomy departments. d'Ettorre and colleagues considered that the safety of anatomy pathology workers who are currently exposed to formaldehyde is still a matter of concern (d'Ettorre, et al. 2017). They found that there is a lack of evidence-based improvement interventions that aimed to control exposure to formaldehyde. They suggested to have more in-depth studies to measure the minimal formaldehyde exposure level in pathology departments. In addition, smoking significantly contributes to indoor formaldehyde level, and thus smoking cessation should be highly encouraged both in working place and at home. Li et al. studied the relationship of occupational exposure to formaldehyde and genetic damage in the peripheral blood lymphocytes of plywood workers. Results showed that there was a dose-response relationship between the current formaldehyde exposure levels and DNA strand breaks and between chromosome damage in peripheral blood lymphocytes of plywood workers (Lin et al. 2013). Sie and colleagues showed that occupational exposure of formaldehyde increases the risk of nasal cancer but not of nasopharyngeal or lung cancer (Sie et al. 2012). Zhang and collaborators examined the ability of formaldehyde to disrupt haematopoiesis in a study of 94 workers in China (43 exposed to formaldehyde and 51 frequency-matched controls); their results showed that among exposed workers, peripheral blood cell counts were significantly reduced in a manner consistent with toxic effects on the bone marrow, and leukaemia-specific chromosome changes were significantly elevated in myeloid blood progenitor cells (Roberts et al. 2016; Zhang et al. 2010b).

The concerns about these adverse effects have attracted the attentions on public health. A decade has passed since formaldehyde was classified as a known human carcinogen by the International Agency for Research on Cancer (World Health Organization 2010). As described by Han and co-workers, the filter media had a crucial role in the performance for formaldehyde abatement (Han et al. 2017). Their work represents a breakthrough in providing experimental clues and key parameters for developing and adopting an effective catalytic oxidation filer for indoor hazard abatement including formaldehyde in real applications.

6 Formaldehyde in Biological Evolution and Metabolism

6.1 The Role of Formaldehyde in Biological Evolution

Kalapos believes that formaldehyde is a component which had a role in evolutionary processes (Kalapos 1998). He raised a hypothesis that the (methyl) glyoxalase path might have bridged the early stage of evolution between formose and archaic reductive citric acid cycles (Kalapos 1999). For instance, formaldehyde acts as one of intermediate products in the Miller-Urey experiment (Hill and Nuth 2003), which implied that reactions of formaldehyde in primaeval aquatic environments could be involved in the abiotic synthesis of complex organic molecules for the origin of life. As described by Roscoe and Miller, CH4 and CO mixture efficiently produced HCN and H_2CO (Stribling and Miller 1987). Gaseous molecules excited by electric discharge led to the production of organic compounds, including amino acids (Parker et al. 2014).

Furthermore, formaldehyde also functions as an energy source in some yeast and bacteria though it has cytotoxicity (Müller and Babel 1991). From prokaryotic to eukaryotic organisms, formaldehyde plays a role in methylation and demethylation of DNA (Crider et al. 2012; Yan and Frey 2011), histone (Li and He 2016; Shi et al. 2004) and RNA (Gerken et al. 2007; Su et al. 2015). As described in a recent report, the life span of Drosophila is related to formaldehyde in a concentration-dependent manner (Li et al. 2014; Li and He 2016). These discoveries demonstrate the importance of formaldehyde in biological processes (see Chap. 3 and 13).

6.2 The Metabolism of Formaldehyde

Formaldehyde is a reactive carbonyl species with strong reducing property, which enables the cross-linking reaction with DNA or proteins (Hoffman et al. 2015). People who stayed in an environment with exceeding indoor formaldehyde will suffer many health problems. Formaldehyde in vivo is accommodated and transported by lysosome; it induces oxidative stress (Evans et al. 2016) and is harmful to the human central nervous system, leading to cognitive impairment (Chen et al. 2017; Hu et al. 2016; Jian and Zhu 2016). Children under the age of 16 are more sensitive to higher concentrations of formaldehyde, asthma and other respiratory system disorders and even leukaemia (Yao et al. 2014; Zaitseva et al. 2013; Zhai et al. 2013; Zhang et al. 2010a). The concentration and exposure time are critical factors to assess the risk of formaldehyde on human health. Generally, a low concentration of formaldehyde has no obvious effects on the tissue and organ. But either long-term or short-term exposure with high doses of formaldehyde will result in lesions to the eyes, skins, respiratory tract, blood system, reproductive system, immune system or even central nervous system in varying degrees (Duong et al. 2011; Tang et al. 2009).

7 Conclusion

In general, formaldehyde is produced naturally in the universe and the upper atmosphere. Formaldehyde can be detected universally, from air to food and in single cells. It will be generated by chemical synthesis or by biological metabolism. The wide usage of formaldehyde-containing products has caused severe formaldehyde pollution and a series of health issues, such as cancer, or neurological disorders. Outdoor formaldehyde levels are usually low, far lower from the toxic dosage set up by national standards, unless in the factory area with massive emission of formaldehyde. Indoor (including residential, office, public place) formaldehyde pollution is the main concern of human health, especially those formaldehyde emitted from the construction and decoration materials, furniture and textiles used. Occupational formaldehyde pollution is still a major risk for employees working in factories or medical-related laboratories. Protection equipment and well-ventilated facilities are necessary. In order to control health issues caused by formaldehyde, it is crucial to control the sources of formaldehyde.

Acknowledgements We would like to thank NCBI for the data on formaldehyde we cited from the website: www.ncbi.nlm.nih.gov. This project was supported by grants from the National Key Research and Development Programme of China (2016YFC1306300), the National Basic Research Programme of China (973 Programme) (2012CB911004), the Beijing Municipal Science and Technology Project (Z161100000217141; Z161100000216137), the Natural Scientific Foundation of China (NSFC 31270868), Foundation of Chinese Academy of Sciences CAS-20140909 and the Queensland-Chinese Academy of Sciences Biotechnology Fund (GJHZ201302). This project was also supported by grants from the National Natural Science Foundation of China (NSFC 81274093), Shandong Province Natural Science Foundation (ZR2015HL128) and Health Department of Shandong Province (2014WS0478).

Competing Financial Interests The authors declare no competing financial interests.

References

Acheson ED, Gardner MJ, Pannett B, Barnes HR, Osmond C, Taylor CP (1984) Formaldehyde in the british chemical industry: an occupational cohort study. Lancet 1:611–616

Agency for Toxic Substances and Disease Registry (1999a) Medical management guidelines for formaldehyde. Toxic Substances Portal. Website: https://www.atsdr.cdc.gov/mmg/mmg.asp?id=216&tid=39

Agency for Toxic Substances and Disease Registry (1999b) Toxicological profile for formaldehyde. Agency for Toxic Substances and Disease Registry, US Department of Health and Human Services

Andersen I, Lundqvist GR, Mølhave L (1975) Indoor air pollution due to chipboard used as a construction material. Atmos Environ 9:1121–1127

ARB (1997) Formaldehyde, toxic air contaminant identification. ARB/SSD/SES, September 1997

Azari MR, Parisa A, Mohammad Javad J, Hamid S, Vajihe H (2012) Occupational exposure of a medical school staff to formaldehyde in tehran. Tanaffos 11:36–41

Base USMC, Lejeune C, Carolina N, Maslia ML, Wre D (1999) Agency for Toxic Substances and Disease Registry toxicological profile information

Binazzi A, Ferrante P, Marinaccio A (2015) Occupational exposure and sinonasal cancer: a systematic review and meta-analysis. BMC Cancer 2015:15–49

Blair SK, Magnani L, Brand J, Wouterloot JG (2008) Formaldehyde in the far outer galaxy: constraining the outer boundary of the galactic habitable zone. Astrobiology 8:59–73

Bosetti C, Mclaughlin JK, Tarone RE, Pira E, La VC (2008) Formaldehyde and cancer risk: a quantitative review of cohort studies through 2006. Ann Oncol 19:29–43

Bradman A, Gaspar F, Castorina R, Williams J, Hoang T, Jenkins PL, Mckone TE, Maddalena R (2016) Formaldehyde and acetaldehyde exposure and risk characterization in California early childhood education environments. Indoor Air 27(1):104–113

Brooks NM (1998) Alexander Butlerov and the professionalization of science in Russia. Russ Rev 57:10–24

Bruckner JV, Warren DA (2001) Toxic effects of solvents and vapors. In Klaassen CD (ed) Casarett Doull's toxiology, the basic science of poisons. 6th edn. The McGraw-Hill Co., Kansas, pp 894–895
Bruinen de Bruin Y, Koistinen K, Kephalopoulos S, Geiss O, Tirendi S, Kotzias D (2008) Characterisation of urban inhalation exposures to benzene, formaldehyde and acetaldehyde in the European Union: comparison of measured and modelled exposure data. Environ Sci Pollut Res 15:417–430
Casanova-Schmitz M, David RM, Heck HD (1984a) Oxidation of formaldehyde and acetaldehyde by NAD+−dependent dehydrogenases in rat nasal mucosal homogenates. Biochem Pharmacol 33:1137–1142
Casanova-Schmitz M, Starr TB, Heck HD (1984b) Differentiation between metabolic incorporation and covalent binding in the labeling of macromolecules in the rat nasal mucosa and bone marrow by inhaled [14C]- and [3H]formaldehyde. Toxicol Appl Pharmacol 76:26–44
Chen XX, Su T, He YG, He RQ (2017) Effect of oxidative stress on accommodation and transportation of formaldehyde by lysosome. Prog Biochem Biophys 44(6):486–494
Cheremisinoff, N. P. (2016). 74. Photochemical smog, John Wiley & Sons, Inc
Chi RH, Wu L (1998) The case of catastrophic producing counterfeit spirits in Shanxi province. Zhi Liang Tian Di 2:13–13
Coggon D, Ntani G, Harris EC, Palmer KT (2014) Upper airway cancer, myeloid leukemia, and other cancers in a cohort of British chemical workers exposed to formaldehyde. Am J Epidemiol 179:1301–1311
Crawford S, Lungu CT (2011) Influence of temperature on styrene emission from a vinyl ester resin thermoset composite material. Sci Total Environ 409:3403–3408
Crider KS, Yang TP, Berry RJ, Bailey LB (2012) Folate and DNA methylation: a review of molecular mechanisms and the evidence for Folate's role. Adv Nutr 3:21–38
d'Ettorre G, Criscuolo M, Mazzotta M (2017) Managing formaldehyde indoor pollution in anatomy pathology departments. Work 56(3):397–402
Domingo JL, Rovira J, Roig N, Schuhmacher M, Nadal M (2014) Risk assessment of indoor exposure to formaldehyde: a potential threat. Toxicol Lett 229:S117–S117
Dreyfuss JH (2010) Occupational formaldehyde exposure linked to increased risk of myeloid leukemia and death. Ca A Cancer J Clinicians 60:135–136
Du Z, Mo J, Zhang Y, Xu Q (2013) Volatile organic compound in newly renovated homes and associated Cancer risk in Guangzhou. A Preliminary Study, China
Du Z, Mo J, Zhang Y (2014) Risk assessment of population inhalation exposure to volatile organic compounds and carbonyls in urban China. Environ Int 73:33–45
Dunky M (2004) Adhesives based on formaldehyde condensation resins. Macromol Symp 217:417–430
Duong A, Steinmaus C, Mchale CM, Vaughan CP, Zhang L (2011) Reproductive and developmental toxicity of formaldehyde: a systematic review. Mut Res Rev Mut Res 728:118–138
European Food Safety Authority (2014) Endogenous formaldehyde turnover in humans compared with exogenous contribution from food sources. Efsa J 12:n/a–n/a
Evans AM, Fameli N, Ogunbayo OA, Duan J, Navarro-Dorado J (2016) From contraction to gene expression: nanojunctions of the sarco/endoplasmic reticulum deliver site- and function-specific calcium signals. Sci China Life Sci 59:749–763
Fox CH, Johnson FB, Whiting J, Roller PP (1985) Formaldehyde fixation. J Histochem Cytochem 33:845–853
Fromme H, Heitmann D, Dietrich S, Schierl R, Körner W, Kiranoglu M, Zapf A, Twardella D (2008) Air quality in schools - classroom levels of carbon dioxide (CO2), volatile organic compounds (VOC), aldehydes, endotoxins and cat allergen. Das Gesundheitswesen 70:88–97
Gerken T, Girard CA, Tung YCL, Webby CJ, Saudek V, Hewitson KS, Yeo GSH, Mcdonough MA, Cunliffe S, Mcneill LA (2007) The obesity-associated FTO gene encodes a 2-Oxoglutarate-dependent nucleic acid demethylase. Science 318:1469–1472

Gustafson P, Barregård L, Lindahl R, Sällsten G (2004) Formaldehyde levels in Sweden: personal exposure, indoor, and outdoor concentrations. J Expo Anal Environ Epidemiol 15:252–260
Han KH, Zhang JS, Guo B (2017) Toward effective design and adoption of catalyst-based filter for indoor hazards: formaldehyde abatement under realistic conditions. J Hazard Mater 331:161–170
He Z, Zhang Y, Wei W (2012) Formaldehyde and VOC emissions at different manufacturing stages of wood-based panels. Build Environ 47:197–204
Heck HD, White EL, Casanovaschmitz M (1982) Determination of formaldehyde in biological tissues by gas chromatography/mass spectrometry. Biomed Mass Spectrom 9:347–353
Hill HG, Nuth JA (2003) The catalytic potential of cosmic dust: implications for prebiotic chemistry in the solar nebula and other protoplanetary systems. Astrobiology 3:291–304
Ho SS, Ip HS, Ho KF, Ng LP, Chan CS, Dai WT, Cao JJ (2013) Hazardous airborne carbonyls emissions in industrial workplaces in China. J Air Waste Manag Assoc 63:864–877
Hoekman SK (1992) Speciated measurements and calculated Reactivities of vehicle exhaust emissions from conventional and reformulated gasolines. Environ Sci Technol 26:1206–1216
Hoffman EA, Frey BL, Smith LM, Auble DT (2015) Formaldehyde crosslinking: a tool for the study of chromatin complexes. J Biol Chem 290:26404–26411
Hu X, Wang T, Jin F (2016) Alzheimer's disease and gut microbiota. Sci China Life Sci 59:1006–1023
Huang L, Mo J, Sundell J, Fan Z, Zhang Y (2013) Health risk assessment of inhalation exposure to formaldehyde and benzene in newly remodeled buildings, Beijing. PLoS One 8:e79553
Huang S, Xiong J, Zhang Y (2015) Impact of temperature on the ratio of initial Emittable concentration to Total concentration for formaldehyde in building materials: theoretical correlation and validation. Environ Sci Technol 49:1537–1544
Huang S, Wei W, Weschler LB, Salthammer T, Kan H, Bu Z, Zhang Y (2017) Indoor formaldehyde concentrations in urban China: preliminary study of some important influencing factors. Sci Total Environ 590–591:394–405
Hult EL, Willem H, Price PN, Hotchi T, Russell ML, Singer BC (2015) Formaldehyde and acetaldehyde exposure mitigation in US residences: inhome measurements of ventilation control and source control. Indoor Air 25:523–535
Institute for Health and Consumer Protection, P. a. C. E. U., I-21020 Ispra (VA) Italys (2005) HEXPOC-human exposure characterization of chemical substances; quantification of exposure routes. Luxembourg: Office for Official Publication of the European Communities. ISBN 92-894-8848-4, http://bookshop.europa.eu/en/hexpoc-human-exposure-characterization-of-chemical-substances-pbLBNA21501/downloads/LB-NA-21-501-EN-C/LBNA21501ENC
Jian X, Zhu MX (2016) Regulation of lysosomal ion homeostasis by channels and transporters. Sci China Life Sci 59:777–791
Johannsen FR, Levinskas GJ, Tegeris AS (1986) Effects of formaldehyde in the rat and dog following oral exposure. Toxicol Lett 30:1–6
Jurvelin J, Edwards R, Saarela K, Laine-Ylijoki J, De Bortoli M, Oglesby L, Schlapfer K, Georgoulis L, Tischerova E, Hanninen O, Jantunen M (2001) Evaluation of VOC measurements in the EXPOLIS study. Air pollution exposure distributions within adult Urban Urban Populations in Europe. J Environ Monit 3:159–165
Kalapos MP (1998) An evolutionary role of formaldehyde. Acta Biol Hung 49:167–171
Kalapos MP (1999) A possible evolutionary role of formaldehyde. Exp Mol Med 31:1–4
Kerns WD, Pavkov KL, Donofrio DJ, Gralla EJ, Swenberg JA (1983) Carcinogenicity of formaldehyde in rats and mice after long-term inhalation exposure. Cancer Res 43:4382–4392
Kim KH, Jahan SA, Lee JT (2011) Exposure to formaldehyde and its potential human health hazards. J Environ Sci Health Part C Environ Carcinogen Ecotoxicol Rev 29:277–299
Kimbell J, Overton J, Subramaniam R, Schlosser P, Morgan K, Conolly R, Miller F (2001) Dosimetry modeling of inhaled formaldehyde: binning nasal flux predictions for quantitative risk assessment. Toxicol Sci 64:111–121

Kirchner S, Derbez M, Duboudin C, Elias P, Gregoire A, Lucas JP, Pasquier N, Ramalho O, Weiss N (2008) Indoor air quality in French dwellings. Dissert Theses Gradw 22:1

Kotzias D (2005) Indoor air and human exposure assessment–needs and approaches. Exp Toxicol Pathol 57(Suppl 1):5–7

Kotzias DK, Koistinen K, Kephalopoulos S, Schlitt C, Carrer P, Maroni M (2005) The INDEX project: critical appraisal of the setting and implementation of indoor exposure limits in the EU, European Commission. JRC, Ispra

Kotzias D, Geiss O, Tirendi S, Barreromoreno J, Reina V, Gotti A, Ciminoreale G, Casati B, Marafante E, Sarigiannis D (2009) Exposure to multiple air contaminants in public buildings, schools and kindergartens–the European indoor air monitoring and exposure assessment (AIRMEX) study. Fresenius Environ Bull 18:670

Ladeira C, Gomes MC, Brito M (2010) 74 genotoxicity biomarkers in occupational exposure to formaldehyde in pathology anatomy laboratories. Eur J Cancer Suppl 8:19–19

Ladeira C, Viegas S, Carolino E, Prista J, Gomes MC, Brito M (2011) Genotoxicity biomarkers in occupational exposure to formaldehyde--the case of histopathology laboratories. Mut Res Fundam Mol Mech Mutagen 721:15–20

Lai HK, Bayer-Oglesby L, Colvile R, Götschi T, Jantunen MJ, Künzli N, Kulinskaya E, Schweizer C, Nieuwenhuijsen MJ (2006) Determinants of indoor air concentrations of PM in six European cities (EXPOLIS study). Atmos Environ 40:1299–1313

Lee EG, Magrm R, Kusti M, Kashon ML, Guffey S, Costas MM, Boykin CJ, Harper M (2017) Comparison between active (pumped) and passive (diffusive) sampling methods for formaldehyde in pathology and histology laboratories. J Occup Environ Hyg 14:31–39

Leon CPD, Pletcher D (1995) Removal of formaldehyde from aqueous solutions via oxygen reduction using a reticulated vitreous carbon cathode cell. J Appl Electrochem 25:307–314

Li YN, He RQ (2016) The effects of formaldehyde on life span and stress resistance in Drosophila melanogaster. Prog Biochem Biophys 43:420–428

Li T, Su T, He Y-G, He R-Q (2014) Formaldehyde, an epigenetic factor for learning and memory participating histone (de)methylation. Acta Neuropharmacologica 4:21–29

Lin CC, Yu KP, Zhao P, Lee WM (2009) Evaluation of impact factors on VOC emissions and concentrations from wooden flooring based on chamber tests. Build Environ 44:525–533

Lin D, Guo Y, Yi J, Kuang D, Li X, Deng H, Huang K, Guan L, He Y, Zhang X (2013) Occupational exposure to formaldehyde and genetic damage in the peripheral blood lymphocytes of plywood workers. J Occup Health 55:284

Liu C, Zhang Y, Benning JL, Little JC (2015) The effect of ventilation on indoor exposure to semivolatile organic compounds. Indoor Air 25:285–296

Miller FP, Vandome AF, Mcbrewster J (2010) August Wilhelm von Hofmann I. Eur J Inorg Chem 19:314–316

Madureira J, Paciencia I, Cavaleiro-Rufo J, de Oliveira Fernandes E (2016) Indoor pollutant exposure among children with and without asthma in Porto, Portugal, during the cold season. Environ Sci Pollut Res Int 23:20539–20552

Mangum JG, Darling J, Menten KM, Henkel C (2007) Formaldehyde densitometry of starburst galaxies. Astrophys J 673:832

Marcon A, Fracasso ME, Marchetti P, Doria D, Girardi P, Guarda L, Pesce G, Pironi V, Ricci P, de Marco R (2014) Outdoor formaldehyde and NO2 exposures and markers of genotoxicity in children living near chipboard industries. Environ Health Perspect 122:639–645

Matsuo Y, Nishino Y, Fukutsuka T, Sugie Y (2008) Removal of formaldehyde from gas phase by silylated graphite oxide containing amino groups. Carbon 46:1162–1163

McLaughlin JK (1994) Formaldehyde and cancer: a critical review. Int Arch Occup Environ Health 66:295–301

Meng F, Wu H (2015) Indoor air pollution by Methylsiloxane in household and automobile settings. PLoS One 10:e0135509

Mirabelli MC, Holt SM, Cope JM (2011) Anatomy laboratory instruction and occupational exposure to formaldehyde. Occupat Environ Med 68:375–378

Mui KW, Wong LT, Hui PS, Chan WY (2009) Formaldehyde exposure risk in air-conditioned offices of Hong Kong. Build Serv Eng 30:279–286
Müller RH, Babel W (1991) Formaldehyde as a carbon and energy source in yeast growth. Zentralbl Mikrobiol 146:25–33
Myers GE (1985) The effects of temperature and humidity on formaldehyde emission from UF-bonded boards: a literature critique. For Prod J 35:20–31
National Research Council (US) Committee to Review EPA's Draft IRIS Assessment of Formaldehyde (2011) Review of the Environmental Protection Agency's draft IRIS assessment of formaldehyde. In: Review of the Environmental Protection Agency's draft IRIS assessment of formaldehyde. National Academies Press (US) https://www.ncbi.nlm.nih.gov/books/NBK208227/
Nielsen GD, Larsen ST, Wolkoff P (2017) Re-evaluation of the WHO (2010) formaldehyde indoor air quality guideline for cancer risk assessment. Arch Toxicol 91:35–61
Cchs SDM, Grotz LDO, Factorine LS, Rodrigues MR, Netto ADP (2012) Occupational exposure to formaldehyde in an institute of morphology in Brazil: a comparison of area and personal sampling. Environ Sci Pollut Res 19:2813–2819
Oglesby L (2000) Population sampling in European air pollution exposure study, EXPOLIS: comparisons between the cities and representativeness of the samples. J Expos Anal Environ Epidemiol 10:355
Parker ET, Cleaves JH, Burton AS, Glavin DP, Dworkin JP, Zhou M, Bada JL, Fernändez FM (2014) Conducting miller-urey experiments. J Vis Exp 83:e51039
Parthasarathy S, Maddalena RL, Russell ML, Apte MG (2011) Effect of temperature and humidity on formaldehyde emissions in temporary housing units. J Air Waste Manage Assoc 61:689–695
Peters TL, Kamel F, Lundholm C, Feychting M, Weibull CE, Sandler DP, Wiebert P, Sparen P, Ye W, Fang F (2017) Occupational exposures and the risk of amyotrophic lateral sclerosis. Occup Environ Med 74:87–92
Pinto JP, Gladstone GR, Yung YL (1980) Photochemical production of formaldehyde in Earth's primitive atmosphere. Science 210:183–185
Pizzi A (2003) Phenolic resin adhesives. In: Handbook of adhesive technology. CRC Press, Boca Raton
Restani P, Galli CL (1991) Oral toxicity of formaldehyde and its derivatives. Crit Rev Toxicol 21:315–328
Reuss G, Disteldorf W, Gamer AO, Hilt A (2000) Formaldehyde. Wiley-VCH Verlag GmbH & Co. KGaA
Roberts AL, Johnson NJ, Cudkowicz ME, Eum KD, Weisskopf MG (2016) Job-related formaldehyde exposure and ALS mortality in the USA. J Neurol Neurosurg Psychiatry 87:786–788
Salonen HJ, Pasanen AL, Lappalainen SK, Riuttala HM, Tuomi TM, Pasanen PO, Back BC, Reijula KE (2009) Airborne concentrations of volatile organic compounds, formaldehyde and ammonia in Finnish office buildings with suspected indoor air problems. J Occup Environ Hyg 6:200–209
Salthammer T (2013) Formaldehyde in the ambient atmosphere: from an indoor pollutant to an outdoor pollutant? ChemInform 44:3320
Salthammer T, Fuhrmann F, Kaufhold S, Meyer B, Schwarz A (1995) Effects of climatic parameters on formaldehyde concentrations in indoor air. Indoor Air 5:120–128
Salthammer T, Mentese S, Marutzky R (2010) Formaldehyde in the indoor environment. Chem Rev 110:2536–2572
Salthammer T (2015) The formaldehyde dilemma. Int J Hyg Environ Health 218(4):433–436
Saowakon N, Ngernsoungnern P, Watcharavitoon P, Ngernsoungnern A, Kosanlavit R (2015) Formaldehyde exposure in gross anatomy laboratory of Suranaree University of Technology: a comparison of area and personal sampling. Environ Sci Pollut Res Int 22:19002–19012
Sarigiannis DA, Karakitsios SP, Gotti A, Liakos IL, Katsoyiannis A (2011) Exposure to major volatile organic compounds and carbonyls in European indoor environments and associated health risk. Environ Int 37:743–765

Seco R, Filella PI (2008) Formaldehyde emission and uptake by Mediterranean trees Quercus Ilex and Pinus Halepensis. Atmos Environ 42:7907–7914
Sekine Y (2005) Sick house/building syndrome in JAPAN – current status and practical research on indoor air quality. Korean J Environ Health 31(3):207–214
Sekine Y, Katori R, Tsuda Y, Kitahara T (2016) Colorimetric monitoring of formaldehyde in indoor environment using built-in camera on mobile phone. Environ Technol 37:1647–1655
Shi Y, Lan F, Matson C, Mulligan P, Whetstine J, Cole P, Casero R (2004) Histone demethylation mediated by the nuclear amine oxidase homolog LSD1. Cell 119:941–953
Sie SS, Timo K, Pentti K, Pirjo H, Eero P (2012) Occupational exposure to wood dust and formaldehyde and risk of nasal, nasopharyngeal, and lung cancer among Finnish men. Cancer Manage Res 4:223–232
Stranger M, Potgieter-Vermaak SS, Van GR (2007) Comparative overview of indoor air quality in Antwerp, Belgium. Environ Int 33(6):789–797
Stribling R, Miller SL (1987) Energy yields for hydrogen cyanide and formaldehyde syntheses: the HCN and amino acid concentrations in the primitive ocean. Orig Life Evol Biosph 17:261–273
Su T, Dan S, Li T, Wang XH, He R-Q (2015) Nucleic acid methylation/demethylation, endogenous formaldehyde and age-related cognitive impairment. Prog Biochem Biophys 42:211–219
Suarez-Bertoa R, Clairotte M, Arlitt B, Nakatani S, Hill L, Winkler K, Kaarsberg C, Knauf T, Zijlmans R, Boertien H (2017) Intercomparison of ethanol, formaldehyde and acetaldehyde measurements from a flex-fuel vehicle exhaust during the WLTC. Fuel 203:330–340
Tang X, Bai Y, Duong A, Smith MT, Li L, Zhang L (2009) Formaldehyde in China: production, consumption, exposure levels, and health effects. Environ Int 36:1210–1224
Til HP, Woutersen RA, Feron VJ, Hollanders VHM, Falke HE, Clary JJ (1989) Two-year drinking-water study of formaldehyde in rats. Food Chem Toxicol 27:77–87
Tohmura S (2001) Influence of the melamine content in melamine-urea-formaldehyde resins on formaldehyde emission and cured resin structure. J Wood Sci 47:451–457
Toth J, Biggin MD (2000) The specificity of protein–DNA crosslinking by formaldehyde: in vitro and in drosophila embryos. Nucleic Acids Res 28:e4
Tulpule K, Dringen R (2013) Formaldehyde in brain: an overlooked player in neurodegeneration? J Neurochem 127:7–21
Uchiyama I (2008) The regulations for indoor air pollution in Japan: a public health perspective. J Risk Res 11:301–314
Vaughan TL, Strader C, Davis S, Daling JR (1986a) Formaldehyde and cancers of the pharynx, sinus and nasal cavity: I. Occupational exposures. Int J Cancer 38:677–683
Vaughan TL, Strader C, Davis S, Daling JR (1986b) Formaldehyde and cancers of the pharynx, sinus and nasal cavity: II. Residential exposures. Int J Cancer 38:685–688
Victorin K (1998) Risk assessment of carcinogenic air pollutants. IMM-report 1998. Institutet for miljo medicin, Stockholm. Available: http://www.imm.ki.se/publ/PDF/Rapp1-98.pdf
Viegas S, Nunes C, Maltavacas J, Gomes M, Brito M, Mendonça P, Prista J (2010) Genotoxic effects in occupational exposure to formaldehyde: a study in anatomy and pathology laboratories and formaldehyde-resins production. J Occupat Med Toxicol 196:1–8
Viegas S, Ladeira C, Gomes M, Nunes C, Brito M, Prista J (2013) Exposure and genotoxicity assessment methodologies - the case of formaldehyde occupational exposure. Curr Anal Chem 9:476–484
Wahed P, Razzaq MA, Dharmapuri S, Corrales M (2016) Determination of formaldehyde in food and feed by an in-house validated HPLC method. Food Chem 202:476–483
Weaver HF (1970) Some characteristics of interstellar gas in the galaxy. Springer, Dordrecht
World Health Organization (2010) WHO guidelines for indoor air quality: selected pollutants. In: WHO guidelines for indoor air quality: selected pollutants, vol. World Health Organization, Geneva. Available at http://www.ncbi.nlm.nih.gov/books/NBK138711/
World Health Organization (2014) WHO indoor air quality guidelines: household fuel combustion. Indoor Air Pollution

Xing C, Riedl B, Cloutier A, Shaler SM (2005) Characterization of urea-formaldehyde resin penetration into medium density fiberboard fibers. Wood Sci Technol 39:374–384
Xiong J, Zhang Y (2010) Impact of temperature on the initial emittable concentration of formaldehyde in building materials: experimental observation. Indoor Air 20:523–529
Xiong J, Wei W, Huang S, Zhang Y (2013) Association between the emission rate and temperature for chemical pollutants in building materials: general correlation and understanding. Environ Sci Technol 47:8540–8547
Yan F, Frey PA (2011) RNA methylation by radical SAM enzymes RlmN and Cfr proceeds via methylene transfer and hydride shift. Proc Natl Acad Sci U S A 108:3930–3934
Yao Y, Liang W, Zhu L, Duan Y, Jin Y, He L (2014) Relationship between the concentration of formaldehyde in the air and asthma in children: a meta-analysis. Int J Clin Exp Med 8:8358–8362
Ye W, Zhang X, Gao J, Cao G., Zhou X, Su X (2017) Indoor air pollutants, ventilation rate determinants and potential control strategies in Chinese dwellings: a literature review. Sci Total Environ 586:696–729
Zaitseva N, Zemlyanova M, Dolgikh O (2013) Formaldehyde exposure and allergic sensitization in children. In: Environment and health – bridging south, north, east and west. MDPI, Basel
Zhai L, Zhao J, Xu B, Deng Y, Xu Z (2013) Influence of indoor formaldehyde pollution on respiratory system health in the urban area of Shenyang, China. Afr Health Sci 13:137–143
Zhang L, Freeman LEB, Nakamura J, Hecht SS, Vandenberg JJ, Smith MT, Sonawane BR (2010a) Formaldehyde and leukemia: epidemiology, potential mechanisms and implications for risk assessment. Environ Mol Mutagen 51:181–191
Zhang L, Tang X, Rothman N, Vermeulen R, Ji Z, Shen M, Qiu C, Guo W, Liu S, Reiss B, Freeman LB, Ge Y, Hubbard AE, Hua M, Blair A, Galvan N, Ruan X, Alter BP, Xin KX, Li S, Moore LE, Kim S, Xie Y, Hayes RB, Azuma M, Hauptmann M, Xiong J, Stewart P, Li L, Rappaport SM, Huang H, Fraumeni JF Jr, Smith MT, Lan Q (2010b) Occupational exposure to formaldehyde, hematotoxicity, and leukemia-specific chromosome changes in cultured myeloid progenitor cells. Cancer Epidemiol Biomark Prev 19:80–88
Zuckerman et al (1970) Formaldehyde appears to be a useful probe for astrochemists due to its low reactivity in the gas. Wikipedia.unicefuganda.org

Chapter 2
Metabolism of Formaldehyde In Vivo

Rong Xiao and Rongqiao He

Abstract Age-related cognitive impairment is regarded as a chronic progression of dysfunction and loss of neurons, accompanied with formation of senile plaque (β-amyloid deposition) and neurofibrillary tangles (paired helical filaments of hyperphosphorylated Tau). Approximately 40% of the cases are suffered from endogenous formaldehyde accumulation. Dysmetabolism of formaldehyde shows a positive relationship with Alzheimer's disease (AD) for about 40% of AD patients who have high levels of urine formaldehyde (see Chap. 8 Cognitive Impairment). To have a convenient and clear explanation, we would like to divide the cellular formaldehyde pathways into two parts: "formaldehyde coming (synthesis or production)" and "formaldehyde going (process or degradation)". The metabolic homeostasis of formaldehyde is maintained between the "coming" and "going" pathways under the physiological state. In fact, formaldehyde comes from multiple pathways related with many compounds such as sugars, lipids, proteins, and nucleic acids, as well as some small molecules and goes away in a relatively narrow lane; besides, it also goes to participate in the other pathways, for instance, as donor to modify histones and DNA. Either "coming" or "going" metabolism dysfunctions would lead to increase or decrease of endogenous formaldehyde. As described in Chap. 8, imbalance (lack or excess) of formaldehyde metabolism induces decline of learning and memory for drosophila and rodents. It has been known that intestinal microbiota is involved in the metabolism of endogenous formaldehyde, and we discuss this topic in details in Chap. 6. In this chapter, we discuss the characteristics of the endogenous formaldehyde metabolism and summarize the pathways, which are responsible for the generation and clearance of endogenous formaldehyde.

Keywords Formaldehyde • Generation • Metabolism • Lysosome • Formaldehyde dehydrogenase

---

R. Xiao
School of Life Sciences, Liaoning Normal University, Dalian 116081, Liaoning, China

R. He (✉)
State Key Laboratory of Brain and Cognitive Science, Institute of Biophysics, Chinese Academy of Sciences, Beijing 100101, China

University of Chinese Academy of Sciences, Beijing 100049, China
e-mail: rongqiaohe@163.com

R. He, *Formaldehyde and Cognition*, DOI 10.1007/978-94-024-1177-5_2

1 Introduction

In the 1960s, Japanese scientists found that some of the trimethylamine oxide (TMAO) in the muscle of gadoid is converted into formaldehyde and dimethylamine when the fish fillets are stored at temperatures above −25 °C (Amano and Yamada 1964a, b; Tokunaga 1964). More studies were carried out in the 1970s on the storage of fish meat under low temperatures concomitant with the production of endogenous formaldehyde which was resulted from TMAO (Castell et al. 1973; Tokunaga 1974). Sotelo's work demonstrates the production of endogenous formaldehyde from protein denaturation during frozen storage (Sotelo et al. 1995). Their work revealed a fact that formaldehyde will be produced and increased in fish fillets while the cells are dead and tissues are degraded.

Endogenous formaldehyde comes from multiple metabolic pathways such as denaturation of protein (Sotelo et al. 1995), oxidation of lipid (Li et al. 2008), decomposition of reducing sugar (our unpublished data), and oxidative stress (Yu 1991). Furthermore, formaldehyde is also produced from the metabolism of methanol, methylamine, glycerol, sarcosine (N-methylglycine), dimethylaminoethanol, dimethylglycine, and other substrates which contain the CH_3N or CH_3O group. Demethylation of DNA, RNA, and histones is also considered as a main way to generate formaldehyde in vivo (see Chap. 3). Recently, intestinal flora has been proved to be involved in age-related cognitive impairment (Hu et al. 2016) and formaldehyde metabolism (Liu et al. 2017), involving synthesis and degradation of formaldehyde (see Chap. 6).

In fact, the whole living things, including humans, produce and process formaldehyde at each moment all the time (The Environmental Protection Agency, USA "Toxicological Review of Formaldehyde Inhalation Assessment". June/18/2010). Metabolism of formaldehyde in vivo is extremely complicated and the pathways process in diversity (Fig. 2.1). In this chapter, to review this complex topic, we would like to discuss the metabolic pathways of formaldehyde by dividing the metabolic pathway into two aspects: "formaldehyde coming" and "formaldehyde going" such as generation and degradation. Excess or lack of either of them will disturb the homeostasis of formaldehyde metabolism, which could lead to disorders of the central nervous system and other systems.

2 Endogenous Formaldehyde Generation

Formaldehyde comes from various sources, such as environment, food, compound degradation, and enzymatic catalysis. A series of enzymes are responsible for formaldehyde production (Table 2.1), including semicarbazide-sensitive amine oxidase (SSAO) (Boor et al. 1992; Yu et al. 1997), alcohol dehydrogenase 1 (ADH1) (MacAllister et al. 2011; Pocker and Li 1991), cytochrome P-450 enzymes (Clejan and Cederbaum 1991, 1992b; Megaraj et al. 2014), serine

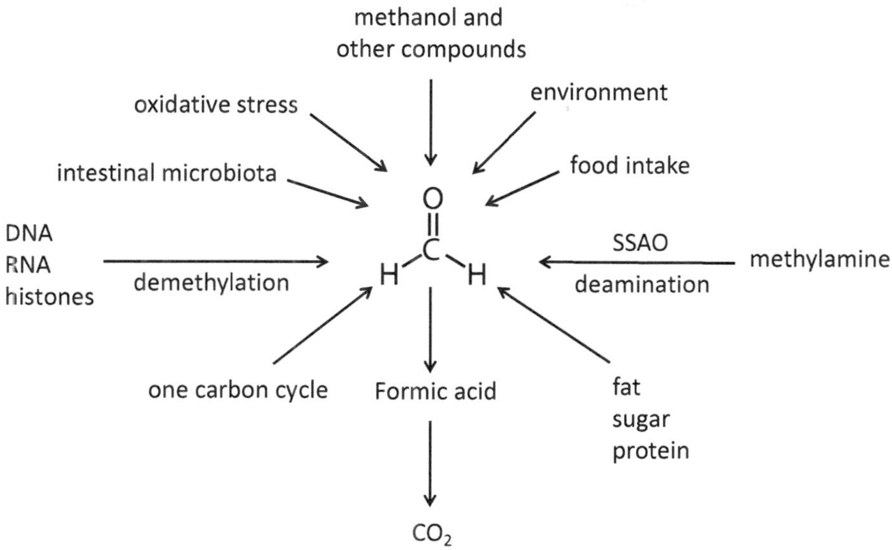

Fig. 2.1 Pathways of formaldehyde generation and degradation. Endogenous formaldehyde comes from different pathways which are diverse and complex as indicated (Source: Adapted from Su et al. 2016)

hydroxymethyltransferase (SHMT) (Renwick et al. 1998), trimethylamine oxide demethylase (TMAO-ase) (Gill and Paulson 1982), sarcosine dehydrogenase (Frisell and Mackenzie 1962; Porter et al. 1985), catalases (MacAllister et al. 2011; Slegers et al. 1984), myeloperoxidase (MPO) (Hazen et al. 1998), dimethylglycine dehydrogenase (Frisell and Mackenzie 1962; Porter et al. 1985), and demethylases (lysine-specific demethylase 1, JmjC domain-containing proteins, and AlkB family members)(Aas et al. 2003; Burg et al. 2016; Chang et al. 2007; Jia et al. 2008; Shi et al. 2004; Trewick et al. 2002; Tsukada et al. 2006).

2.1 *From Environment*

Formaldehyde inters the respiratory system when individuals breathe polluted air which contains high levels of formaldehyde from various sources such as some vehicle exhaust, insecticides, insulation, embalming fluids, plywood adhesives, and abrasive materials ("Toxicological Review of FA Inhalation Assessment" reposted by Environmental Protection Agency USA, June/18/2010). Manufacturing plants and power plants are major sources of anthropogenic emissions of formaldehyde which could generate or consume formaldehyde or materials that contain formaldehyde including petroleum refineries, wood burning, coking operations, incinerating, as well as tobacco smoke (see Chap. 1 for this topic).

Table 2.1 Enzymes in production and degradation of formaldehyde

Formaldehyde producing enzyme	Superfamily	Substrate	References
Catalases		Methanol	MacAllister et al. (2011) and Slegers et al. (1984)
ADH1	Dehydrogenases	Methanol	MacAllister et al. (2011) and Pocker and Li (1991)
CYP4502E1	Cytochrome P-450 enzymes	Methanol, glycerol and drugs	Clejan and Cederbaum (1991, 1992b) and Megaraj et al. (2014)
SHMT		Serine	Renwick et al. (1998)
TMAO-ase	Lyases	Trimethylamine oxide	Gill and Paulson (1982)
Dimethylglycine dehydrogenase	Oxidoreductases	Dimethylglycine	Frisell and Mackenzie (1962) and Porter et al. (1985)
Sarcosine dehydrogenase	Oxidoreductases	Sarcosine (N-methylglycine)	Frisell and Mackenzie (1962) and Porter et al. (1985)
SSAO	Amine oxidases	Methylamine	Boor et al. (1992) and Yu et al. (1997)
MPO	Peroxidases	Glycine	Hazen et al. (1998)
LSD-1 (KIAA0601 or KDM1A)	KDM	Lysines in histones	Burg et al. (2016) and Shi et al. (2004)
Jmjc domain-containing proteins	KDM	Lysines or arginines in histones	Chang et al. (2007) and Tsukada et al. (2006)
AlkB (hABH or FTO)	AlkB	Methyl group in adenine, cytosine, uracil and thymine of DNA and RNA	Aas et al. (2003), Jia et al. (2008) andTrewick et al. (2002)
Formaldehyde degrading enzyme	**Superfamily**	**Product**	**References**
ALDH2	Aldehyde dehydrogenase	Formic acid	Dicker and Cederbaum (1986), Dorokhov et al. (2015) and MacAllister et al. (2011)
ADH3	Dehydrogenase	Formic acid	Koivusalo et al. (1989) and MacAllister et al. (2011)

The abbreviations used are listed as followed. *ADH1* alcohol dehydrogenase 1, *ADH3* alcohol dehydrogenase 3, *ALDH2* aldehyde dehydrogenase, *FTO* fat mass and obesity-associated protein, *KDMs* lysine demethylases, *LSD-1* lysine-specific demethylase 1, *MPO* myeloperoxidase, *SHMT* serine hydroxymethyl transferase, *SSAO* semicarbazide-sensitive amine oxidase, *TMAO-ase* trimethylamine oxide demethylase

2.2 From Food

Foods including fruits, vegetables, and alcoholic beverages could generate methanol in the healthy human body (Trézl et al. 1997; Burbacher et al. 1999). Usually, methanol is chemically locked to pectin from fruits and vegetables and can pass digestion without absorption by the gut. In 2014, Shindyapina and colleagues found that when the human subjects consumed citrus pectin, ethanol, and red wine, the concentrations of methanol and formaldehyde in these volunteers' blood plasma would be increased (Shindyapina et al. 2014). During heat and packaged processes, methanol would also be released from fruits, vegetables, or other juices. Furthermore, recent report also showed that fermentation of gut microbiota is also an important pathway to metabolize methanol into formaldehyde (Kostov et al. 1983). Although human diet is usually composed of a very tiny component of methanol, human began to take aspartame over the last 40 years which would also be converted into methanol and might associate with the increase in AD incidence at the same time (Monte 2010). This might be attributed to gut esterases and peptidases which could metabolize aspartame into three common chemicals: aspartic acid, phenylalanine, and methanol (Choudhary and Sheela Devi 2015).

As early as 1953, Mackenzie and colleagues found that formaldehyde was isolated when methanol was incubated with an unfractionated liver homogenate (Mackenzie et al. 1953). With the combination of a great number of studies, the detailed mechanism which accounts for the production of endogenous formaldehyde from methanol (CH_3OH) has been extensively studied in vivo. Although the physical and chemical properties of methanol are similar to that of alcohol, methanol is not suitable for drinking or inhalation due to the toxic products generated during its metabolism (Jacobsen and McMartin 1986; Pohanka 2016). Among the toxic products, the newly formed formaldehyde could lead to the cytotoxicities or protein aggregations which would finally cause cancer or AD, and so forth. According to the summarization of previous studies, there are mainly three pathways which account for the generation of formaldehyde from methanol in vivo (Dorokhov et al. 2015; MacAllister et al. 2011). Firstly, methanol in the presence of H_2O_2 could be oxidized to formaldehyde and H_2O by catalase (MacAllister et al. 2011; Slegers et al. 1984). During this reaction cycle, both the peroxidatic activities of catalase and hydroxyl radicals generated from H_2O_2 are crucial for the oxidation of methanol (Cederbaum and Qureshi 1982). Secondly, methanol could be converted to hydroxyl radicals (•OH) by Fe^{2+} and H_2O_2. At the same time, the hydroxyl radicals (•OH) are able to react with methanol, releasing formaldehyde and O_2 (MacAllister et al. 2011). Thirdly, ADH1 can oxidize the methanol to formaldehyde with the help of NAD^+ (MacAllister et al. 2011; Pocker and Li 1991). Besides, cytochrome P-450 monooxygenase might also participate in the generation of formaldehyde (Guo et al. 2010; Wang et al. 2012). In vivo, endogenous methanol could be formed through carboxyl methyl esters' hydrolyzation, which are formed by the methylation of the -COOH group of amino acids in proteins with endogenous S-adenosylmethionine (SAM) as the methyl donor (Diliberto and Axelrod 1976;

Lee et al. 2008). In 2008, Lee et al. reported that the level of methanol, formaldehyde, or formic acid was increased when the striatal homogenates of male Sprague-Dawley (SD) rats were incubated with SAM (Lee et al. 2008). This suggested that the endogenously formed methanol is the source of formaldehyde in vivo.

2.3 Formaldehyde in One-Carbon Circle

Formaldehyde plays a role in one-carbon circle. One-carbon units which are in the form of methyl ($-CH_3$), methylene ($-CH_2$-), methenyl ($-CH=$), formyl ($-CHO$), and formimino ($-CH = NH$) groups can be transferred from one compound to another (Sun et al. 2011). This circle provides cellular components including proteins, lipids, and nucleic acids for cell growth and proliferation (Locasale 2013). In vivo, the donors for one-carbon units are glycine, histidine, serine, tryptophan, and methionine. In the presence of tetrahydrofolate (THF), SHMT catalyzes the serine to produce formaldehyde and glycine. Subsequently, formaldehyde reacts with THF to form 5,10-methylene-THF (Renwick et al. 1998). Furthermore, plants might remove the formaldehyde through C1 metabolism. As described by Zhang and colleagues, petunia plants use ^{13}C-labeled formaldehyde (2 mM) to synthesize [5-^{13}C] methionine via C1 metabolism probably in the cytoplasm (Zhang et al. 2014). When the concentration of ^{13}C formaldehyde increased to 4–6 mM, $H^{13}COOH$ and [2-^{13}C] glycine were also generated in *Petunia* plant through C1 metabolism probably in chloroplasts and peroxisomes (Zhang et al. 2014). During this process, a small amount of [U-^{13}C] glucose was also generated via the Calvin cycle.

2.4 Formaldehyde from Biochemical Compounds

2.4.1 From Protein Denaturation

TMAO whose chemical formula is $(CH_3)_3NO$ is composed of a small hydrophilic group (N^+O^-) and three methyl groups (Mondal et al. 2013). As early as 1956, Vaisey EB firstly reported that TMAO could be reduced to trimethylamine, dimethylamine, and formaldehyde in a nonenzymatic process (Vaisey 1956). Subsequently, Parkin and colleagues found that the microsomal fraction from skeletal muscle of red hake (*Urophycis chuss*) was able to reduce TMAO to dimethylamine and formaldehyde in the presence of $Fe^{2+/3+}$ and either cysteine or ascorbat (Parkin and Hultin 1982). At the same time, an enzyme named as TMAO-ase which accounts for the reduction of TMAO to formaldehyde and dimethylamine was purified from cod kidney (Gill and Paulson 1982). During this process, the generated formaldehyde was found to trigger protein denaturation in the fish muscle which would finally result in the bad flavor, odor, and color of fish (Sotelo et al. 1995). Although TMAO is also known as a chemical chaperone which could help protein folding in denaturing conditions, the conversion of TMAO to formaldehyde would make TMAO lose the protective property (Tatzelt et al. 1996). Furthermore, the levels of TMAO, dimethylamine, trimethylamine, and formaldehyde were studied in 266 different types of fishes (Chung and Chan 2009). Regrettably, the reduction of TMAO to formaldehyde has not been reported in human being, and its relation to cognitive impairment has not been reported yet.

2.4.2 From Lipid Oxidation

During the oxidative degradation of polyunsaturated lipids, malondialdehyde (MDA) is released (Stancliffe et al. 2011). Thus, MDA is usually considered as a biomarker for oxidative stress (Del Rio et al. 2005). As one of the major products of lipid peroxidation, accumulation of MDA in vivo can be observed in many pathophysiological processes (Antus et al. 2014). In fact, the patients with diabetes (Su and He 2014), cardiovascular diseases (Bagatini et al. 2011), and neurodegenerative disorders have higher concentration of MDA in their blood (Sanyal et al. 2009). MDA which acts as a reactive electrophile species is cytotoxic probably due to the two aldehyde groups. Therefore, to trace the changes of MDA and its products that resulted from proteins modified with MDA (both in vivo and in vitro studies) is conducive in understanding their effects on oxidative stress and the related diseases.

In 2008, Li and his colleagues incubated N-Acyl-4-sphingoine-1-phosphocholine (1 mg/ml myelin) with 50 mM phosphate buffer (pH 7.2) containing 1 mM H_2O_2 at 37 °C for 24 h. They measured formaldehyde with the high-performance liquid chromatography (HPLC) coupled with dinitrophenylhydrazine (DNPH) absorption measurement as described previously and found the production of formaldehyde in the oxidation of myelin (Li et al. 2008) (Fig. 2.2). The reaction between 10 mM

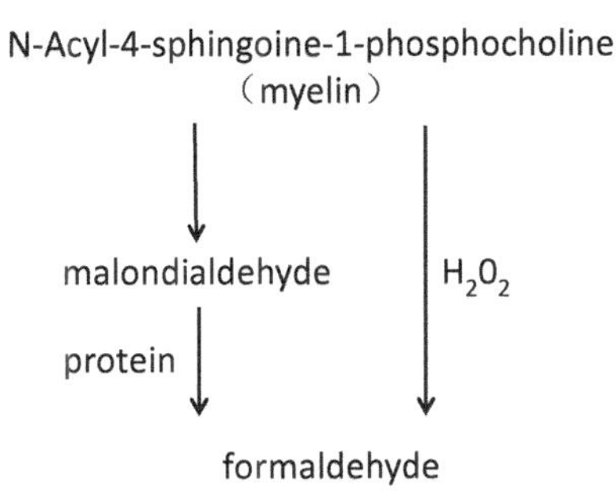

Fig. 2.2 Formation of formaldehyde from malondialdehyde and myelin oxidation. Formaldehyde is one of the products from decomposed malondialdehyde. Incubation of N-Acyl-4-sphingoine-1-phosphocholine (1 mg/ml myelin) with 50 mM phosphate buffer (pH 7.2) containing 1 mM H_2O_2 at 37 °C for 24 h produces formaldehyde and malondialdehyde in the oxidation of myelin (Li et al. 2008)

MDA and 0.05 mM calf serum albumin also produced formaldehyde. These data demonstrated that oxidation of cellular lipid can produce formaldehyde.

2.4.3 From Reducing Sugar Decomposition

As described by Pinto and collaborators, formaldehyde plays as a basic block for many compounds in organisms, for instance, the reducing sugar D-ribose (Pinto et al. 1980; Shapiro 1988). In the open-chain form, D-ribose has an active aldehyde group and constitutes the structural backbone of purines and pyrimidines (Shapiro 1988; Sutherland 2010). Usually, riboflavin and nucleic acids are the two major sources of D-ribose from the human diet. In healthy individuals, the concentration of D-ribose in the blood and the cerebrospinal fluid is about 0.01–0.1 mM (Cai et al. 2005; Seuffer 1977). In vivo, hexose monophosphate shunt and nucleotide metabolism could also generate endogenous D-ribose (Lager et al. 2003).

In 2013, Su and his collaborators found the abnormally high levels of D-ribose in the urine of type 2 diabetic patients compared with the normal participants (Su et al. 2013). Three years later, Chen and colleagues demonstrated the high level of D-ribose in the blood of diabetic patients (Chen et al. 2016). These cases may support that diabetic patients suffer from not only dysmetabolism of glucose but also of ribose. Wang and her colleagues observed that formaldehyde is produced from the decomposition of ribose (our unpublished data). Ribose is first decomposed into glyceraldehyde, followed by the production of formaldehyde under alkaline condition. Yield of formaldehyde was also observed from D-ribose in the presence of lysine. Experiments on D-ribose incubated with glycine further demonstrated similar results. Intraperitoneal injection of ribose into C57BL/6 J mice increased the concentration of formaldehyde in the blood compared with the control. However, whether D-ribose can produce formaldehyde in vivo needs to be further investigated.

2.4.4 Formaldehyde from Dimethylaminoethanol, Dimethylglycine, and Sarcosine (N-Methylglycine)

In the 1950s, formaldehyde was observed to be formed during the oxidation reaction of dimethylaminoethanol, dimethylglycine, sarcosine (N-methylglycine), or methanol catalyzed by liver extractions without fraction (Johnston and Mackenzie 1956; Mackenzie et al. 1953). Later, mitochondria were found to be responsible for the formation of formaldehyde through the conversion of dimethylglycine or sarcosine (Mackenzie et al. 1953). After separation through DEAE-cellulose and $Ca_3(Po_4)_2$-cellulose chromatography, Frisell and Mackenzie obtained the purified dimethylglycine dehydrogenase and sarcosine dehydrogenase from mitochondria, which could act on the CH-NH group of dimethylglycine and sarcosine to generate formaldehyde, respectively (Frisell and Mackenzie 1962). Based on the enzymological characterizations, both the dehydrogenases belong to oxidoreductases family, which require a flavin adenine dinucleotide (FAD) as a cofactor (Frisell and Mackenzie 1962). In the dimethylglycine dehydrogenase-catalyzed reaction, N, N-dimethylglycine, acceptor, and H_2O were converted to sarcosine, formaldehyde, and reduced acceptor (Porter et al. 1985). As for the sarcosine dehydrogenase-catalyzed reaction, sarcosine, H_2O, and O_2 were converted to glycine, formaldehyde, and H_2O_2 with the help of FAD (Porter et al. 1985).

2.4.5 From Substrates Containing the CH_3N or CH_3O Group Mediated by Cytochrome P-450 System

A great number of studies have shown that cytochrome P-450 in microsome or mitochondria is closely related to the formaldehyde liberation by glycerol oxidation. In 1988, Winters and colleagues found that glycerol could be oxidized to a Nash-reactive material by rat liver microsomes in a NADPH-dependent manner (Winters et al. 1988). This Nash-reactive material was proved to be formaldehyde by the positive reaction of glutathione-dependent formaldehyde dehydrogenase (Winters et al. 1988). Due to inhibition of carbon monoxide on glycerol oxidation reaction, the authors speculated that cytochrome P-450 might participate in this oxidative process (Winters et al. 1988). Two years later, the same authors used cytochrome P-450 isozyme inducers such as pyrazole, ethanol, or acetone to treat rats and then isolated their microsomes (Winters and Cederbaum 1990). When these microsomes were incubated with glycerol, the level of formaldehyde was increased, suggesting that glycerol might be an effective substrate for cytochrome P-450 (Winters and Cederbaum 1990). In the 1990s, the mechanism of glycerol oxidation mediated by cytochrome P-450 has been gradually figured out. The oxidation of glycerol catalyzed by microsomes or purified cytochrome P-450 (P4502E1) also required the involvement of H_2O_2 and nonheme iron except the NAPDH (Clejan and Cederbaum 1991; Wang et al. 2012). During this reaction cycle, the authors hold that cytochrome P-450 catalyzed iron with H_2O_2 to generate an oxidant which could further metabolize glycerol to formaldehyde (Clejan and Cederbaum 1992a). And glycerol could be cleaved by the oxidant to form formaldehyde with breakage of a carbon-hydrogen bond (Rashba-Step et al. 1994). Besides glycerol, vicinal diols such as 1,2-propanediol and 1,2-butanediol or the charged glycol such as alpha-glycerophosphate could also liberate formaldehyde with the interaction of an oxidant derived from H_2O_2 plus nonheme iron (Clejan and Cederbaum 1992b). With the summarization of the studies about cytochrome P-450-catalyzed formaldehyde generation, cytochrome P-450 was proved to act as a terminal oxidase in the electron transfer chains which could participate in the demethylation process of a variety of substrates containing the CH_3N or CH_3O group (Kobayashi et al. 2000; Vaz et al. 1996). Recently, more than 21 cytochrome P-450 isoenzymes have been reported. Importantly, cytochrome P-450 could participate in the metabolism of various drugs and procarcinogens in vivo. During this process, formaldehyde could also be released. Very recently, Megaraj and colleagues found that cytochrome P-450 enzymes, especially P-450 2E1 in the microsome of liver and intestine, were responsible for catalyzing azoxymethane which is a colon carcinogen to produce formaldehyde in vivo (Megaraj et al. 2014). In this reaction, azoxymethane is hydroxylated to form methylazoxymethanol which subsequently converts into formaldehyde and methyldiazonium ion (Megaraj et al. 2014).

2.5 *Oxidative Stress Produces Formaldehyde*

2.5.1 SSAO Catalyzes Methylamine to Generate Formaldehyde

SSAO is also known as an adhesion molecule (VAP-1) and requires topaquinone as the cofactor (Jakobsson et al. 2005). Besides in the blood plasma, SSAO is found to distribute in many cells such as vascular smooth muscle cells, adipocytes, chondrocytes, and odontoblasts (Boomsma et al. 1997; Moldes et al. 1999; Filip et al. 2016; O'Sullivan et al. 2002). It is also localized in the cytoplasm of endothelial cells and at the outer membrane surface (Jaakkola et al. 1999). During the past decades, a great number of studies have shown that the expression level of SSAO would be increased during the pathological progresses such as vascular disorders, chronic heart failure, diabetic complications, and AD (Boomsma et al. 1997; Obata 2006; Ullah et al. 2013; Yu 2001; Yu et al. 2003). In 1992, SSAO was firstly reported to catalyze the deamination reaction of endogenous methylamine to produce the toxic formaldehyde, hydrogen peroxide, and ammonia (Boor et al. 1992). Therefore, the released formaldehyde may be the crucial factor which leads to the lesions in endothelial cells, and plaques formed from aberrant protein aggregation (Obata 2006; Yu 2001; Yu et al. 2003; Yu and Zuo 1993).

In 1997, Yu et al. found that endogenous adrenaline is also an important source for the formation of formaldehyde in vivo (Yu et al. 1997). During the stress conditions, adrenaline could be released into the circulatory system and deaminated by type A monoamine oxidase (MAO-A) to form 2-(3,4-dihydroxyphenyl)-2-hydroxyacetaldehyde, hydrogen peroxide, and methylamine (Yu et al. 1997). Subsequently, the generated methylamine could be further converted to toxic formaldehyde, hydrogen peroxide, and ammonia through deamination induced by SSAO which is distributed in the cardiovascular smooth muscles and the circulatory system (Yu et al. 1997). The authors hold that the formed formaldehyde may be responsible for the initiation of endothelial injury (Yu et al. 1997). In 2004, Gubisne-Haberle et al. proved that the formaldehyde released from methylamine deamination which is catalyzed by SSAO in human brain meninges, could cause cross-linkage between formaldehyde and proteins effectively (Gubisne-Haberle et al. 2004). In 2014, Qiang and her colleagues found that both formaldehyde and SSAO levels were significantly elevated in the brain of senescence-accelerated mouse-prone 8 strain (SAMP8) mice, which are classic AD models, at the age of 3 months compared with the age-matched SAMR1 mice (Qiang et al. 2014). The authors speculated that the elevated levels of SSAO might account for the high concentration of formaldehyde in the brain of SAMP8 mice, which probably leads to the Tau hyperphosphorylation and protein aggregation with cognitive impairment as an ultimate result (Qiang et al. 2014).

2.5.2 Formaldehyde from Glycine Oxidation by MPO

MPO is a heme peroxidase which is found to express in neutrophils, monocytes, and macrophages (Bos et al. 1978; Daugherty et al. 1994; Klebanoff and Clark 1978). During pathogen invading process, activated neutrophils would release MPO to produce hypohalous acids from reacting with H_2O_2 and halide at inflammatory sites (Kato 2016). The generated hypohalous acids would account for much of the antibactericidal activity of neutrophils (Davies 2011). As early as 1968, Zgliczynski and colleagues found that MPO could not only catalyze the classic substrates for peroxidase but also α-amino acids, which provided clues for studies on aldehyde generation through MPO-H_2O_2-chloride system (Zgliczynski et al. 1968). In the 1990s, lots of reports found that human neutrophils could employ this MPO-H_2O_2-Cl^- system to oxidize α-amino acids to reactive aldehydes (Anderson et al. 1997). And their structures were analyzed by a variety of mass spectrometric methods (Hazen et al. 1998). Among the amino acid-derived aldehydes, formaldehyde could be released by oxidation of aliphatic glycine in the presence of MPO, H_2O_2, and Cl^- in vitro experiments (Hazen et al. 1998). And cytotoxicity assays showed that the aldehyde derivatives including formaldehyde, 2-hydroxyproprianaldehyde, and acrolein were found more toxic than the other α-amino acid-derived aldehydes in HA1 fibroblasts survival assays (Vasilyev et al. 2005). In order to determine whether amino acid-derived aldehydes are also formed in vivo, both wild-type and $MPO^{-/-}$ mice were used. After acute myocardial infarction (AMI) induction, the aldehydes including formaldehyde in ventricular tissues of wild-type mice were generated significantly compared with that in the $MPO^{-/-}$ mice (Vasilyev et al. 2005). This means that the formaldehyde could be generated through the MPO-H_2O_2-chloride system in vivo.

2.5.3 Cells Produce Reactive Oxygen Species (ROS) in the Presence of Formaldehyde and Vice Versa

Breakthroughs in biochemistry have furthered our understanding of the onset and progression of various diseases, for instance, AD, and have advanced the development of new therapeutics (Evans et al. 2016; Umeno et al. 2017). Abnormally high level of formaldehyde induced the cellular oxidative stress, resulting in abnormal modification and accumulation of proteins, neuron death, and cognitive impairment. Jung and colleagues observed that when the concentration of formaldehyde increased, the level of intracellular ROS was also elevated at the same time (Jung et al. 2007). Formaldehyde-induced ROS can be released from eosinophils after they migrated to the inflammatory sites of the airways (Jung et al. 2007). Recently, Chen and her colleagues added formaldehyde to N2a cells and observed the yield of

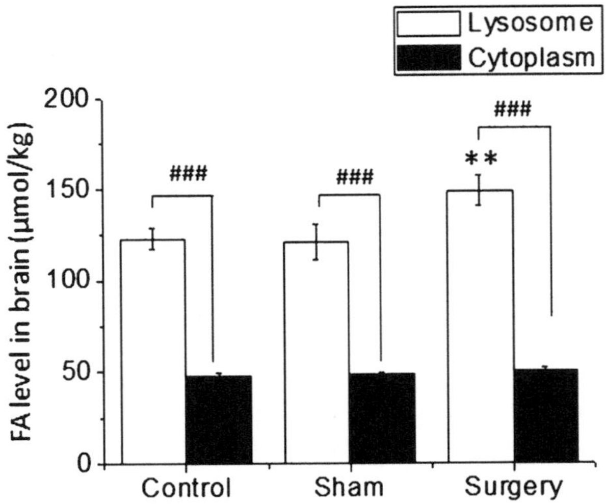

Fig. 2.3 Formaldehyde levels in the SD rats with bilateral carotid occlusion. Levels of formaldehyde (FA) in the lysosome and cytoplasm of brain cells of SD rats ($n = 12$) which were operated with bilateral carotid occlusion. Data are expressed as the mean ± s.d.; $^{**}P < 0.01$; $^{###}P < 0.001$ (Chen et al. 2017)

ROS. This is to say, formaldehyde is able to induce N2a cells to produce ROS through triggering cellular oxidative stress during the culture (Chen et al. 2017).

To investigate excess formaldehyde induced by the oxidative stress which leads to neuron death and cognitive dysfunction, the technique for monitoring the dynamic distribution of formaldehyde in living cells has been performed by Chen and coworkers (Chen et al. 2017). In the cellular experiment, they added hydrogen peroxide (H_2O_2) to bEed.3 cells and found that formaldehyde is produced and localized in the lysosome, detected by the formaldehyde molecular probe FAP-1. They added formaldehyde to the cells and also observed that the signal of formaldehyde markedly increases inside lysosome and then decreases with time. Furthermore, they employed SD rats which were operated with bilateral carotid occlusion and isolated lysosomes from rat brains. Then formaldehyde was detected with the fluorescent probe. The results showed that formaldehyde is located in the lysosomes but not much in the cytoplasm (Fig. 2.3). These results suggested that ROS is able to trigger the yield of formaldehyde in cells during the oxidative stress (Chen et al. 2017).

Get together with these two paragraphs, ROS is able to promote cellular oxidative stress and the production of formaldehyde (Chen et al. 2017). In contrast, formaldehyde also triggers the oxidative stress and induces the production of ROS. Thus, ROS and formaldehyde may form positive circle in the cells during oxidative stress as shown in Fig. 2.4. Because lysosome is an important cellular organelle for endogenous formaldehyde (Jian and Zhu 2016), we hypothesized here that endogenous formaldehyde could be used as a biomarker for lysosome function and dysfunction (He 2016).

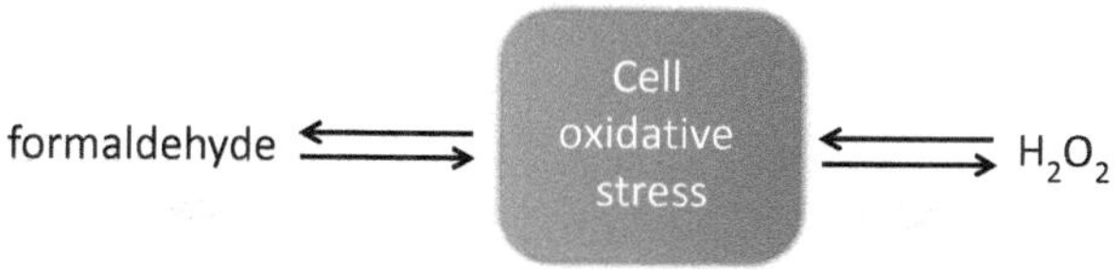

Fig. 2.4 Circle of hydrogen peroxide and formaldehyde. Addition of hydrogen peroxide promotes oxidative stress for cells followed by the production of formaldehyde. Cells produce hydrogen peroxide in the presence of formaldehyde (Chen et al. 2017)

2.6 *From Intestinal Flora*

The microbiota-gut-brain axis is composed of the neural, immune, endocrine, and metabolic pathways, which is a bidirectional communication system. Gut microbiota also called "the second brain" can modulate brain function and regulate pathogenesis of neurodegeneration, including age-related AD (Hu et al. 2016). It has been found that *Salmonella enterica subsp. enterica serovar Typhimurium* possesses the genes encoding methanol dehydrogenase (https://www.ncbi.nlm.nih.gov/protein). Furthermore, recent studies showed that gut microbiota is closely associated with host cognition, as well as AD-related pathogenesis (Jiang et al. 2017). It was reported that colonization of intrinsic pathogens is able to disturb the gut-brain axis, which finally causes the inflammatory host responses and pathogen-mediated diseases (Stecher 2015). Furthermore, highly elevated levels of formaldehyde were observed in the appendix of the transgenic mouse APP+/PS1+, compared with the age-match wild-type C57BL/6 mouse (Liu et al. 2017). These data suggest that intestinal microbiota is involved in the metabolic pathways for endogenous formaldehyde in the transgenic mouse models for AD (see Chap. 6).

2.7 *From Oxidative Demethylation*

In the live organisms, methylation is one of the most important modifications for histones, which could regulate a variety of biological processes including transcription activation and repression. To date, the methyl groups in the lysines of histones could be removed by lysine demethylases (KDMs), a family of enzymes including the flavin-dependent KDM1 enzymes and the 2-oxoglutarate- and oxygen-dependent JmjC KDMs (Burg et al. 2016; Chang et al. 2007; Shi et al. 2004; Tsukada et al. 2006). During this oxidative demethylation processes, formaldehyde could be released. Furthermore, the methyl group of the adenine as well as cytosine in DNA and RNA could also be demethylated by a type of enzymes named as AlkB, hABH2, or fat mass and obesity-associated protein (FTO), concomitant with the production of formaldehyde (Aas et al. 2003; Jia et al. 2008; Trewick et al. 2002). The detailed processes and mechanisms for the production of formaldehyde that resulted from the reaction about methylation and demethylation are discussed in Chap. 3.

3 Formaldehyde Metabolic Pathways

To date, a great number of studies have shown that formaldehyde is a cancerigenic molecule which would bring a variety of serious diseases and even cause human being death. Thus, the scientific reports have recommended that we should try our best to avoid inhalation of the toxic formaldehyde as much as possible. Even we are successfully protected from the adverse effects induced by exogenous formaldehyde, the endogenous formaldehyde could also be formed by a series of pathways which have been mentioned above. How could the endogenous formaldehyde be effectively metabolized? Fortunately, mitochondrial aldehyde dehydrogenase 2 (ALDH2) or glutathione (GSH)-dependent alcohol dehydrogenase 3 (ADH3) was found to be in charge of the metabolism of endogenous formaldehyde in vivo (Table 2.1). Furthermore, intestinal microbiota and tetrahydrofolic acid could also consume excess formaldehyde. However, the formaldehyde would react with biomacromolecules including protein, DNA, and RNA if it is accumulated in excess.

3.1 Mitochondrial ALDH2 Pathway

In our daily lives, methanol from fruits, vegetables, fermented beverages, soft drinks, as well as aspartame-sweetened foods is usually metabolized by multiple reactions in vivo (Burbacher et al. 1999). As we have described above, the ingested methanol at low dose could be converted to formaldehyde through oxidation in three different pathways (Dorokhov et al. 2015; MacAllister et al. 2011). Noticeably, the generated formaldehyde could be further oxidized to formic acid catalyzed by ALDH2 in the presence of NAD^+ (Dicker and Cederbaum 1986; Dorokhov et al. 2015; MacAllister et al. 2011). Under physiological conditions, formic acid is usually dissociated to release formate and hydrogen ions. Importantly, the neurotoxic formate would be converted to carbon dioxide (CO_2) in a tetrahydrofolate-dependent multistep pathway (Noker and Tephly 1980). During this reaction cycle, formate could firstly bind to THF to form [^{10}N] formyl-THF, which is catalyzed by 10-formyl-THF synthetase. Subsequently, NADP oxidoreductase catalyzes the oxidation of the formyl group on [^{10}N] formyl-THF directly to release CO_2 (Brosnan et al. 2015).

3.2 Glutathione (GSH)-Dependent Pathway

GSH is a thiol-containing tripeptide which is able to act as an important antioxidant to prevent carbonyl derivatives generated through lipid peroxidation (Palamanda and Kehrer 1992). During the detoxification process, GSH requires to cooperate with a series of enzymes including GSH reductase (GSHRx), GSH peroxidase

(GSHPx), and GSH S-transferase (GSHST). Koivusalo and colleagues found that formaldehyde could be consumed by formaldehyde dehydrogenase which is also known as class III alcohol dehydrogenase (ADH3 or FDH) in a GSH-dependent pathway (Koivusalo et al. 1989). Firstly, formaldehyde reacts with GSH to form S-hydroxymethyl GSH in a nonenzymatic reaction. Secondly, S-hydroxymethyl GSH is transformed into S-formyl GSH catalyzed by ADH3 in the presence of NAD^+. Finally, S-formyl GSH hydrolase irreversibly catalyzes the S-formyl GSH to form GSH and formate (MacAllister et al. 2011). Similarly, the formate could be metabolized to CO_2 in the multistep reactions which have been mentioned above. Hedberg and coworkers analyzed the mRNA or protein level of ADH3 in oral mucosa and cell lines, as well as its activity (Hedberg et al. 2000). Their results indicated that ADH3 is a major enzyme for the prevention of formaldehyde toxicity in human oral mucosa (Hedberg et al. 2000). Recently, Qiang and her colleagues reported that the mRNA or protein level, as well as enzyme activity of ADH3, which is responsible for the clearance of endogenous formaldehyde, was also lower in the brain of SAMP8 mice than that in SAMR1 mice (Qiang et al. 2014). This might lead to the relatively higher level of formaldehyde in the brains of SAMP8 than that in the SAMR1 (Qiang et al. 2014). Taken together, the GSH-dependent ADH3 is an important enzyme to eliminate the damage brought by endogenous formaldehyde.

3.3 *Intestinal Microbiota*

The gut/gastrointestinal tracts, which contain the environment of microbes, generate, process, and scavenge formaldehyde (see Chap. 6). The assimilation and dissimilation pathways of formaldehyde in gut bacteria could contribute to the metabolism of endogenous formaldehyde in physiology and pathology for humans (Li et al. 2016a; b; Marx et al. 2003; Mei et al. 2015; Vorholt et al. 2000).

3.4 *Synthesis of THF*

As previously described by Kallen and Jencks, the formaldehyde generated from one-carbon circle could also react with THF to form 5, 10-methylene-THF (Kallen and Jencks 1966). As we all know, THF is an important cofactor during the synthesis of amino acids and nucleic acids. Thus, the interaction between formaldehyde and THF might also be another pathway which decreases the level of free formaldehyde in vivo (Kallen and Jencks 1966). The authors put forward that thiols could inhibit the condensation of formaldehyde with THF since thiols react with formaldehyde with the formation of hemithioacetal at neutral and alkaline pH. Furthermore, THF could compete with thiols to bind to unhydrated formaldehyde (Kallen and Jencks 1966).

3.5 *Reaction with Biomacromolecules Such as Protein, DNA, and RNA*

In the live organisms, the formaldehyde is usually produced within the cells, and this toxic molecule could be metabolized through GSH-dependent pathways, aldehyde dehydrogenases, intestinal microbiota, or THF. If the dynamic equilibrium between formaldehyde generation and clearance was destroyed, the excess formaldehyde would be accumulated within the cells, which could further react with biomacromolecules including protein, DNA, and RNA and finally bring harmful impacts on their biological functions.

3.5.1 The Reaction Between Formaldehyde and Proteins

In daily experiments, formaldehyde is usually used as a fixing factor or cross-linking reagent to study the distribution of biomacromolecules in the cells and tissues. In the 1930–1950s, formaldehyde was reported to react with proteins with forming methylene cross-links (Clark and Rowan 1938; Fraenkel-Conrat and Olcott 1948). With the many efforts from the scientists all over the world, the mechanisms of the cross-links generated between formaldehyde and proteins have been discovered and summarized in details. Firstly, the α/ε-amino group and thiol of amino acid residues could react with formaldehyde to form methylol adducts (Metz et al. 2004). Next, the methylol adducts of the amino groups would be dehydrated to generate an imine, which is also named as Schiff base (Metz et al. 2004). Finally, labile imine could further form cross-links with a battery of residues including arginine, asparagine, cysteine, glutamine, histidine, tryptophan, as well as tyrosine (Metz et al. 2004). Therefore, formaldehyde could induce intramolecular and intermolecular cross-linkages between protein residues. In 2004, Metz and colleagues analyzed the products from peptides modified by formaldehyde through tandem reversed-phase liquid chromatography and electrospray ionization mass spectrometry (LC/MS) (Metz et al. 2004). Their study tried to observe the different modifications of every individual amino acid residue induced through formaldehyde and found that peptides (or proteins)' different sequences decided the formaldehyde modification types including methylol groups, Schiff bases, and methylene bridges (Metz et al. 2004). In 1996, Yu and his colleagues found that formaldehyde from deamination of methylamine mediated by SSAO might lead to the formation of irreversible adducts of proteins in vivo (Yu and Zuo 1996). They speculated that the generated formaldehyde, hydrogen peroxide, and formaldehyde-modified proteins might be risk factors for the formation of endothelial injury and initiation of atherosclerosis (Yu and Zuo 1996). In 2004, Gubisne-Haberle et al. did further study to discuss the products after formaldehyde modification (Gubisne-Haberle et al. 2004). They observed that formaldehyde generated from methylamine deamination could effectively induce bipeptide (H-Lys-Leu-OH) cross-linkage, which is not stable in six N HCl. This finding suggested that Schiff bases between two free ε-amino groups of lysine residues were formed in the presence of formaldehyde (Gubisne-Haberle et al. 2004). After reducing by sodium cyanoborohydride, the Schiff bases are converted to N-methyl-lysine and formyl-lysine (Gubisne-Haberle et al. 2004). Besides the bipeptides, the authors also used the mouse intestines which are rich in SSAO activity to produce formaldehyde to further modify BSA (Gubisne-Haberle et al. 2004). Similar to the data from bipeptide, N-methyl-lysine that originated from the hydrolyzed proteins was also found by HPLC. Importantly, the proteins after modification by formaldehyde might lose their original functions and could become pathogenic agents which are related to the various diseases including vascular disorders, chronic heart failure, diabetic complications, and AD (Gubisne-Haberle et al. 2004). In 2006, Chen and colleagues proved that formaldehyde could significantly trigger the β-sheet, oligomers, protofibrils, and even aggregate formation from seed-free $A\beta_{1-40}$ in vitro (Chen et al. 2006). Recently, endogenous formaldehyde was also proved to participate in the formation of N^6-formyl-lysine which could disturb the functions of other posttranslational modifications (Edrissi et al. 2013).

3.5.2 The Reaction Between Formaldehyde and Nucleic Acids

In addition to proteins, formaldehyde could also trigger the inactivation of nucleic acids including DNA and RNA. Similar to the protein adducts modified by formaldehyde, labile aminomethylol ($-NHCH_2OH$), or a Schiff base ($-N = CH_2$), as well as stable methylene ($R\text{-}CH_2\text{-}R'$) derivatives are also generated when formaldehyde is reacted with nucleic acids (Feldman 1967; Penniston and Doty 1963). Among the two types of derivatives, the NCH_2OH derivatives are formed through the conversion of the –NH groups and amino groups in a rapid and reversible reaction which could release formaldehyde after dilution. Thus, the derivatives are usually undetectable due to the unstable property, while the R-CH2-R' derivatives need two steps to form: The first stage is the formation of methylol derivatives ($RH + CH_2O \rightleftharpoons R\text{-}CH_2OH$), and the second stage is the formation of methylene derivatives ($R\text{-}CH_2OH + RH \rightleftharpoons R\text{-}CH_2\text{-}R + H_2O$) (Feldman 1967). At this stage, the generated methylene derivatives are more stable. In 1980, Chaw and colleagues purified the cross-links generated between formaldehyde and exocyclic amino groups of the bases in the DNA or RNA through reverse-phase high-pressure liquid chromatography (Chaw et al. 1980). After analysis by a series of methods including ultraviolet or nuclear magnetic resonance (NMR) spectra, both cytosines and purines were found to be involved in the methylene-bridged products (Chaw et al. 1980). Furthermore, formaldehyde was also found to attack the "imino" groups of thymine and uracil derivatives (McGhee and von Hippel 1975). During this reaction, N-3 proton is converted to a hydroxymethyl or methylol group (McGhee and von Hippel 1975). Interestingly, even though the secondary structure from both RNA and DNA is different, the cross-links generated between formaldehyde and RNA or DNA were similar (Chaw et al. 1980). Except the multiple studies on the structure of adducts formed from formaldehyde and DNA or RNA, the side effects of these adducts were also focused for a very long time. To date, we can confirm that DNA or RNA after aberrant modification by formaldehyde could be a potential risk for gene mutation and genome instability, which finally cause cancer and even death.

3.5.3 The Cross-Links of Proteins and Nucleic Acids Induced by Formaldehyde

Although reaction with formaldehyde would lead to the inactivation of proteins, DNA, or RNA, formaldehyde which acts as an ideal fixer is usually used to stabilize and characterize the complex generated between proteins and DNA in vitro studies. Therefore, it is likely that excess of endogenous formaldehyde might have the potential to induce the aberrant DNA-protein cross-links (DPX or DPCs), which would make protein or DNA lose their original functions, and finally cause cancer. As early as 1983, Casanova-Schmitz and Heck observed that formaldehyde could induce the cross-links between proteins and nucleic acids in vitro experiments (Casanova-Schmitz and Heck 1983). Furthermore, they also treated rats with formaldehyde and found that the content of DPX was increased in respiratory mucosa of the rats (Casanova-Schmitz and Heck 1983). Therefore, the newly formed DPX with genotoxicity might be one of the pathogenic factors to trigger cancer. In 2010, Lu and colleagues characterized the structure of DPCs formed from Lysine, Cysteine, Histidine, and Tryptophan with deoxyguanosine, deoxyadenosine, and deoxycytosine (Lu et al. 2010). Among the cross-links, Lysine is liable to react with deoxyguanosine (Lu et al. 2010). In general, there are two routes accounting for the generation of DPCs. One route is that electrophilic carbon of formaldehyde attacks the nucleophilic site of proteins, and subsequently, the methylol adduct of proteins reacts with the nucleophilic site of DNA to form DPCs (Kennedy-Darling and Smith 2014). The other route is that formaldehyde initially attacks the nucleophilic site of DNA, and its methylol adduct further forms cross-links with the residue in protein (Lu et al. 2010).

4 Conclusion

Formaldehyde comes from many different metabolic pathways but goes away through a narrow degradation lane in human cells (Fig. 2.5). Among the "coming" pathways, formaldehyde resulted from methanol is one of the main ways in the metabolism because humans take methanol from food. However, cellular oxidative stress is also one of the most common pathways to produce endogenous formaldehyde, such as oxidation of proteins and lipids as well as monoamine catalyzed by SSAO. Hydrogen peroxide triggers the production of formaldehyde through cellular oxidative stress. Besides these pathways, DNA, RNA, and protein demethylation also produce plenty of formaldehyde. It should never be forgotten that intestinal flora regulates formaldehyde generation, process, and clearance. In the "going" pathways, the conversion of formaldehyde to formic acid then to carbon dioxide and

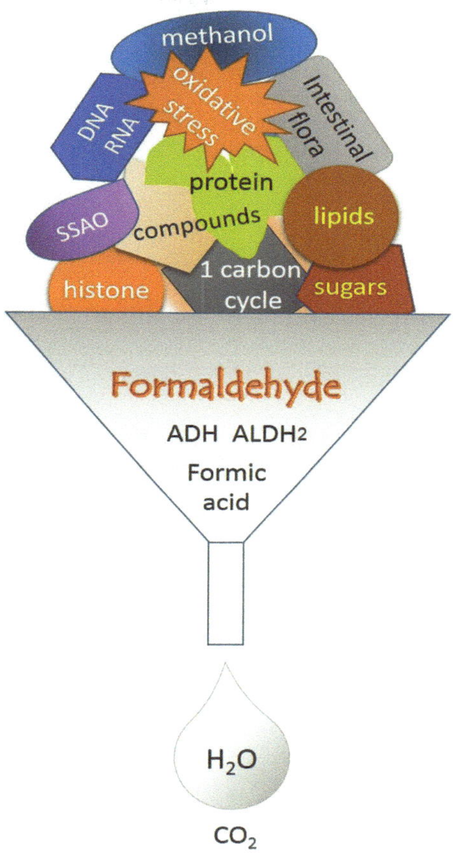

Fig. 2.5 Formaldehyde comes from many pathways but goes away through a narrow lane (diagrammatic drawing). The ability of scavenging endogenous formaldehyde for humans is weaker than that for rodents (Bruckner and Warren 2001). Humans face with so many endogenous metabolic products including formaldehyde daily, which needs to be processed and scavenged. However, the ability of scavenging becomes weaker as aging (Tong et al. 2013), resulting in chronic accumulation of metabolic products, including formaldehyde

water is the most important way in degradation. The metabolic homeostasis of endogenous formaldehyde is maintained by the activities of generation and degradation pathways, which are complicated and diversified. Either the pathway of generation or degradation dysfunctions, which will result in lack or excess of formaldehyde, leads to imbalance of the metabolic formaldehyde. Dysmetabolism is one of the risk factors related to the onset of age-cognitive impairment. The ability of scavenging endogenous formaldehyde in humans is much weaker than that in rodents (Bruckner and Warren 2001). Humans have to face with so many endogenous metabolic products including formaldehyde daily, which need to be processed and scavenged. Lysosome acts as a centre to accommodate and transport formaldehyde for a cell. Dysfunction of lysosome could lead to imbalance of formaldehyde metabolism (Fig. 2.6). However, the ability of processing and scavenging becomes weaker as aging (Tong et al. 2013), resulting in chronic accumulation of metabolic products in vivo, including formaldehyde at last.

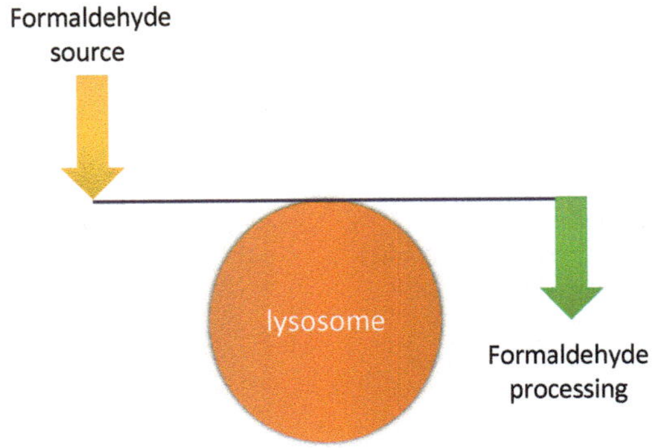

Fig. 2.6 Lysosome is able to accommodate and transport formaldehyde for a cell (diagrammatic drawing). Formaldehyde homeostasis could depend upon its source and degradation as well as the function of lysosome (Chen et al. 2017)

Acknowledgment This project was supported by grants from the Beijing Municipal Science and Technology Project (Z161100000217141; Z161100000216137), the National Key Research and Development Program of China (2016YFC1306300), the National Basic Research Program of China (973 Program) (2012CB911004), the National Natural Science Foundation of China (NSFC 31301880, NSFC 31270868), the Foundation of Chinese Academy of Sciences (CAS-20140909), the Queensland-Chinese Academy of Sciences Biotechnology Fund (GJHZ201302), the Program for Liaoning Excellent Talents in University (LJQ2015057), and the Dalian High-Level Talent Innovation Support Plan (2015R067).

Competing Financial Interests The authors declare no competing financial interests.

References

Aas PA, Otterlei M, Falnes PO, Vagbo CB, Skorpen F, Akbari M, Sundheim O, Bjoras M, Slupphaug G, Seeberg E, Krokan HE (2003) Human and bacterial oxidative demethylases repair alkylation damage in both RNA and DNA. Nature 421(6925):859–863

Amano K, Yamada K (1964a) A biological formation of formaldehyde in the muscle tissue of gadoid fish. Bull Jap Soc Fish 30(5):430–435

Amano K, Yamada K (1964b) Formaldehyde formation from trimethylamine oxide by the action of pyloric caeca of cod. Bull Jap Soc Sci Fish 30:639–645

Anderson MM, Hazen SL, Hsu FF, Heinecke JW (1997) Human neutrophils employ the myeloperoxidase-hydrogen peroxide-chloride system to convert hydroxy-amino acids into glycolaldehyde, 2-hydroxypropanal, and acrolein. A mechanism for the generation of highly reactive alpha-hydroxy and alpha, beta-unsaturated aldehydes by phagocytes at sites of inflammation. J Clin Invest 99(3):424–432

Antus B, Harnasi G, Drozdovszky O, Barta I (2014) Monitoring oxidative stress during chronic obstructive pulmonary disease exacerbations using malondialdehyde. Respirology 19(1):74–79

Bagatini MD, Martins CC, Battisti V, Gasparetto D, da Rosa CS, Spanevello RM, Ahmed M, Schmatz R, Schetinger MR, Morsch VM (2011) Oxidative stress versus antioxidant defenses in patients with acute myocardial infarction. Heart Vessel 26(1):55–63

Boomsma F, van Veldhuisen DJ, de Kam PJ, Man in't Veld AJ, Mosterd A, Lie KI, Schalekamp MA (1997) Plasma semicarbazide-sensitive amine oxidase is elevated in patients with congestive heart failure. Cardiovasc Res 33(2):387–391

Boor PJ, Trent MB, Lyles GA, Tao M, Ansari GA (1992) Methylamine metabolism to formaldehyde by vascular semicarbazide-sensitive amine oxidase. Toxicology 73(3):251–258

Bos A, Wever R, Roos D (1978) Characterization and quantification of the peroxidase in human monocytes. Biochim Biophys Acta 525(1):37–44
Brosnan ME, MacMillan L, Stevens JR, Brosnan JT (2015) Division of labour: how does folate metabolism partition between one-carbon metabolism and amino acid oxidation? Biochem J 472(2):135–146
Bruckner JV, Warren DA (2001) Toxic effects of solvents and vapors. In: Klaassen CD (ed) Casarett Doull's toxicology, the basic science of poisons, 6th edn. The McGraw-Hill Co., Kansas, pp 894–895
Burbacher T, Shen D, Grant K, Sheppard L, Damian D, Ellis S, Liberato N (1999) Reproductive and offspring developmental effects following maternal inhalation exposure to methanol in nonhuman primates. Res Rep Health Eff Inst i–ii:1–117. discussion 119–133
Burg JM, Gonzalez JJ, Maksimchuk KR, McCafferty DG (2016) Lysine-specific demethylase 1A (KDM1A/LSD1): product recognition and kinetic analysis of full-length histones. Biochemistry 55(11):1652–1662
Cai Y, Liu J, Shi Y, Liang L, Mou S (2005) Determination of several sugars in serum by high-performance anion-exchange chromatography with pulsed amperometric detection. J Chromatogr A 1085(1):98–103
Casanova-Schmitz M, Heck HD (1983) Effects of formaldehyde exposure on the extractability of DNA from proteins in the rat nasal mucosa. Toxicol Appl Pharmacol 70(1):121–132
Castell CH, Smith B, Dyer WJ (1973) Effect of formaldehyde on salt extractable proteins of gadoid muscle. J Fish Res Board Can 30(8):1205–1213
Cederbaum AI, Qureshi A (1982) Role of catalase and hydroxyl radicals in the oxidation of methanol by rat liver microsomes. Biochem Pharmacol 31(3):329–335
Chang B, Chen Y, Zhao Y, Bruick RK (2007) JMJD6 is a histone arginine demethylase. Science 318(5849):444–447
Chaw YF, Crane LE, Lange P, Shapiro R (1980) Isolation and identification of cross-links from formaldehyde-treated nucleic acids. Biochemistry 19(24):5525–5531
Chen K, Maley J, Yu PH (2006) Potential implications of endogenous aldehydes in beta-amyloid misfolding, oligomerization and fibrillogenesis. J Neurochem 99(5):1413–1424
Chen L, Cheng CY, Choi H, Ikram MK, Sabanayagam C, Tan GS, Tian D, Zhang L, Venkatesan G, Tai ES, Wang JJ, Mitchell P, Cheung CM, Beuerman RW, Zhou L, Chan EC, Wong TY (2016) Plasma metabonomic profiling of diabetic retinopathy. Diabetes 65(4):1099–1108
Chen XX, Su T, He YG, He RQ (2017) Spatial cognition decline caused by excess formaldehyde in the lysosome. Prog Biochem Biophys 44(6):486–494
Choudhary AK, Sheela Devi R (2015) Longer period of oral administration of aspartame on cytokine response in Wistar albino rats. Endocrinol Nutr 62(3):114–122
Chung SW, Chan BT (2009) Trimethylamine oxide, dimethylamine, trimethylamine and formaldehyde levels in main traded fish species in Hong Kong. Food Addit Contam Part B Surveill 2(1):44–51
Clark GL, Rowan R (1938) The formaldehyde-protein reaction. Science 87(2265):487
Clejan LA, Cederbaum AI (1991) Role of iron, hydrogen peroxide and reactive oxygen species in microsomal oxidation of glycerol to formaldehyde. Arch Biochem Biophys 285(1):83–89
Clejan LA, Cederbaum AI (1992a) Role of cytochrome P450 in the oxidation of glycerol by reconstituted systems and microsomes. FASEB J 6(2):765–770
Clejan LA, Cederbaum AI (1992b) Structural determinants for alcohol substrates to be oxidized to formaldehyde by rat liver microsomes. Arch Biochem Biophys 298(1):105–113
Daugherty A, Dunn JL, Rateri DL, Heinecke JW (1994) Myeloperoxidase, a catalyst for lipoprotein oxidation, is expressed in human atherosclerotic lesions. J Clin Invest 94(1):437–444
Davies MJ (2011) Myeloperoxidase-derived oxidation: mechanisms of biological damage and its prevention. J Clin Biochem Nutr 48(1):8–19
Del Rio D, Stewart AJ, Pellegrini N (2005) A review of recent studies on malondialdehyde as toxic molecule and biological marker of oxidative stress. Nutr Metab Cardiovasc Dis 15(4):316–328

Dicker E, Cederbaum AI (1986) Inhibition of CO_2 production from aminopyrine or methanol by cyanamide or crotonaldehyde and the role of mitochondrial aldehyde dehydrogenase in formaldehyde oxidation. Biochim Biophys Acta 883(1):91–97

Diliberto EJ, Axelrod J (1976) Regional and subcellular distribution of protein carboxymethylase in brain and other tissues. J Neurochem 26(6):1159–1165

Dorokhov YL, Shindyapina AV, Sheshukova EV, Komarova TV (2015) Metabolic methanol: molecular pathways and physiological roles. Physiol Rev 95(2):603–644

Edrissi B, Taghizadeh K, Dedon PC (2013) Quantitative analysis of histone modifications: formaldehyde is a source of pathological n(6)-formyllysine that is refractory to histone deacetylases. PLoS Genet 9(2):e1003328

Evans AM, Fameli N, Ogunbayo OA, Duan J, Navarro-Dorado J (2016) From contraction to gene expression: nanojunctions of the sarco/endoplasmic reticulum deliver site- and function-specific calcium signals. Sci China Life Sci 59(8):749–763

Feldman MY (1967) Reaction of formaldehyde with nucleotides and ribonucleic acid. Biochim Biophys Acta 149(1):20–34

Filip A, Pinzano A, Bianchi A, Feve B, Jalkanen S, Gillet P, Mainard D, Lacolley P, Magdalou J, Mercier N (2016) Expression of the semicarbazide-sensitive amine oxidase in articular cartilage: its role in terminal differentiation of chondrocytes in rat and human. Osteoarthr Cartil 24(7):1223–1234

Fraenkel-Conrat H, Olcott HS (1948) The reaction of formaldehyde with proteins; cross-linking between amino and primary amide or guanidyl groups. J Am Chem Soc 70(8):2673–2684

Frisell WR, Mackenzie CG (1962) Separation and purification of sarcosine dehydrogenase and dimethylglycine dehydrogenase. J Biol Chem 237:94–98

Gill TA, Paulson AT (1982) Localization, characterization and partial purification of TMAO-ase. Comp Biochem Physiol B 71(1):49–56

Gubisne-Haberle D, Hill W, Kazachkov M, Richardson JS, Yu PH (2004) Protein cross-linkage induced by formaldehyde derived from semicarbazide-sensitive amine oxidase-mediated deamination of methylamine. J Pharmacol Exp Ther 310(3):1125–1132

Guo L, Zeng XY, Wang DY, Li GQ (2010) Methanol metabolism in the Asian corn borer, Ostrinia furnacalis (Guenée) (Lepidoptera: Pyralidae). J Insect Physiol 56(3):260–265

Hazen SL, Hsu FF, d'Avignon A, Heinecke JW (1998) Human neutrophils employ myeloperoxidase to convert alpha-amino acids to a battery of reactive aldehydes: a pathway for aldehyde generation at sites of inflammation. Biochemistry 37(19):6864–6873

He RQ (2016) Abnormal lysosome, formaldehyde dysmetabolism and age-related cognitive impairment. Prog Biochem Biophys 43(12):1197

Hedberg JJ, Hoog JO, Nilsson JA, Xi Z, Elfwing A, Grafstrom RC (2000) Expression of alcohol dehydrogenase 3 in tissue and cultured cells from human oral mucosa. Am J Pathol 157(5):1745–1755

Hu X, Wang T, Jin F (2016) Alzheimer's disease and gut microbiota. Sci China Life Sci 59(10):1006–1023

Jaakkola K, Kaunismaki K, Tohka S, Yegutkin G, Vanttinen E, Havia T, Pelliniemi LJ, Virolainen M, Jalkanen S, Salmi M (1999) Human vascular adhesion protein-1 in smooth muscle cells. Am J Pathol 155(6):1953–1965

Jacobsen D, McMartin KE (1986) Methanol and ethylene glycol poisonings. Mechanism of toxicity, clinical course, diagnosis and treatment. Med Toxicol 1(5):309–334

Jakobsson E, Nilsson J, Ogg D, Kleywegt GJ (2005) Structure of human semicarbazide-sensitive amine oxidase/vascular adhesion protein-1. Acta Crystallogr D Biol Crystallogr 61(Pt11):1550–1562

Jia G, Yang CG, Yang S, Jian X, Yi C, Zhou Z, He C (2008) Oxidative demethylation of 3-methylthymine and 3-methyluracil in single-stranded DNA and RNA by mouse and human FTO. FEBS Lett 582(23–24):3313–3319

Jian X, Zhu MX (2016) Regulation of lysosomal ion homeostasis by channels and transporters. Sci China Life Sci 59:777–791

Jiang C, Li G, Huang P, Liu Z, Zhao B (2017) The gut microbiota and Alzheimer's disease. J Alzheimers Dis 58(1):1–15

Johnston JM, Mackenzie CG (1956) Isolation of radioformaldehyde in the metabolism of dimethylaminoethanol-C H. J Biol Chem 221(1):301–305

Jung WW, Kim EM, Lee EH, Yun HJ, Ju HR, Jeong MJ, Hwang KW, Sul D, Kang HS (2007) Formaldehyde exposure induces airway inflammation by increasing eosinophil infiltrations through the regulation of reactive oxygen species production. Environ Toxicol Pharmacol 24(2):174–182

Kallen RG, Jencks WP (1966) The mechanism of the condensation of formaldehyde with tetrahydrofolic acid. J Biol Chem 241(24):5851–5863

Kato Y (2016) Neutrophil myeloperoxidase and its substrates: formation of specific markers and reactive compounds during inflammation. J Clin Biochem Nutr 58(2):99–104

Kennedy-Darling J, Smith LM (2014) Measuring the formaldehyde protein-DNA cross-link reversal rate. Anal Chem 86(12):5678–5681

Klebanoff SJ, Clark RA (1978) The neutrophil: function and clinical disorders. Elsevier/North Holland Biomedical Press, Amsterdam, pp 447–451

Kobayashi K, Yamamoto T, Taguchi M, Chiba K (2000) High-performance liquid chromatography determination of N- and O-demethylase activities of chemicals in human liver microsomes: application of postcolumn fluorescence derivatization using Nash reagent. Anal Biochem 284(2):342–347

Koivusalo M, Baumann M, Uotila L (1989) Evidence for the identity of glutathione-dependent formaldehyde dehydrogenase and class III alcohol dehydrogenase. FEBS Lett 257(1):105–109

Kostov V, Dimov T, Kominkova V, Taseva S (1983) Effect of certain factors on the growth of bacteria contaminating hydrolysis yeast production and of yeasts assimilating and not assimilating methanol. Acta Microbiol Bulg 13:46–55

Lager I, Fehr M, Frommer WB, Lalonde S (2003) Development of a fluorescent nanosensor for ribose. FEBS Lett 553(1–2):85–89

Lee ES, Chen H, Hardman C, Simm A, Charlton C (2008) Excessive S-adenosyl-L-methionine-dependent methylation increases levels of methanol, formaldehyde and formic acid in rat brain striatal homogenates: possible role in S-adenosyl-L-methionine-induced Parkinson's disease-like disorders. Life Sci 83(25–26):821–827

Li FX, Jing L, Xu YJ, Tong ZQ, Nie CL, He RQ (2008) Formaldehyde-mediated chronic damage may be related to sporadic neurodegeneration. Prog Biochem Biophys 35(4):393–400

Li Y, Song Z, Ding Y, Xin Y, Wu T, Su T, He R, Tai F, Lian Z (2016a) Effects of formaldehyde exposure on anxiety-like and depression-like behavior, cognition, central levels of glucocorticoid receptor and tyrosine hydroxylase in mice. Chemosphere 144:2004–2012

Li T, Su T, He Y, Lu J, Mo W, Wei Y, He R (2016b) Brain formaldehyde is related to water intake behavior. Aging Dis 7:561–584

Liu KL, He YG, Yu LX, He RQ (2017) Elevated formaldehyde in the cecum of APP/PS1 mouse. Microbiol China 44(8):1761–1766

Locasale JW (2013) Serine, glycine and one-carbon units: cancer metabolism in full circle. Nat Rev Cancer 13(8):572–583

Lu K, Ye W, Zhou L, Collins LB, Chen X, Gold A, Ball LM, Swenberg JA (2010) Structural characterization of formaldehyde-induced cross-links between amino acids and deoxynucleosides and their oligomers. J Am Chem Soc 132(10):3388–3399

MacAllister SL, Choi J, Dedina L, O'Brien PJ (2011) Metabolic mechanisms of methanol/formaldehyde in isolated rat hepatocytes: carbonyl-metabolizing enzymes versus oxidative stress. Chem Biol Interact 191(1–3):308–314

Mackenzie CG, Johnston JM, Frisell WR (1953) The isolation of formaldehyde from dimethylaminoethanol, dimethylglycine, sarcosine, and methanol. J Biol Chem 203(2):743–753

Marx CJ, Chistoserdova L, Lidstrom ME (2003) Formaldehyde-detoxifying role of the tetrahydromethanopterin-linked pathway in methylobacterium extorquens AM1. J Bacteriol 185(24):7160–7168

McGhee JD, von Hippel PH (1975) Formaldehyde as a probe of DNA structure. II. Reaction with endocyclic imino groups of DNA bases. Biochemistry 14(6):1297–1303

Megaraj V, Ding X, Fang C, Kovalchuk N, Zhu Y, Zhang QY (2014) Role of hepatic and intestinal p450 enzymes in the metabolic activation of the colon carcinogen azoxymethane in mice. Chem Res Toxicol 27(4):656–662

Mei Y, Jiang C, Wan Y, Lv J, Jia J, Wang X, Yang X, Tong Z (2015) Aging-associated formaldehyde-induced norepinephrine deficiency contributes to age-related memory decline. Aging Cell 14(4):659–668

Metz B, Kersten GF, Hoogerhout P, Brugghe HF, Timmermans HA, de Jong A, Meiring H, ten Hove J, Hennink WE, Crommelin DJ, Jiskoot W (2004) Identification of formaldehyde-induced modifications in proteins: reactions with model peptides. J Biol Chem 279(8):6235–6243

Moldes M, Feve B, Pairault J (1999) Molecular cloning of a major mRNA species in murine 3T3 adipocyte lineage differentiation-dependent expression, regulation, and identification as semicarbazide-sensitive amine oxidase. J Biol Chem 274(14):9515–9523

Mondal J, Stirnemann G, Berne BJ (2013) When does TMAO fold a polymer chain and urea unfold it? J Phys Chem B 117(29):8723–8732

Monte WC (2010) Methanol: a chemical Trojan horse as the root of the inscrutable U. Med Hypotheses 74(3):493–496

Noker PE, Tephly TR (1980) The role of folates in methanol toxicity. Adv Exp Med Biol 132:305–315

Obata T (2006) Diabetes and semicarbazide-sensitive amine oxidase (SSAO) activity: a review. Life Sci 79(5):417–422

O'Sullivan M, Tipton KF, McDevitt WE (2002) Immunolocalization of semicarbazide-sensitive amine oxidase in human dental pulp and its activity towards serotonin. Arch Oral Biol 47(5):399–406

Palamanda JR, Kehrer JP (1992) Inhibition of protein carbonyl formation and lipid peroxidation by glutathione in rat liver microsomes. Arch Biochem Biophys 293(1):103–109

Parkin KL, Hultin HO (1982) Fish muscle microsomes catalyze the conversion of trimethylamine oxide to dimethylamine and formaldehyde. FEBS Lett 139(1):61–64

Penniston JT, Doty P (1963) Reaction of formaldehyde with soluble ribonucleic acid. Biopolymers 1:145–163

Pinto JP, Gladstone GR, Yung YL (1980) Photochemical production of formaldehyde in earth's primitive atmosphere. Science 210(4466):183–185

Pocker Y, Li H (1991) Kinetics and mechanism of methanol and formaldehyde interconversion and formaldehyde oxidation catalyzed by liver alcohol dehydrogenase. Adv Exp Med Biol 284:315–325

Pohanka M (2016) Toxicology and the biological role of methanol and ethanol: current view. Biomed Pap Med Fac Univ Palacky Olomouc Czech Repub 160(1):54–63

Porter DH, Cook RJ, Wagner C (1985) Enzymatic properties of dimethylglycine dehydrogenase and sarcosine dehydrogenase from rat liver. Arch Biochem Biophys 243(2):396–407

Qiang M, Xiao R, Su T, Wu BB, Tong ZQ, Liu Y, He RQ (2014) A novel mechanism for endogenous formaldehyde elevation in SAMP8 mouse. J Alzheimers Dis 40(4):1039–1053

Rashba-Step J, Step E, Turro NJ, Cederbaum AI (1994) Oxidation of glycerol to formaldehyde by microsomes: are glycerol radicals produced in the reaction pathway? Biochemistry 33(32):9504–9510

Renwick SB, Snell K, Baumann U (1998) The crystal structure of human cytosolic serine hydroxymethyltransferase: a target for cancer chemotherapy. Structure 6(9):1105–1116

Sanyal J, Bandyopadhyay SK, Banerjee TK, Mukherjee SC, Chakraborty DP, Ray BC, Rao VR (2009) Plasma levels of lipid peroxides in patients with Parkinson's disease. Eur Rev Med Pharmacol Sci 13(2):129–132

Seuffer R (1977) A new method for the determination of sugars in cerebrospinal fluid (author's transl). J Clin Chem Clin Biochem 15(12):663–668

Shapiro R (1988) Prebiotic ribose synthesis: a critical analysis. Orig Life Evol Biosph 18(1–2):71–85

Shi Y, Lan F, Matson C, Mulligan P, Whetstine JR, Cole PA, Casero RA, Shi Y (2004) Histone demethylation mediated by the nuclear amine oxidase homolog LSD1. Cell 119(7):941–953

Shindyapina AV, Petrunia IV, Komarova TV, Sheshukova EV, Kosorukov VS, Kiryanov GI, Dorokhov YL (2014) Dietary methanol regulates human gene activity. PLoS One 9(7):e102837

Slegers G, Lambrecht RH, Vandewalle T, Meulewaeter L, Vandecasteele C (1984) Enzymatic synthesis of C-11 formaldehyde: concise communication. J Nucl Med 25(3):338–342

Sotelo CG, Piñeiro C, Pérez-Martín RI (1995) Denaturation of fish proteins during frozen storage: role of formaldehyde. Z Lebensm Unters Forsch 200(1):14–23

Stancliffe RA, Thorpe T, Zemel MB (2011) Dairy attenuates oxidative and inflammatory stress in metabolic syndrome. Am J Clin Nutr 94(2):422–430

Stecher B (2015) The roles of inflammation, nutrient availability and the commensal microbiota in enteric pathogen infection. Microbiol Spectr 3(3):1–17

Su T, He R (2014) D-ribose, an overlooked player in type 2 diabetes mellitus? Sci China Life Sci 57(3):361

Su T, Xin L, He YG, Wei Y, Song YX, Li WW, Wang XM, He RQ (2013) The abnormally high level of uric D-ribose for type-2 diabetics. Prog Biochem Biophys 40(9):816–825

Su T, Monte WC, Hu X, He Y, He R (2016) Formaldehyde as a trigger for protein aggregation and potential target for mitigation of age-related, progressive cognitive impairment. Curr Top Med Chem 16(5):472–484

Sun J, Steindler L, Thrash JC, Halsey KH, Smith DP, Carter AE, Landry ZC, Giovannoni SJ (2011) One carbon metabolism in SAR11 pelagic marine bacteria. PLoS One 6(8):e23973

Sutherland JD (2010) Ribonucleotides. Cold Spring Harb Perspect Biol 2(4):a005439

Tatzelt J, Prusiner SB, Welch WJ (1996) Chemical chaperones interfere with the formation of scrapie prion protein. EMBO J 15(23):6363–6373

Tokunaga T (1964) Studies on the development of dimethylamine and formaldehyde in Alaska pollack muscle during frozen storage. Bull Hokkaido Reg Fish Res Lab 29:108–122

Tokunaga T (1974) The effect of decomposed products of trimethylamine oxide on quality of frozen Alaska pollock fillet. Nippon Suisan Gakkaishi 40(2):167–174

Tong Z, Han C, Luo W, Li H, Luo H, Qiang M, Su T, Wu B, Liu Y, Yang X, Wan Y, Cui D, He R (2013) Aging-associated excess formaldehyde leads to spatial memory deficits. Sci Rep 3:1807

Trewick SC, Henshaw TF, Hausinger RP, Lindahl T, Sedgwick B (2002) Oxidative demethylation by *Escherichia coli* AlkB directly reverts DNA base damage. Nature 419(6903):174–178

Trézl L, Csiba A, Juhász S, Szentgyörgyi M, Lombai G, Hullán L, Juhász A (1997) Endogenous formaldehyde level of foods and its biological significance. Zeitschrift für Lebensmitteluntersuchung und-Forschung A 205(4):300–304

Tsukada Y, Fang J, Erdjument-Bromage H, Warren ME, Borchers CH, Tempst P, Zhang Y (2006) Histone demethylation by a family of JmjC domain-containing proteins. Nature 439(7078):811–816

Ullah K, Xie B, Iqbal J, Rasool A, Qing H, Deng Y (2013) Arterial vascular cell line expressing SSAO: a new tool to study the pathophysiology of vascular amine oxidases. J Neural Transm (Vienna) 120(6):1005–1013

Umeno A, Biju V, Yoshida Y (2017) *In vivo* ROS production and use of oxidative stress-derived biomarkers to detect the onset of diseases such as Alzheimer's disease, Parkinson's disease, and diabetes. Free Radic Res 25:1–15

Vaisey EB (1956) The non-enzymic reduction of trimethylamine oxide to trimethylamine, dimethylamine, and formaldehyde. Can J Biochem Physiol 34(6):1085–1090

Vasilyev N, Williams T, Brennan ML, Unzek S, Zhou X, Heinecke JW, Spitz DR, Topol EJ, Hazen SL, Penn MS (2005) Myeloperoxidase-generated oxidants modulate left ventricular remodeling but not infarct size after myocardial infarction. Circulation 112(18):2812–2820

Vaz AD, Pernecky SJ, Raner GM, Coon MJ (1996) Peroxo-iron and oxenoid-iron species as alternative oxygenating agents in cytochrome P450-catalyzed reactions: switching by threonine-302 to alanine mutagenesis of cytochrome P450 2B4. Proc Natl Acad Sci U S A 93(10):4644–4648

Vorholt JA, Marx CJ, Lidstrom ME, Thauer RK (2000) Novel formaldehyde-activating enzyme in methylobacterium extorquens AM1 required for growth on methanol. J Bacteriol 182(23):6645–6650

Wang SP, He GL, Chen RR, Li F, Li GQ (2012) The involvement of cytochrome P450 monooxygenases in methanol elimination in Drosophila melanogaster larvae. Arch Insect Biochem Physiol 79(4–5):264–275

Winters DK, Cederbaum AI (1990) Oxidation of glycerol to formaldehyde by rat liver microsomes. Effects of cytochrome P-450 inducing agents. Biochem Pharmacol 39(4):697–705

Winters DK, Clejan LA, Cederbaum AI (1988) Oxidation of glycerol to formaldehyde by rat liver microsomes. Biochem Biophys Res Commun 153(2):612–617

Yu PH (1991) Oxidative deamination of aliphatic amines by rat aorta semicarbazide-sensitive amine oxidase. J Pharm Pharmacol 42(12):882–884

Yu PH (2001) Involvement of cerebrovascular semicarbazide-sensitive amine oxidase in the pathogenesis of Alzheimer's disease and vascular dementia. Med Hypotheses 57(2):175–179

Yu PH, Zuo DM (1993) Oxidative deamination of methylamine by semicarbazide-sensitive amine oxidase leads to cytotoxic damage in endothelial cells. Possible consequences for diabetes. Diabetes 42(4):594–603

Yu PH, Zuo DM (1996) Formaldehyde produced endogenously via deamination of methylamine. A potential risk factor for initiation of endothelial injury. Atherosclerosis 120(1–2):189–197

Yu PH, Lai CT, Zuo DM (1997) Formation of formaldehyde from adrenaline in vivo; a potential risk factor for stress-related angiopathy. Neurochem Res 22(5):615–620

Yu PH, Wright S, Fan EH, Lun ZR, Gubisne-Harberle D (2003) Physiological and pathological implications of semicarbazide-sensitive amine oxidase. Biochim Biophys Acta 1647(1–2):193–199

Zgliczynski JM, Stelmaszynska T, Ostrowiski W, Naskalski J, Sznajd J (1968) Myeloperoxidase of human leukaemic leucocytes. Oxidation of amino acids in the presence of hydrogen peroxide. Eur J Biochem 4(4):540–547

Zhang W, Tang L, Sun H, Han S, Wang X, Zhou S, Li K, Chen L (2014) C1 metabolism plays an important role during formaldehyde metabolism and detoxification in petunia under liquid HCHO stress. Plant Physiol Biochem 83:327–336

Chapter 3
Formaldehyde Playing a Role in (De)methylation for Memory

Tao Su and Rongqiao He

Abstract Methylation of DNA, RNA, and histone is one of the mechanisms for epigenetic regulation that occurs by the addition of a methyl (CH_3) group to DNA, RNA, and histone, thereby often modifies the function of the genes through chromatin remodeling. As described by recent studies, methylation of DNA or histone is regarded as a critical step in the process of memory formation. Formaldehyde as a methyl donor for the methylation of DNA, RNA, and histone acts as an epigenetic factor participating in the reversible and dynamic methylation. DNA demethylation elicits formaldehyde generation in the dividing cells and post-mitotic neurons. Endogenous formaldehyde is observably increased in aging population, but DNA methylation decreases, which closely links learning-responsive DNA methylation and memory formation. Dysmetabolism of endogenous formaldehyde, which affects DNA and histone methylation, is involved in age-related cognitive impairment. The level of formaldehyde is positively correlated with cognitive impairment, such as Alzheimer's disease (AD) and post-stroke dementia (PSD). Enhancement of DNA demethylation or block of DNA re-methylation using 5-aza-2-deoxycytidine (an inhibitor of DNA methyltransferase, DNMT) in rats leads to spatial memory deficits during spatial memory formation. Scavenging the elevated formaldehyde effectively relieves memory loss for rats. Here, we discuss the role of endogenous formaldehyde in methylation and demethylation of DNA, RNA, and histone as well as the formation or loss of memory.

Keywords Formaldehyde • Methylation • Demethylation • DNA • RNA • Histones • Memory • Cognitive impairment

---

T. Su
State Key Laboratory of Brain and Cognitive Science, Institute of Biophysics, Chinese Academy of Sciences, Beijing 100101, China

R. He (✉)
State Key Laboratory of Brain and Cognitive Science, Institute of Biophysics, Chinese Academy of Sciences, Beijing 100101, China

University of Chinese Academy of Sciences, Beijing 100049, China
e-mail: rongqiaohe@163.com

R. He, *Formaldehyde and Cognition*, DOI 10.1007/978-94-024-1177-5_3

1 Introduction

Epigenetic mechanisms, including a diverse range of histone, DNA, and RNA modifications, specifically turn genes "on" or "off" without a change in the nucleotide sequence (Berger et al. 2009). Methylation and demethylation of DNA (Bochtler et al. 2016), RNA (Dezi et al. 2016), and histone (Gupta et al. 2010) are involved in regulating many cellular processes, such as genomic imprinting, chromatin conformation, transcription, chromosome stability, and embryonic development in all cell types (Bergman and Cedar 2013). The study of methylation and demethylation is beneficial to the development of new drugs (Mirfattah et al. 2016). Of course, the functionally relevant changes are closely associated with brain functions (Martinowich et al. 2003). The establishment of persistent epigenetic states in nerve cells is dramatically related with the neural plasticity and memory formation (Reik et al. 2001). Learning, which is a complex process involving a variety of proteins and epigenetic regulatory enzymes, induces DNA (de)methylation (Ooi and Bestor 2008). As recent studies show, formaldehyde acting as a significant factor participates in reversible and dynamic methylation of the biomacromolecules (Bird 1992; Shi et al. 2004).

In this laboratory, Tong and his colleagues have observed that age-related cognitive impairment is closely correlated with endogenous formaldehyde metabolism and DNA methylation (Tong et al. 2012). Dysmetabolism of endogenous formaldehyde causes memory loss through interfering with DNA methyltransferase (DNMT) in age-related cognitive impairment (Tong et al. 2013). Consistent with the roles in epigenetic modification and regulation, formaldehyde could be called an epigenetic factor, for which a growing number of fundamental insights were provided to discuss the function of endogenous formaldehyde in memory formation, cognitive ability, and impairment as well as neurodegenerative disease in this chapter.

2 Formaldehyde Plays a Role in DNA Methylation and Demethylation

DNA methylation/demethylation was commonly found on CpG dinucleotide (CpG islands), which is located in 40% gene promoters (Jabbari and Bernardi 2004). Other positions of DNA methylation/demethylation without CpG dinucleotide were usually observed in embryonic stem cell (Abdelfatah et al. 2016; Li et al. 2016). 5-Methylcytosine (5mC) plays a role in gene expression, genomic imprinting, and suppression of transposable elements. Thus, the most widely characterized DNA methylation process is the covalent addition of the methyl group at the 5-carbon of the cytosine ring resulting in 5mC. These methyl groups project into the major groove of DNA and inhibit transcription (Lister et al. 2009). DNA methyltransferase (DNMT) participates in the generation of 5mC in an enzymatic activity dependent manner. Through the "one-carbon pool" of the human body, DNMT uses

S-adenosyl-L-methionine (SAM) as the methyl donor in the methylation of DNA/RNA/histone (Yan and Fujimori 2011; Crider et al. 2012). Formaldehyde serves as a methyl donor involving in the formation of SAM. Formaldehyde reacts with tetrahydrofolic acid (THF) to form 5,10-methylene tetrahydrofolic acid, which gives the methyl group in the formation of methionine.

Three kinds of DNA methyltransferases (DNMTs), DNMT1, DNMT3a, and DNMT3b, play important roles in methylating the vertebrate genomes (Law and Jacobsen 2010). DNMT1 is a component of the DNA replication complex, and it is able to maintain the level of methylated DNA, generated during DNA replication (Pollema-Mays et al. 2014). DNMT1 is present both in the germ cells and somatic cells, but it cannot turn on the methylation solely (Wang and Shen 2004). DNMT3a and DNMT3b are largely expressed in the early development stages and embryonic stem cells, which are able to turn on the DNA methylation process (Metivier et al. 2008). Tong and his colleagues have reported that the brain formaldehyde level influences the activity of DNMT and the degree of DNA methylation (Tong et al. 2013).

Formaldehyde not only serves as methyl donor in DNA methylation process, but also as one of the products in the process of DNA demethylation (Kress et al. 2001). The biochemical mechanisms include passive demethylation and active demethylation. Passive demethylation of CpG dinucleotides, which is the conserved methylation pattern, occurs in the DNA replication. The replication machinery incorporates unmethylated 5mC in the newly synthesized strand. The activity of DNMT1 was suppressed by chromatin structure and transcription complex after replication, which resulted in the low degree of DNA methylation. The demethylation of deoxycytidine and 5-azacytidine simply results from an absence of methylation maintenance after replication, following the formation of endogenous formaldehyde (Venturelli et al. 2013). So it is the passive process of demethylation and independent of any demethylating enzyme. With the progression of aging, the degree of DNA methylation decreases, and the accumulation of endogenous formaldehyde may occur if the elder people got dysmetabolism of formaldehyde.

An active demethylation involves many enzymes responsible for the modification, such as direct demethylation, nucleotide excision repair (NER), base excision repair (BER), 5mC deamination-dependent BER, 5mC oxidative demethylation, elongator complex protein 3, and catalyzed 5mC demethylation (Niehrs 2009; Wu and Zhang 2010). The active process does not related with replication, but accompanies with the formation of formaldehyde.

The modification of 5mC includes deamination, methyl dehydrogenation, and oxidative demethylation (Tahiliani et al. 1999). The main pattern of 5mC deamination is oxidative demethylation (Smith and Meissner 2013). The family of ten-eleven translocated oncogene (TET), including of TET1, TET2, and TET3, mainly participates in the oxidative demethylation (Wu and Zhang 2011). TET is a 2-oxoglutarate (2OG) and Fe^{2+}-dependent enzyme that catalyzes conversion of 5mC to 5-hydroxymethylcytosine (5hmC), 5-formylcytosine (5fC), and 5′-carboxylcytosine (5caC) through three consecutive oxidation reactions in cultured cells and in vitro (Wu and Ling 2014). The endogenous formaldehyde is produced in the process of oxidative demethylation of 5fC into cytosine. Additionally, the oxidative methyl of cytosine can be demethylated by glucosyltransferase through BER pathway, following which the endogenous formaldehyde and cytosine are produced (Su et al. 2015).

3 Formaldehyde and RNA Methylation/Demethylation

In the central dogma of molecular biology, genetic information flows from DNA to RNA and then to protein. RNA participates in biological systems not only by passing genetic information from DNA to protein but also by regulating various biological processes. The reversible RNA modifications play crucial roles in posttranscriptional regulation of gene expression. N6-Methyladenosine (m6A) is investigated as the most prevalent modification in mRNAs and long noncoding RNAs (lncRNAs) including ribosomal RNA (rRNA), small nuclear RNA (snRNA), and tRNA in the eukaryotes (Wang and He 2014). Several lncRNAs contain m6A which indicates that certain ncRNAs transcribed by RNA polymerase II are also subject to m6A methylation.

Like the process of active DNA demethylation, the balance of reversible RNA demethylation needs several crucial RNA demethylases, such as fat mass and obesity-associated gene (FTO) and alkB homolog 5 (ALKBH5) (Meyer et al. 2012). FTO and ALKBH5 are the members of the α-ketoglutarate and Fe^{2+}-dependent AlkB family of proteins that catalyze oxidative demethylation (Sanchez-Pulido and Andrade-Navarro 2007). M6A on nuclear RNA (including mRNA, lncRNA, and other types of RNA) is the main substrate of FTO (Zheng et al. 2012). M6A is oxidized at the methyl group to produce N6-hydroxymethyladenosine (hm6A), or undergo further oxidation to produce N6-formyladenosine (f6A), which leads to the demethylated products with the release of formaldehyde or formic acid (Gerken et al. 2007).

The noticeable phenotypes of both FTO and ALKBH5 mutations indicate the functional importance of reversible m6A methylation on RNA. FTO gene is associated with body mass index and predisposing to obesity and type 2 diabetes (Field et al. 2007; Frayling et al. 2007). FTO is also highly expressed in human brain and relates with cognitive change in a prospective cohort study (Bressler et al. 2012). ALKBH5 is important in meiosis, and the mice lacking of ALKBH5 are infertile (Klungland and Dahl 2014). However, the more detailed information on the demethylation to produce endogenous formaldehyde and its roles in the cognitive impairment should be pursued.

4 Formaldehyde Is Also Active in Histone Methylation/Demethylation

Histone modifications are thought to contribute to the control of gene expression through influencing chromatin compaction or other protein complexes. Reversible histone methylation/demethylation plays an important role in regulation of chromatin dynamics and gene transcription (Cheung et al. 2010). Histone methylation occurs on all basic residues: arginine, lysine, and histidine (Borde et al. 2009). Lysine can be monomethylated (me1), dimethylated (me2), or trimethylated (me3) on their ε-amine group, while arginines can be monomethylated (me1) (De Santa et al. 2007). Depending on the methyl of SAM supported by formaldehyde, the lysine and arginine of histone (H3/H4) can be methylated by the methylase (Pedersen and Helin 2010). On the other hand, the methylated histone can also be catalyzed by demethylase to generate formaldehyde, participating in the "one-carbon pool" cycle (Shi et al. 2004). The metabolic cycle of reversible histone methylation involves formaldehyde, SAM, and histone (Luka et al. 2014).

The most extensively studied histone methylation sites include histone H3 lysine 4 (H3K4), H3K9, H3K27, H3K36, H3K79, and H4K20. The discovery of an H3K4 demethylase, lysine-specific demethylase 1A (KDM1A), revealed that histone methylation is reversible. Lysine-specific demethylase 1 (LSD1) is a flavin-containing protein based on its ability to bind FAD. Its family includes KDM1, KDM2, KDM3, KDM4, KDM5, and KDM6 (Pedersen and Helin 2010; Cheng et al. 2014; Kumarasinghe and Woster 2014; Zhang et al. 2014). Different locations of N-methylated histones and amounts of methyl catalyzed by KDM result in different epigenetic effects (Li et al. 2014).

LSD1 belongs to the flavin adenine dinucleotide (FAD)-dependent amine oxidase family of demethylases and demethylates mono- and dimethylated histone H3 lysine 4 and H3 lysine 9, which is hydrolyzed to form formaldehyde and other aldehydes via a nonenzymatic process (Neelamegam et al. 2012). Semicarbazide-sensitive amine oxidase (SSAO), which catalyzes methylamine to produce formaldehyde, shares the same cofactor (FAD) and mechanism as LSD1 in the cleavage of the inactivated carbon–nitrogen bonds from their substrates (Zhou et al. 2014). LSD1 may catalyze the conversion of one molecule of H3K4 to generate nonmethylated H3K4 and two molecules of formaldehyde. Lysine of me1 and me2 can be methylated, because the oxidation of methylated amine needs at least one proton (Zhou et al. 2014). In a complete catalytic cycle, the cofactor FAD is reduced to FADH2 and then is likely to be reoxidized by oxygen to produce hydrogen peroxide and formaldehyde.

Methylation/demethylation of DNA, RNA, and histone plays crucial roles in the development, behavior, cognition, and other processes in mammals (Klose and Zhang 2007; Smith and Meissner 2013). The dysfunction of reversible DNA, RNA, and histone methylation participates in many diseases, such as cancer, development malformation, and cognitive impairments (Wang et al. 2016). Excess endogenous formaldehyde is also reported as one of the important risks of cancer and cognitive dysfunction (He et al. 2010). Formaldehyde participates in the reversible methylation of DNA, RNA, and histone, which could regulate the epigenetic memory in chromosome segregation, transcriptional regulation, and DNA repair. Considering DNA methylation is one of the key factors in the memory formation in the mammal; formaldehyde can influence the learning and memory process. As the previous study reported, the excess formaldehyde in the brain has repressive effect on the DNMT activities, resulting in decreased level of DNA methylation (Liu et al. 2011; Tong et al. 2014).

Formaldehyde is the intermediate product of methanol metabolism. The methanol is catalyzed to generate formaldehyde by ethanol dehydrogenase (ADH) and further to produce formic acid in the oxidation of formaldehyde dehydrogenase (FDH) (Staab et al. 2009). In the transgenic model which knocks down the FDH, the rat performed spatial learning and memory impairment compared with the control group (Qiang et al. 2014; Wu et al. 2014). Formaldehyde can pass blood–brain barrier (BBB) as a small molecule compound (Shcherbakova et al. 1986). Formaldehyde metabolism declines gradually with aging (Tong et al. 2013). The accumulation of formaldehyde in the central nervous system could result in brain damage and cognitive dysfunction.

The reversible methylation of DNA, RNA, and histone may be one of the sources of endogenous formaldehyde in the body. Formaldehyde participates in the epigenetic modification as the methyl donor. The important role of endogenous formaldehyde in reversible methylation and the mechanism of endogenous formaldehyde influencing the cognition need further investigation in the future.

5 Methylation and Demethylation and Memory

In 1969, Griffith and Mahler put forward the DNA ticketing theory of memory, which regards DNA as one kind of information storage molecules with all lifelong maintenance (Griffith and Mahler 1969). Afterward, Crick hypothesized that memory is stored by reversible modifications to DNA and protein by some kinds of enzymes (Crick 1984). DNA methyltransferase is reported as such an enzyme. Holliday and colleagues indicated that methylation of DNA cytosine residues is important for memory formation (Holliday 1999). According to the time of information storage, memory is divided into short-term memory (STM) and long-term memory (LTM). As previously reported (Fiumara et al. 2007), STM formation usually requires modification of the existing synapses in the absence of changes in protein expression. LTM formation requires activation and inactivation of transcription factors and the induction of complicated synapse-to-nuclear signal transduction cascades that lead to stable changes in gene expression (Morris and Monteggia 2014; Heyward and Sweatt 2015).

Alzheimer's disease (AD) is a chronic neurodegenerative disease that causes 60% to 70% of cases of dementia. AD is a polygenetic disease whose characteristic symptom is memory loss (Lippa 1999). The genetic research identifies that mutations in presenilin 1 (PS1) and amyloid precursor protein (APP) linked to the early onset of familial Alzheimer's disease (FAD) (Hsiao et al. 1996). However, the most AD patients are with sporadic AD which the risk is believed to be polygenetic (Roger et al. 1995). Pollwein and colleagues found that the expression of APP is regulated by its promoter rich in CpG islands (Pollwein et al. 2016). The methylation of upstream promoter element was observed with high tissue specificity (Rogaev et al. 1994). Furthermore, the methylation of APP promoter is highly correlated with age (Tahiliani Met al. 1999). The mRNA level of APP in the brain of female rats is lower than that of male rats (Mani and Thakur 2006). In addition to APP, other genes such as BACE and PS1 are also associated with DNA methylation (Fuso et al. 2005).

So far, abundant investigations have been carried out on the relation between (de) methylation and memory formation (Jarome and Lubin 2014). DNA or histone methylation, which is a stable epigenetic modification in regulating transcription and imprinting and silencing of transposons, plays central roles in memory formation and maintenance in synapse development (Dias et al. 2015). In this topic, Heyward and Sweatt emphasized the hypothesis on epigenetics in memory (Heyward and Sweatt 2015). Methylation-mediated alterations in gene expression drive memory formation have adopted a long-standing conceptual framework in which genes are either classified as being permissive for memory (i.e., memory promoters) or disruptive toward memory formation (i.e., memory suppressor) Lister and Mukamel thought dynamics of DNA methylation in the brain is important to learning and memory for human. DNA methylation featured with different spatio-temporal characteristics may facilitate distinct permissive or instructive roles in brain function, memory, and learning (Lister and Mukamel 2015). Of course, DNA methyltransferases play a critical role in DNA methylation and in learning and memory (Morris and Monteggia 2014). Since formaldehyde plays a role in methylation and demethylation, we have carried out some work on formaldehyde to test its roles with memory and the function of DNMT (Tong et al. 2013).

Regulation of chromatin structure through posttranslational modification of histone proteins, primarily histone H3 phosphorylation and acetylation, is an important early step in the induction of synaptic plasticity and formation of long-term memory (Ng et al. 2003; Sims et al. 2003). Gupta and colleagues found that trimethylation of histone H3 at lysine 4 (H3K4), an active mark for transcription, is upregulated in the hippocampus one hour following contextual fear conditioning (Gupta et al. 2010). They believe that histone methylation is actively regulated in the hippocampus and facilitates long-term memory formation. Ashley and colleagues provided the initial demonstration that histone modifications and DNA methylation play vital roles in the estrogenic regulation of memory (Fortress and Frick 2014). These investigations show the close association of demethylation/methylation of DNA and histones with memory formation and suppression.

Although international colleagues pay much attention to the relationship between epigenetics and memory, and try to clarify the different mechanisms of methylation and demethylation undergoing the formation and suppression of memory, very few have got down to investigate where methyl groups come and go and what the effect the product (formaldehyde) of released methyl group has on cognitive ability and impairment in vivo. The formaldehyde acts as a donor as well as a participant in the methylation and demethylation. We would like to have a paragraph to discuss the topic.

6 Formaldehyde Acting as an Epigenetic Factor in Formation and Loss of Memory

Histone methylation or demethylation is a process by which methyl groups are transferred onto or taken off the amino acids of histone proteins. Formaldehyde, a methyl group donor, participates in the modification of DNA and histone. In fact, cytosine methylation of genomic DNA decreases with age in different tissues of mammals (Qazi et al. 2017). On the other hand, endogenous formaldehyde increases with aging, which may result from the demethylation of DNA, RNA, and histones. The role of epigenetic factors, for instance, formaldehyde in developing neurodegenerative disorders in aging, should be focused. Formaldehyde is involved in different epigenetic mechanisms including DNA and histone (de)methylation in Alzheimer's disease (Tong et al. 2013). This suggests that imbalance of formaldehyde metabolism interferes with DNA and histone methylation and demethylation, which is hypothesized as a risk factor involved in age-related cognitive impairment including memory loss.

To investigate the intrinsic linkage of endogenous formaldehyde to learning-responsive DNA methylation and memory formation, Tong and colleagues trained Sprague–Dawley rats in spatial learning and observed whether the training induces changes in DNA methylation and demethylation. During the training, an initial global DNA demethylation occurs, followed by a subsequent re-methylation associated with hippocampal formaldehyde elevation, and then decreased to baseline level (Tong et al. 2013). They intrahippocampally injected normal adult rats with excess formaldehyde which can imitate the global DNA methylation decline associated with age-related spatial memory deficits. To further investigate the relation of formaldehyde, DNA methylation, and memory, they scavenged the elevated formaldehyde by formaldehyde-degrading enzyme (FDH), enhanced DNA demethylation by a DNA-demethylating agent, or blocked DNA re-methylation using 5-aza-2-deoxycytidine, a DNA-demethylating agent by inhibiting DNMT activity (Yamagata et al. 2012; Hamm et al. 2009) in rats. They observed that all the three processes led to spatial memory deficits during spatial memory formation. That is to say, aging-associated excess formaldehyde may contribute to cognitive decline during aging.

DNA methyltransferase (DNMT) plays a role in memory formation through its regulation of DNA methylation (Day and Sweatt 2010; Levenson et al. 2006). Reduction in global DNA methylation by simultaneous knockout of DNMT1 and DNMT3a results in deficits in both memory acquisition and retrieval in mice (Feng et al. 2010; Miller et al. 2010). As described by Shi and colleagues, memory reconsolidation for a cocaine-paired stimulus depends critically on DNMT activity in the bilateral intra-basolateral amygdala (Shi et al. 2015). The spatial memory deficits of aged mice could be reversed by overexpression of DNMT3a in the hippocampus (Su and Tsai 2012). On the other hand, activity and expression of DNMT as well as levels of global DNA methylation observably decrease in brains of the aged mice, aged rats and in autopsied brain from AD patients. Hippocampus-related topographic amnesia is the most common symptom of memory disorders in Alzheimer's disease (AD) patients. Therefore, the misregulated DNMT function associated with defects in global DNA methylation may be one of the patho-mechanisms underlying hippocampal-related spatial memory deficits.

Noticing that endogenous formaldehyde is involved in global DNA methylation and demethylation, Tong and coworkers have investigated the relation between endogenous formaldehyde and DNMT activity to understand the experience-mediated DNA methylation, which is required for the formation of recent memory as well as the maintenance of remote memory. In their experiments, they found a marked increase in endogenous formaldehyde levels associated with a decline in global DNA methylation in the autopsied hippocampus from AD patients. Formaldehyde in excess of normal physiological levels decreased the DNA methylation by disturbance of the activities of DNMTs in vitro and in vivo. As described by Tong and coworkers, intrahippocampal injection of excess formaldehyde before spatial learning in healthy adult wild-type rats results in learning difficulties similar to the patients at early stage of age-related cognitive impairment (Tong et al. 2015). After injection of excess formaldehyde after spatial learning, the rats suffered from the loss of remote spatial memory as observed in late stage of AD. These data support the contribution of aging-associated formaldehyde to topographic amnesia in age-related cognitive impaired patients through DNA methylation and demethylation. Whether formaldehyde, associated with histone methylation and demethylation, is also directly or indirectly involved in learning and memory needs further investigation.

7 Conclusion

DNA methylation, histone methylation, and histone acetylation, as the three major epigenetic processes, are involved in memory regulation. Chromatin remodeling through epigenetic controls such as methylation and demethylation of DNA or histones is thought as one of the mechanisms for learning and memory including long-term memory storage. In the methylation and demethylation, we supposed that endogenous formaldehyde acts as an epigenetic factor involved in the formation or loss of memory (Su et al. 2015). This viewpoint is supported by these observations as follows:

1. As described previously, endogenous formaldehyde participates in DNA and histone methylation (Su et al. 2015; Li et al. 2014).
2. Formaldehyde participates in demethylation and methylation cycle of DNA or histone, in which demethylation of DNA or histone is associated with the release of formaldehyde (Shi et al. 2004). DNA demethylation elicits formaldehyde generation in the dividing cells and post-mitotic neurons (Tong et al. 2012).
3. Endogenous formaldehyde is increased with aging and positively correlated with cognitive impairment, such as AD and post-stroke dementia (Tong et al. 2009, 2016).
4. Enhancement of DNA demethylation or block of DNA re-methylation using 5-aza-2-deoxycytidine in rats leads to spatial memory deficits during spatial memory formation (Tong et al. 2011).
5. Changes in levels of formaldehyde are associated with the expression and activity of DNMTs in vivo and in vitro (Tong et al. 2015).
6. High concentration of formaldehyde inhibited the N-methyl-D-aspartate receptor, which impaired the learning and memory in rats (Yu et al. 2011; Taylor et al. 2014).
7. Epigenetic mechanisms can contribute to the pathologies of neurological disorders and cause memory-related symptoms (Kim and Kaang 2017).

Methylation of DNA or histone is a critical step for memory formation (Gupta et al. 2010). All these results suggest that homeostasis of endogenous formaldehyde metabolism benefited the cognitive behaviors, and dysmetabolism of endogenous formaldehyde may interfere with DNA or histone methylation and leads to memory loss. Whether formaldehyde involved in RNA methylation is also related to formation and loss of memory should be investigated further.

Acknowledgment This project was supported by grants from the Beijing Municipal Science and Technology Project (Z161100000217141; Z161100000216137), the National Key Research and Development Program of China (2016YFC1306300), the National Basic Research Program of China (973 Program) (2012CB911004), the Natural Scientific Foundation of China (NSFC 31301880, NSFC 31270868), the Foundation of Chinese Academy of Sciences CAS-20140909, the Queensland-Chinese Academy of Sciences Biotechnology Fund (GJHZ201302), the Program for Liaoning Excellent Talents in University (LJQ2015057), and the Dalian High Level Talent Innovation Support Plan (No. 2015R067).

Competing Financial Interests The authors declare no competing financial interests.

References

Abdelfatah E, Kerner Z, Nanda N, Ahuja N (2016) Epigenetic therapy in gastrointestinal cancer: the right combination. Therap Adv Gastroenterol 9(4):560–579

Berger SL, Kouzarides T, Shiekhattar R, Shilatifard A (2009) An operational definition of epigenetics. Genes Dev 23(7):781–783

Bergman Y, Cedar H (2013) DNA methylation dynamics in health and disease. Nat Struct Mol Biol 20(3):274–281

Bird A (1992) The essentials of DNA methylation. Cell 70(1):5–8
Bochtler M, Kolano A, Xu GL (2016) DNA demethylation pathways: additional players and regulators. BioEssays 39(1):1–13
Borde V, Robine N, Lin W, Bonfils S, Géli V, Nicolas A (2009) Histone H3 lysine 4 trimethylation marks meiotic recombination initiation sites. EMBO J 28(2):99–111
Bressler J, Fornage M, Demerath EW, Knopman DS, Monda KL, North KE, Penman A, Mosley TH, Boerwinkle E (2012) Fat mass and obesity gene and cognitive decline: the atherosclerosis risk in communities study. Neurology 80(1):92–99
Cheng MB, Zhang Y, Cao CY, Zhang WL, Zhang Y, Shen YF (2014) Specific phosphorylation of histone demethylase KDM3A determines target gene expression in response to heat shock. PLoS Biol 12(12):e1002026
Cheung VG, Sherman SL, Feingold E (2010) Genetic control of hotspots. Science 327(5967):791–792
Chouliaras L, Mastroeni D, Delvaux E, Grover A, Kenis G, Hof PR, Steinbusch HWM, Coleman PD, Rutten BPF, van den Hove DLA (2013) Consistent decrease in global DNA methylation and hydroxymethylation in the hippocampus of Alzheimer's disease patients. Neurobiol Aging 34(9):2091–2099
Crick F (1984) Memory and molecular turnover. Nature 312(5990):101
Crider KS, Yang TP, Berry RJ, Bailey LB (2012) Folate and DNA methylation: a review of molecular mechanisms and the evidence for folate's role. Adv Nutr 3(1):21–38
Day JJ, Sweatt JD (2010) DNA methylation and memory formation. Nat Neurosci 13(11):1319–1323
Dezi V, Ivanov C, Haussmann IU, Soller M (2016) Nucleotide modifications in messenger RNA and their role in development and disease. Biochem Soc Trans 44(5):1385–1393
Dias BG, Maddox S, Klengel T, Ressler KJ (2015) Epigenetic mechanisms underlying learning and the inheritance of learned behaviors. Trends Neurosci 38(2):96–107
Feng J, Zhou Y, Campbell SL, Le T, Li E, Sweatt JD et al (2010) Dnmt1 and Dnmt3a maintain DNA methylation and regulate synaptic function in adult forebrain neurons. Nat Neurosci 13:423–430
Field SF, JMM H, Walker NM, Dunger DB, Todd JA (2007) Analysis of the obesity gene FTO in 14,803 type 1 diabetes cases and controls. Diabetologia 50(10):2218–2220
Fiumara F, Milanese C, Corradi A, Giovedì S, Leitinger G, Menegon A, Montarolo PG, Benfenati F, Ghirardi M (2007) Phosphorylation of synapsin domain a is required for post-tetanic potentiation. J Cell Sci 120(Pt 18):3228–3237
Fortress AM, Frick KM (2014) Epigenetic regulation of estrogen-dependent memory. Front Neuroendocrinol 35(4):530–549
Frayling TM, Timpson NJ, Weedon MN, Zeggini E, Freathy RM, Lindgren CM, Perry JRB, Elliott KS, Lango H, Rayner NW, Shields B, Harries LW, Barrett JC, Ellard S, Groves CJ, Knight B, Patch AM, Ness AR, Ebrahim S, Lawlor DA, Ring SM, Ben-Shlomo Y, Jarvelin MR, Sovio U, Bennett AJ, Melzer D, Ferrucci L, Loos RJF, Barroso I, Wareham NJ, Karpe F, Owen KR, Cardon LR, Walker M, Hitman GA, Palmer CNA, Doney ASF, Morris AD, Smith GD, Hattersley AT, McCarthy MI (2007) A common variant in the FTO gene is associated with body mass index and predisposes to childhood and adult obesity. Science 316(5826):889–894
Fuso A, Seminara L, Cavallaro RA, D'Anselmi F, Scarpa S (2005) S-adenosylmethionine/homocysteine cycle alterations modify DNA methylation status with consequent deregulation of PS1 and BACE and beta-amyloid production. Mol Cell Neurosci 28(1):195–204
Gerken T, Girard CA, Tung YCL, Webby CJ, Saudek V, Hewitson KS, Yeo GSH, McDonough MA, Cunliffe S, McNeill LA, Galvanovskis J, Rorsman P, Robins P, Prieur X, Coll AP, Ma M, Jovanovic Z, Farooqi IS, Sedgwick B, Barroso I, Lindahl T, Ponting CP, Ashcroft FM, O'Rahilly S, Schofield CJ (2007) The obesity-associated FTO gene encodes a 2-oxoglutarate-dependent nucleic acid demethylase. Science 318(5855):1469–1472
Griffith JS, Mahler HR (1969) DNA ticketing theory of memory. Nature 223(5206):580–582
Gupta S, Kim SY, Artis S, Molfese DL, Schumacher A, Sweatt JD, Paylor RE, Lubin FD (2010) Histone methylation regulates memory formation. J Neurosci 30(10):3589–3599

Hamm CA, Xie HH, Costa FF, Vanin EF, Seftor EA, Sredni ST, Bischof J, Wang DL, Bonaldo MF, Hendrix MJC, Soares MB (2009) Global demethylation of rat chondrosarcoma cells after treatment with 5-aza-29-deoxycytidine results in increased tumorigenicity. PLoS One 4:e8340
He RQ, Lu J, Miao JY (2010) Formaldehyde stress. Sci China Life Sci 53(12):1399–1404
Heyward FD, Sweatt JD (2015) DNA methylation in memory formation: emerging insights. Neuroscientist 21(5):475–489
Holliday R (1999) Is there an epigenetic component in long-term memory? J Theor Biol 200(3):339–341
Hsiao K, Chapman P, Nilsen S, Eckman C, Harigaya Y, Younkin S, Yang FS, Cole G (1996). Correlative memory deficits, Abeta elevation, and amyloid plaques in transgenic mice. Science 274(5284):99–102
Jabbari K, Bernardi G (2004) Cytosine methylation and CpG, TpG (CpA) and TpA frequencies. Gene 333:143–149
Jarome TJ, Lubin FD (2014) Epigenetic mechanisms of memory formation and reconsolidation. Neurobiol Learn Mem 115:116–127
Kim S, Kaang BK (2017) Epigenetic regulation and chromatin remodeling in learning and memory. Exp Mol Med 49(1):e281
Klose RJ, Zhang Y (2007) Regulation of histone methylation by demethylimination and demethylation. Nat Rev Mol Cell Biol 8(4):307–318
Klungland A, Dahl JA (2014) Dynamic RNA modifications in disease. Curr Opin Genet De 26:47–52
Kress C, Thomassin H, Grange T (2001) Local DNA demethylation in vertebrates: how could it be performed and targeted? FEBS Lett 494(3):135–140
Kumarasinghe IR, Woster PM (2014) Synthesis and evaluation of novel cyclic peptide inhibitors of lysine-specific demethylase 1. ACS Med Chem Lett 5(1):29–33
Law JA, Jacobsen SE (2010) Establishing, maintaining and modifying DNA methylation patterns in plants and animals. Nat Rev Genet 11(3):204–220
Levenson CW (2006) Regulation of the NMDA receptor: implications for neuropsychological development. Nutr Rev 64(9):428–432
Li T, Su T, He YG, He RQ (2014) Formaldehyde, an epigenetic factor for learning and memory participating histone (de)methylation. Acta Neuropharmacologica 57(4):403–411
Li LL, Li C, Mao HT, Du ZF, Chan WY, Murray P, Luo B, Chan ATC, Mok TSK, Chan FKL, Ambinder RF, Tao Q (2016) Epigenetic inactivation of the CpG demethylase TET1 as a DNA methylation feedback loop in human cancers. Sci Rep 6:26591
Lippa CF (1999) Familial Alzheimer's disease: genetic influences on the disease process. Int J Mol Med 4(5):529–536
Lister R, Mukamel EA (2015) Turning over DNA methylation in the mind. Front Neurosci 9:252
Lister R, Pelizzola M, Dowen RH, Hawkins DR, Hon G, Tonti-Filippini J, Nery JR, Lee L, Ye Z, Ngo Q-M, Edsall L, Antosiewicz-Bourget J, Stewart R, Ruotti V, Millar HA, Thomson JA, Ren B, Ecker JR (2009) Human DNA methylomes at base resolution show widespread epigenomic differences. Nature 462:315–322
Liu QC, Yang LQ, Gong CM, Tao GH, Huang HY, Liu JJ, Zhang HM, Wu DS, Xia B, Hu GH, Wang KP, Zhuang ZX (2011) Effects of long-term low-dose formaldehyde exposure on global genomic hypomethylation in 16HBE cells. Toxicol Lett 205(3):235–240
Luka Z, Pakhomova S, Loukachevitch LV, Calcutt MW, Newcomer ME, Wagner C (2014) Crystal structure of the histone lysine specific demethylase LSD1 complexed with tetrahydrofolate. Protein Sci 23(7):993–998
Mani ST, Thakur MK (2006) In the cerebral cortex of female and male mice, amyloid precursor protein (APP) promoter methylation is higher in females and differentially regulated by sex steroids. Brain Res 1067(1):43–47
Martinowich K, Hattori D, Wu H, Fouse S, He F, Hu Y, Fan G, Sun YE (2003) DNA methylation-related chromatin remodeling in activity-dependent BDNF gene regulation. Science 302(5646):890–893

Mastroeni D, Grover A, Delvaux E, Whiteside C, Coleman PD, Rogers J (2010) Epigenetic changes in Alzheimer's disease: decrements in DNA methylation. Neurobiol Aging 31(12):2025–2037
Mastroeni D, Grover A, Delvaux E, Whiteside C, Coleman PD, Rogers J (2011) Epigenetic mechanisms in Alzheimer's disease. Neurobiol Aging 32(7):1161–1180
Métivier R, Gallais R, Tiffoche C, Le Péron C, Jurkowska RZ, Carmouche RP, Ibberson D, Barath P, Demay F, Reid G, Benes V, Jeltsch A, Gannon F, Salbert G (2008) Cyclical DNA methylation of a transcriptionally active promoter. Nature 452(7183):45–50
Meyer KD, Saletore Y, Zumbo P, Elemento O, Mason CE, Jaffrey SR (2012) Comprehensive analysis of mRNA methylation reveals enrichment in 3′ UTRs and near stop codons. Cell 149(7):1635–1646
Miller CA, Gavin CF, White JA, Parrish RR, Honasoge A, Yancey CR et al (2010) Cortical DNA methylation maintains remote memory. Nat Neurosci 13:664–666
Mirfattah B, Herring J, Tang H, Zhang K (2016) Probes and targets of DNA methylation and demethylation in drug development. Curr Top Med Chem 17(15):1727–1740
Morris MJ, Monteggia LM (2014) Role of DNA methylation and the DNA methyltransferases in learning and memory. Dialogues Clin Neurosci 16(3):359–371
Neelamegam R, Ricq EL, Malvaez M, Patnaik D, Norton S, Carlin SM, Hill IT, Wood MA, Haggarty SJ, Hookerl JM (2012) Brain-penetrant LSD1 inhibitors can block memory consolidation. ACS Chem Neurosci 3(2):120–128
Ng HH, Robert F, Young RA, Struh K (2003) Targeted recruitment of Set1 histone methylase by elongating Pol II provides a localized mark and memory of recent transcriptional activity. Mol Cell 11(3):709–719
Niehrs C (2009) Active DNA demethylation and DNA repair. Differentiation 77(1):1–11
Oliveira AMM, Hemstedt TJ, Bading H (2012) Rescue of aging-associated decline in Dnmt3a2 expression restores cognitive abilities. Nat Neurosci 15(8):1111–1113
Ooi SK, Bestor TH (2008) The colorful history of active DNA demethylation. Cell 133(7):1145–1148
Pogribny IP, Vanyushin BF (2010) Age-related genomic hypomethylation. Epigen Aging:11e27
Pollema-Mays SL, Centeno MV, Apkarian AV, Martina M (2014) Expression of DNA methyltransferases in adult dorsal root ganglia is cell-type specific and up regulated in a rodent model of neuropathic pain. Front Cell Neurosci 8:217
Pollwein P, Masters CL, Beyreuther K (2016) The expression of the amyloid precursor protein (APP) is regulated by two GC-elements in the promoter. Nucleic Acids Res 20(1):63–68
Qazi TJ, Quan Z, Mir A, Qing H (2017) Epigenetics in Alzheimer's disease: perspective of DNA methylation. Mol Neurobiol. doi:10.1007/s12035-016-0357-6
Qiang M, Xiao R, Su T, Wu BB, Tong ZQ, Liu Y, He RQ (2014) A novel mechanism for endogenous formaldehyde elevation in SAMP8 mouse. J Alzheimers Dis 40(4):1039–1053
Reik W, Dean W, Walter J (2001) Epigenetic reprogramming in mammalian development. Science 293(5532):1089–1093
Rogaev EI, Lukiw WJ, Lavrushina O, Rogaeva EA, St George-Hyslop PH (1994) The upstream promoter of the beta-amyloid precursor protein gene (APP) shows differential patterns of methylation in human brain. Genomics 22(2):340–347
Roger L, WestJohn M, LeeLeonard E, Maroun (1995) Hypomethylation of the amyloid precursor protein gene in the brain of an Alzheimer's disease patient. J Mol Neurosci 6(2):141–146
Sanchez-Pulido L, Andrade-Navarro MA (2007) The FTO (fat mass and obesity associated) gene codes for a novel member of the non-heme dioxygenase superfamily. BMC Biochem 8:23
Santa FD, Totaro MG, Prosperini E, Notarbartolo S, Testa G, Natoli G (2007) The histone H3 lysine-27 demethylase Jmjd3 links inflammation to inhibition of polycomb-mediated gene silencing. Cell 130(6):1083–1094
Shcherbakova LN, Tel'pukhov VI, Trenin SO, Bashilov IA, Lapkina TI (1986) Permeability of the blood-brain barrier to intra-arterial formaldehyde. Biull Eksp Biol Med 102(11):573–575

Shi YJ, Lan F, Matson C, Mulligan P, Whetstine JR, Cole PA, Casero RA, Shi Y (2004) Histone demethylation mediated by the nuclear amine oxidase homolog LSD1. Cell 119(7):941–953

Shi HS, Luo YX, Yin X, Wu HH, Xue G, Geng XH, Hou YN (2015) Reconsolidation of a cocaine associated memory requires DNA methyltransferase activity in the basolateral amygdala. Sci Rep 5:13327

Sims RJ, Nishioka K, Reinberg D (2003) Histone lysine methylation: a signature for chromatin function. Trends Genet 19(11):629–639

Smith ZD, Meissner A (2013) DNA methylation: roles in mammalian development. Nat Rev Genet 14(3):204–220

Staab CA, Ålander J, Morgenstern R, Grafström RC, Höög JO (2009) The Janus face of alcohol dehydrogenase 3. Chem Biol Interact 178(1–3):29–35

Su SC, Tsai LH (2012) DNA methylation in cognition comes of age. Nat Neurosci 2012 Jul 26;15(8):1061–1062

Su T, Song D, Li T, Wang XH, He RQ (2015) Nucleic acid methylation/demethylation, endogenous formaldehyde and age-related cognitive impairment. Prog Biochem Biophys 43(3):211–219

Tahiliani M, Koh KP, Shen YH, Pastor WA, Bandukwala H, Brudno Y, Agarwa S, Iyer LM, Liu DR, Aravind L, Tohgi H, Utsugisawa K, Nagane Y, Yoshimura M, Genda Y, Ukitsu M (1999) Reduction with age in methylcytosine in the promoter region −224 approximately −101 of the amyloid precursor protein gene in autopsy human cortex. Brain Res Mol Brain Res 70(2):288–292

Taylor CJ, He RQ, Bartlett PF (2014) The role of the N-methyl-D-aspartate receptor in the proliferation of adult hippocampal neural stem and precursor cells. Sci China Life Sci 57(4):403–411

Tohgi H, Utsugisawa K, Nagane Y, Yoshimura M, Genda Y, Ukitsu M (1999) Reduction with age in methylcytosine in the promoter region -224 approximately -101 of the amyloid precursor protein gene in autopsy human cortex. Brain Res Mol Brain Res 70(2):288–292

Tong ZQ, Zhang JL, Luo WH, Wang WS, Li FX, Li H, Luo HJ, Lu J, Zhou JN, Wan Y, He RQ (2009) Urine formaldehyde level is inversely correlated to mini mental state examination scores in senile dementia. Neurobiol Aging 32(1):31–41

Tong ZQ, Han CS, Miao JY, Lu J, He RQ (2011) Excess endogenous formaldehyde induces memory decline. Prog Biochem Biophys 38(6):575–579

Tong ZQ, Han CS, Luo WH, Wang XH, Li H, Luo HJ, Zhou JN, Qi JS, He RQ (2012) Accumulated hippocampal formaldehyde induces age-dependent memory decline. Age (Dordr) 35(3):583–596

Tong ZQ, Han CS, Luo WH, Li H, Luo HJ, Qiang M, Su T, Wu BB, Liu Y, Yang X, Wan Y, Cui DH, He RQ (2013) Aging-associated excess formaldehyde leads to spatial memory deficits. Sci Rep 3:1807

Tong ZQ, Han CS, Qiang M, Wang WS, Lv JH, Zhang SZ, Luo WH, Lie H, Luo HJ, Zhou JN, Wu BB, Su T, Yang X, Wang XM, Liu Y, He RQ (2015) Age-related formaldehyde interferes with DNA methyltransferase function, causing memory loss in Alzheimer's disease. Neurobiol Aging 36(1):100–110

Tong ZQ, Wang WS, Luo WH, Lv JH, Li H, Luo HJ, Jia JP, He RQ (2016) Urine formaldehyde predicts cognitive impairment in post-stroke dementia and Alzheimer's disease. J Alzheimers Dis 55(3):1031–1038

Venturelli S, Berger A, Weiland T, Essmann F, Waibel M, Nuebling T, Häcker S, Schenk M, Schulze-Osthoff K, Salih HR, Fulda S, Sipos B, Johnstone RW, Lauer UM, Bitzer M (2013) Differential induction of apoptosis and senescence by the DNA methyltransferase inhibitors 5-azacytidine and 5-aza-2′-deoxycytidine in solid tumor cells. Mol Cancer Ther 12(10):2226–2236

Wang X, He C (2014) Dynamic RNA modifications in posttranscriptional regulation. Mol Cell 56(1):5–12

Wang KY, Shen CKJ (2004) DNA methyltransferase Dnmt1 and mismatch repair. Oncogene 23(47):7898–7902

Wang ZQ, Tang BS, He YQ, Jin P (2016) DNA methylation dynamics in neurogenesis. Epigenomics 8(3):401–414

West RL, Lee JM, Maroun LE (1995) Hypomethylation of the amyloid precursor protein gene in the brain of an Alzheimer's disease patient. J Mol Neurosci 6(2):141–146

Wu YC, Ling ZQ (2014) The role of TET family proteins and 5-hydroxymethylcytosine in human tumors. Histol Histopathol 29(8):991–997

Wu SC, Zhang Y (2010) Active DNA demethylation: many roads lead to Rome. Nat Rev Mol Cell Biol 11(9):607–620

Wu H, Zhang Y (2011) Mechanisms and functions of Tet protein-mediated 5-methylcytosine oxidation. Genes Dev 25(23):2436–2452

Wu KY, Ren RT, Su WT, Wen B, Zhang YY, Yi F, Qiao XH, Yuan TT, Wang JH, Liu LM, Belmonte JCI, Liu GH, Chen C (2014) A novel suppressive effect of alcohol dehydrogenase 5 in neuronal differentiation. J Biol Chem 289(29):20193–20199

Yamagata Y, Szabó P, Szüts D, Bacquet C, Arányi T, Páldi A (2012) Rapid turnover of DNA methylation in human cells. Epigenetics 7:141–145

Yan F, Fujimori DG (2011) RNA methylation by radical SAM enzymes RlmN and Cfr proceeds via methylene transfer and hydride shift. Proc Natl Acad Sci 108(10):3930–3934

Yu NK, Baek SH, Kaang BK (2011) DNA methylation-mediated control of learning and memory. Mol Brain 4:5

Zhang LP, Wang J, Pan YQ, Jin J, Sang JQ, Huang P, Shao GB (2014) Expression of histone H3 lysine 4 methylation and its demethylases in the developing mouse testis. Cell Tissue Res 358(3):875–883

Zheng G, Dah JA, Niu YM, Fedorcsak P, Huang CM, Li CJ, Vågbø CB, Shi Y, Wang WL, Song SH, Lu ZK, Bosmans RPG, Dai Q, Hao YJ, Yang X, Zhao WM, Tong WM, Wang XJ, Bogdan F, Furu K, Fu Y, Jia GF, Zhao X, Liu J, Krokan HE, Klungland A, Yang YG, He C (2012) ALKBH5 is a mammalian RNA demethylase that impacts RNA metabolism and mouse fertility. Mol Cell 49(1):18–29

Zhou C, Kang D, Xu YG, Zhang LY, Zha XM (2014) Identification of novel selective lysine-specific demethylase 1 (LSD1) inhibitors using a pharmacophore-based virtual screening combined with docking. Chem Biol Drug Des 85(6):659–671

Chapter 4
Dehydration, Formaldehyde, and Age-Related Cognitive Impairment

Ting Li, Tong Ge, and Rongqiao He

Abstract Chronic dehydration is regarded as a common symptom of patients with age-related cognitive impairment, particularly those with Alzheimer's disease. It has been known that chronic dehydration causes not only serum hyperosmotic pressure increase but also metabolic dysfunction of the central nervous system and cognitive impairment. Recently, imbalance of anabolism and catabolism of endogenous formaldehyde has been considered as one of the risk factors involving age-related cognitive impairment. Formaldehyde at low concentration induces the senile plaques (amyloid β deposition), neurofibrillary-like tangles (containing tau hyperphosphorylation), and cognitive impairment in the brain of monkeys. The levels of endogenous formaldehyde of inpatients were markedly higher than those of the age-matched participants. Thus, we are concerned about the relationship between cognitive impairment and water intake behavior as well as that between dehydration and dysmetabolism of endogenous formaldehyde with aging. As a result of aging, serum AVP concentration and brain endogenous formaldehyde levels increase. Excess formaldehyde could activate ANG II to promote AVP expression. Meanwhile, elevation of AVP activates semicarbazide-sensitive amine oxidase (SSAO) to produce more endogenous formaldehyde. These pathways could create a "vicious cycle" worsening water intake behavior that both drinking frequency and quantity decrease. That is, at least, dysmetabolism of endogenous formaldehyde can affect water intake behaviors. The highest concentration of endogenous formaldehyde is found in the morning, followed by a decrease around noon and an increase in the evening. Therefore, regular water intake including tea or other beverages containing resveratrol is recommended to intervene and reduce the accumulation of formaldehyde in the human body, especially when it is taken in the morning.

Keywords Dehydration • Thirst • Sensation • Respond

T. Li • T. Ge
State Key Laboratory of Biomacromolecules, Institute of Biophysics, Chinese Academy of Sciences, Beijing 100101, China

R. He (✉)
State Key Laboratory of Brain and Cognitive Science, Institute of Biophysics, Chinese Academy of Sciences, Beijing 100101, China

University of Chinese Academy of Sciences, Beijing 100049, China
e-mail: rongqiaohe@163.com

R. He, *Formaldehyde and Cognition*, DOI 10.1007/978-94-024-1177-5_4

1 Introduction

According to Thales (636–546 B.C.), everything arises out of water and returns to it (Thomson 1959). Living things arose in the water, and each human individual starts his/her life in a watery environment in utero. However, human creatures have to carry the watery environment inside their body living on land. Water is the basic and indispensable molecule for every activity of human cells. Therefore, the aqueous environment is essential for intracellular, intercellular biochemical and physiological reactions. Water is central to maintain the most important processes in the human body, such as the circulation (blood and lymph), cerebral spinal fluid (CSF) flowing, removal of soluble waste products, facilitating ingestion and digestion, and flushing out the urinary tract and other crucial organs (Hooper et al. 2014). Thus, death could occur under the absence of fluid intake or serious dehydration, far more quickly than due to lack of any other nutrient.

However, after H_2O has got a carbon atom and converted into CH_2O, it becomes seriously toxic. We also carry formaldehyde in the watery environment inside our body. The balance of anabolism and metabolism of endogenous formaldehyde is maintained by various pathways. Thus, formaldehyde is involved in many aspects of our body. Many colleagues have been absorbed to study the toxicity of formaldehyde on human health. However, the relationship between dehydration and dysmetabolism of endogenous formaldehyde involved in age-related cognitive impairment has not been deeply investigated and discussed. Thus, we would like to discuss the relation of formaldehyde with dehydration and cognitive impairment in this chapter.

2 A Progressive Dehydration as Aging

Over 2000 years ago, Aristotle said: "Old age is dry and cold" (Aristotle 350 B.C.). The Chinese traditional medicine (Plain Questions, published 99–26 B.C.) described (Hou 2009): "Kidney essence withered first, accompanied by weakness of the spleen and the stomach, leading to disability of body fluid processing." In other words, loss of fluid related with imbalance of body water homeostasis had been noticed in the far ancient period. Human beings consume adequate volumes of fluids on a daily basis under physiological condition (Kenney and Chiu 2001). Many elderly people cannot maintain body water homeostasis in the progression of aging (Luckey and Parsa 2003). Aging is a natural progress, which is followed by a progressive decrease or disturbance in the water metabolic balance (Stout et al. 1999). As described by many laboratories (Hooper et al. 2014), in the progress of aging, the proportion of human body fluid compared to body weight reduces from over 70% in newborn babies to 60% in childhood and about 50% in older people (Altman and Katz 1961; Friis-Hansen et al. 1951; Greenleaf 1998; Olde Rikkert et al. 1997). As shown in Fig. 4.1, Li and her collaborators demonstrated a decrease in both the

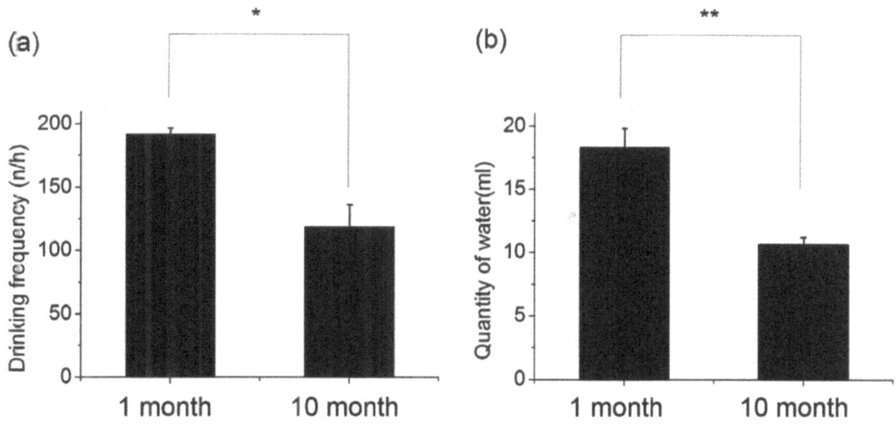

Fig. 4.1 A decrease in frequency and quantity of drinking water as aging. The drinking frequency per hour of mice (C57 BL/6) was monitored with an infrared CCD camera (panel **a**). The drinking frequency (counts per hour) of 10-month-old mice was significantly lower than 1-month-old as control (*$P < 0.05$). The drinking quantity of water was recorded for 72 h (panel **b**). The drinking quantity in 72 h for 10-month-old mice was much less than that for 1-month-old as control (**$P < 0.01$). The usage of this figure was authorized by the authors and Acta Neuropharmacologica

frequency and quantity of water intake of 10-month-old mice (C57BL/6j wild-type) compared with those of 1-month-old ones (Li et al. 2012). This natural reducing is suspected and considered to be a cause of a reduction in general body hydration in non-pathological conditions.

3 Diminished Thirst Sensation as Aging

Dehydration is primarily attributable to inadequate water intake, which could be caused by dysfunction of the central nervous system (CNS) mechanisms controlling thirst. The phenomenon of a reduced thirst in response to dehydration in aging was first observed decades ago and has been examined extensively since then. The reduced thirst and ingestive behavior have been reported consistently in response to hyperosmotic stimuli, hypovolemic stimuli, and dehydration in both elderly humans and animal models of aging (Begg 2017). Elderly people (over the age of 65 years) fail to perceive thirst, and thus they have a chance to face dehydration risk. It is a fact that the elderly progressively diminish thirst sensation followed by decreasing fluid intake as they age (Kenney and Chiu 2001). Especially under a challenge situation, for instance, exercising in a warm environment, a hyperosmotic stimulus, or fluid deprivation, the elderly adults could exhibit diminished thirst sensation and reduce fluid intake, compared with young adults.

During the long process, aging alters all physiological control systems, including the regulation system associated with thirst and satiety. For the elderly, full fluid restoration eventually occurs though full restoration of fluid balance (Kenney and Chiu 2001). In elderly population, depression is regarded as the most common pathologic cause for loss of body weight. The reduction in the feeling of thirst and water intake gradually causes chronic dehydration. On one hand, diminished responses to sodium may be another reason to decrease intake of water. As described by Begg, the reduced thirst and ingestive behaviors respond to hyperosmotic stimuli, hypovolemic stimuli, and dehydration in both elderly humans and animal models of aging (Begg 2017). On the other hand, decreased water intake could be resulted from a physiologic anorexia (Morley 2002). In fact, anorexia for elderly individuals again is caused by multiple reasons and alterations, such as increased leptin levels, fundal compliance, and hedonic qualities of food. So far, the precise etiology and mechanism of reduced thirst still remain to be investigated, although there has been evidence to suggest that aging disturbs the physiological response and regulation of body fluid and sodium homeostasis.

4 Dehydration and Age-Related Cognitive Impairment

4.1 Dehydration as a Predictor for Cognitive Status

Chronic dehydration is one of the risk factors involving age-related cognitive impairment. Archer and colleagues have examined differences, interactions, and associations among cognition, fluid intake, and demographic variables for psychogeriatric inpatients aged 65 years or older. As shown in their results, elderly adults with and without cognitive impairment had different fluid intake over time. They emphasized that the fluid intake of elderly patients with or without cognitive impairment must be monitored throughout their hospitalization (Archer et al. 2015).

Dehydration can be also used as a reliable predictor for impaired cognitive status besides other complications (Warren et al. 1994; Miller et al. 1998). According to Wilson and Morley, dehydration frequently results in delirium as a manifestation of cognitive dysfunction. The cognitive impairment could not be completely reversible, even though the occurrence of delirium appears transient acute global cerebral dysfunction (Wilson and Morley 2003). The activity of nitric oxide synthase (NOS) is reduced in the cortex and striatum of rats with aging (Calka et al. 1994). The decrease of activity of NOS occurring with aging may blunt the increase of NOS in dehydration and possibly interfere with memory processing and cognitive function (Salemme et al. 1996). However, the cause of cognitive impairment that resulted from dehydration is definitely complex. As described by Koch and Fulop, various hormonal and kidney changes occur, both affecting water homeostasis with aging. Aging is a risk factor for chronic kidney disease (CKD), and many features of CKD are reproduced in the aging kidney (Koch and Fulop 2017). Imbalance of ions also affects our brain and cognitive ability during dehydration. We should consider more effects resulted from different factors when using dehydration as a predictor for cognitive status.

4.2 *Dehydration and Alzheimer's Disease (AD)*

In 1989, Albert and colleagues considered that elderly individuals with Alzheimer's disease (AD) are at increased risk of dehydration during periods of fluid restriction due to loss of normal physiological responses of thirst (Albert et al. 1989). Mueller and Boisen paid attention to dehydration and claimed: "Keeping your patient's water level up" (Mueller and Boisen 1989). In fact, weight loss can be observed in elderly patients in clinics, which is a major clinical feature of AD (Koopmans et al. 1992). A great weight loss in a short period is often related with increased disease severity and even mortality (Luckey and Parsa 2003; Gillette et al. 2007). Chronic dehydration has been regarded as a common symptom of patients with age-related cognitive impairment, particularly those with AD (Buffa et al. 2010). The slow, progressive weight loss associated with aging may speed up prior to the onset of age-related dementia according to Purdy (Purdy 2006). Hoseini has had a hypothesis that weight loss primarily results from chronic dehydration, which may be one of the preventable risk factors for Alzheimer's disease (Gharibzadeh and Hoseini 2007). Amyloid β deposition and tau hyperphosphorylation in paired helical filaments are two of the most typical pathological features. Dehydration and hyperosmotic stress trigger tau phosphorylation in human neuroblastoma cells and induce apoptosis (Tanii et al. 2000; Stoothoff and Johnson 2001; Saikia et al. 2014). According to Buffa and collaborators, malnutrition is present even in the mild-moderate stages of AD and a tendency to exhibit dehydration that appears in the severe stage (Buffa et al. 2010). Therefore, it is necessary to explore the relationship between age-related cognitive impairment and dehydration, in particular of chronic dehydration.

5 Formaldehyde and Age-Related Cognitive Impairment

5.1 *Dysmetabolism of Formaldehyde Is Related with Aβ Deposition and Tau Hyperphosphorylation*

Recently, formaldehyde is considered as a major risk factor involved in the progression of age-related cognitive impairment (Su et al. 2016). Formaldehyde at low concentration induced tau protein aggregation (Nie et al. 2007a), which was implicated in cytotoxicity and neural cell apoptosis (Nie et al. 2007b; Chi et al. 2012). Oral administration of methanol (the metabolic precursor of formaldehyde) promoted the formation of senile plaques (SPs) and tau hyperphosphorylation in the brains of monkeys, accompanied with decline in working memory (Yang et al. 2014a, 2014b). Furthermore, intravenous injection of formaldehyde through the tail vein activated glycogen synthase kinase-3β (GSK-3β) and tau hyperphosphorylation in the brain of C57BL/6j wild-type mouse (Lu et al. 2013; He et al. 2016).

Emotional behaviors of animals including depression and anxiety were affected by formaldehyde steam (Leitl et al. 2014; Li et al. 2016a). Although the effect of formaldehyde on brain dysfunction has been widely studied (Mei et al. 2015; Su et al. 2016), the relation of formaldehyde with dehydration and age-related cognitive impairment needs to be deeply investigated.

5.2 Dehydration Leading to High Levels of Formaldehyde

To investigate whether dehydration is involved in formaldehyde metabolism, Li and her colleagues employed C57BL/6j mice and deprived their water intake. As shown in Fig. 4.2, the brain formaldehyde concentration of the mice significantly increased in 72 h during the water deprivation (Li et al. 2012). Consequently, Li and her colleagues established a mouse model to simulate chronic dehydration. They administrated C57 BL/6j mice with 4% NaCl instead of water intake for 3 months. In the 4% NaCl-fed group, the mice showed chronic dehydrated symptoms in which their serum osmotic pressure and sodium concentration significantly increased, accompanied with their decreased body weight (Li et al. 2016b). The levels of formaldehyde were markedly elevated in the brain of the 4% NaCl-fed mice. As shown by these data, both acute dehydration and chronic dehydration can induce dysmetabolism and elevate levels of brain formaldehyde.

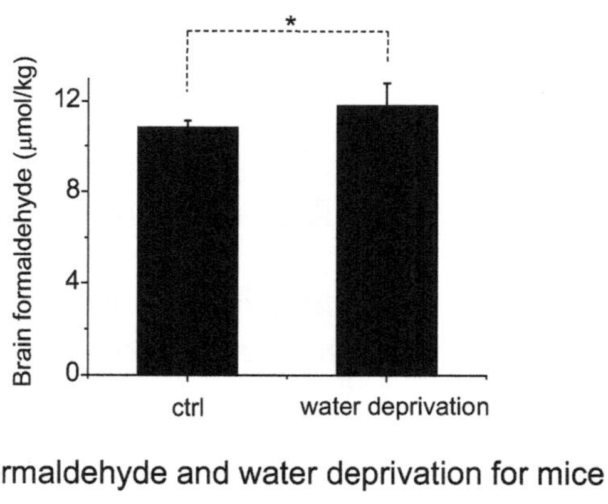

Fig. 4.2 An increase of mouse brain formaldehyde in limit of drinking water. 2-month-old C57 BL/6 mice were deprived of drinking water for 72 h and the concentration of their brain formaldehyde was measured with DNPH coupled with HPLC as described. The age-matched C57 BL/6 mice were employed as control. The concentration of brain formaldehyde of the water-deprived mice is significantly higher than that of control ($*P = 0.037, n = 8$). The usage of this figure was authorized by the authors and Acta Neuropharmacologica

5.3 Levels of Formaldehyde Increases with Aging

As mentioned above, dehydration either physiologically or pathologically occurs in the elderly population (over the age of 65 years) with aging. Similarly, according to Tong and his colleagues, a progressive increase in levels of endogenous formaldehyde could be observed in the elderly individuals (over the age of 75 years) in epidemiological investigation (Tong, et al., 2013). They found that the concentration of hippocampal formaldehyde of rats gets higher with their age. The concentration of brain formaldehyde of 10-month-old mice was higher than that of 3-month-old ones (Li et al. 2016c). The brain formaldehyde of senescence-prone strain 8 (SAMP8) was markedly higher than that of senescence-resistant strain 1 (SAMR1) as control (Qiang et al. 2014). Very recently, Liu and his colleagues revealed that the transgenic APP/PS1 mouse (a typical AD animal model) is suffered from abnormally high levels of formaldehyde from the intestinal microbiota (Liu et al. 2017). All these data suggest that the metabolic rate of endogenous formaldehyde is dependent upon human age.

5.4 Patients with Cognitive Impairment Suffer from Formaldehyde Dysmetabolism

Formaldehyde is produced and processed in all human cells including neural cells, thus, dysmetabolism of formaldehyde is involved in multiple pathways. However, the most sensitive organ or cell prone to dysmetabolism of formaldehyde remains unknown. According to our investigation, we would like to conclude that the brain is the most sensitive. The concentrations of urine formaldehyde are positively related to the decline of cognitive ability not only in AD inpatients but also in elderly individuals living in the communities in Beijing (Tong et al. 2011; Yu et al. 2014). As shown in Chap. 8, even education levels are closely correlated with the levels of urine formaldehyde in the elderly individuals in Beijing communities (Yu et al. 2014). In fact, formaldehyde is associated with the dementia of the patients with either AD or VD (vascular dementia) (Qiang et al. 2014; Chen et al. 2014; Tong et al. 2017). In other words, similar to examination of cognitive ability such as the mini-mental state examination (MMSE) and Montreal Cognitive Assessment (MoCA), endogenous formaldehyde could not be used as a biomarker to identify AD from VD. As a non-invasion biomarker, endogenous formaldehyde can be employed for age-related cognitive impairment in preclinical stage, clinical stage, and epidemiological investigation for AD.

5.5 *Increasing Formaldehyde Concentration Decreases Water Intake*

As mentioned above, both frequency and quantity of water intake progressively decrease with aging, followed by an increase of brain formaldehyde. Brain formaldehyde of 10-month-old mice was significantly higher than young ones (Li et al. 2012). These data alone could not clarify either formaldehyde or dehydration as a cause to interfere with the water intake behaviors. In order to clarify whether formaldehyde can affect water intake, Li and her collaborators administrated C57BL/6j mice with formaldehyde (0.5 mg/kg, once daily) through intraperitoneal injection. As their result shows, the frequency and quantity of water intake markedly decreased after a week the mice were injected with formaldehyde (Li et al. 2016b).

5.6 *Vicious Circle of Dehydration, Formaldehyde, and Cognitive Impairment*

As shown in Fig. 4.3, the interaction between dehydration and cognitive impairment creates a vicious cycle. On one hand, we have known that dehydration affects semicarbazide-sensitive amine oxidase (SSAO) and formaldehyde dehydrogenase 3 (ADH3), leading to an imbalance of the activities between the two enzymes and the elevation of brain formaldehyde (Li et al. 2016b). Water deprivation in 72 hours

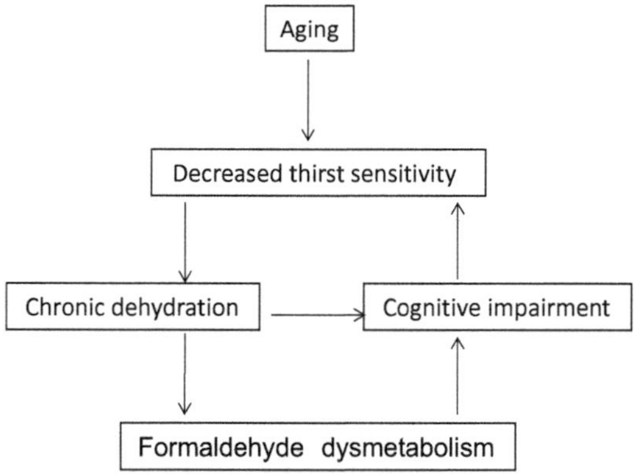

Fig. 4.3 Diagram for a putative "vicious circle" of dehydration, formaldehyde, and cognitive impairment. Chronic dehydration may be a preventable risk factor for Alzheimer's disease (Hoseini 2007). Aging diminishes the sensation of thirst (Kenney and Chiu 2001), leading to decreasing water intake and chronic dehydration (Begg 2017). Dysmetabolism of formaldehyde could be resulted from chronic dehydration that also affects cognitive ability (Li et al. 2012, 2016b). Dysmetabolism of formaldehyde could be observed in approximately 40% of patients with age-related cognitive impairment (Wang et al. 2017). Cognitive impairment worsens the diminished thirst sensation (Constans 2005)

enhances production of endogenous formaldehyde, especially in the brain (Li et al. 2012). On the other hand, injection with formaldehyde decreases the frequency and volume of water intake (Li et al. 2016b). Elevation of serum arginine vasopressin (AVP) levels could be detected when there is a distinct increase in the brain formaldehyde level after the injection. The concentration of serum angiotensin II (ANG II) significantly increased after administration of formaldehyde. A significant increase of the serum aldosterone (ALD) was detected as a result of the increased angiotensin levels. These data indicated that formaldehyde promoted AVP and ALD expression through increasing the level of serum ANG II.

6 Changes in Catecholamine and Behaviors

Liu and colleagues showed that the rats treated with inhalation formaldehyde (monitored to be 13.5+/−1.5 ppm) showed more aggressive behavior. At the same time, they observed the elevation of dopamine and decrease of 5-hydroxltryptamine (5-HT) significantly in the frontal cortex synaptosome (Liu et al. 2009). To study whether dehydration can disturb the anabolism and metabolism of catecholamine, Li and her colleagues employed the chronic dehydrated mouse model (C57 BL/6j mice fed with 4% NaCl instead of water intake for 3 months) (Li et al. 2016b). With a high level of formaldehyde, the chronic dehydrated (4% NaCl-fed) mice have got their brain 5-HT markedly decreased, acting as slow learners in the "shuttle box" behavior assay compared with those fed with water as control (Li et al. 2016b). They intraperitoneally injected C57BL/6j mouse with formaldehyde for a week and found a lower level of 5-HT and slow learning of the animal in the shuttle box (Fig. 4.4). As a control, 4% NaCl injection did not result in any significant differences in 5-HT levels and behavior assays. Their data indicated that chronic dehydration leads to dysregulation of brain formaldehyde, accompanied with a decrease of 5-HT levels in the mouse brain and decline of learning in the shuttle box.

As we know, dopamine (DA) is one of the catecholamine neurotransmitters playing an important role in the control of motor activity and emotional behavior. Tyrosine hydroxylase (TH) is a rate-limiting enzyme for DA synthesis. After the mice had been exposed in 2 ppm formaldehyde for one week, their immobility time increased significantly in the forced swimming test, indicative of depression-like behavior. Noticeably, the numbers of TH-immunoreactive neurons had a significant decrease under the same conditions (Li et al. 2016a). More researches on the relationship between depression behaviors and formaldehyde should be carried out to clarify the pathomechamisms that underlie. Mei and coworkers observed norepinephrine deficiency induced by formaldehyde played a role in age-related memory loss (Mei et al. 2015).

Except for "shuttle box" learning deficiency, long-term 4% NaCl-fed mice did not show significant differences in other behavioral tests such as Y maze, radial arm maze, elevated plus maze, and the open field test. No noticeable results were observed for grooming and the time spent in the central square in open field test

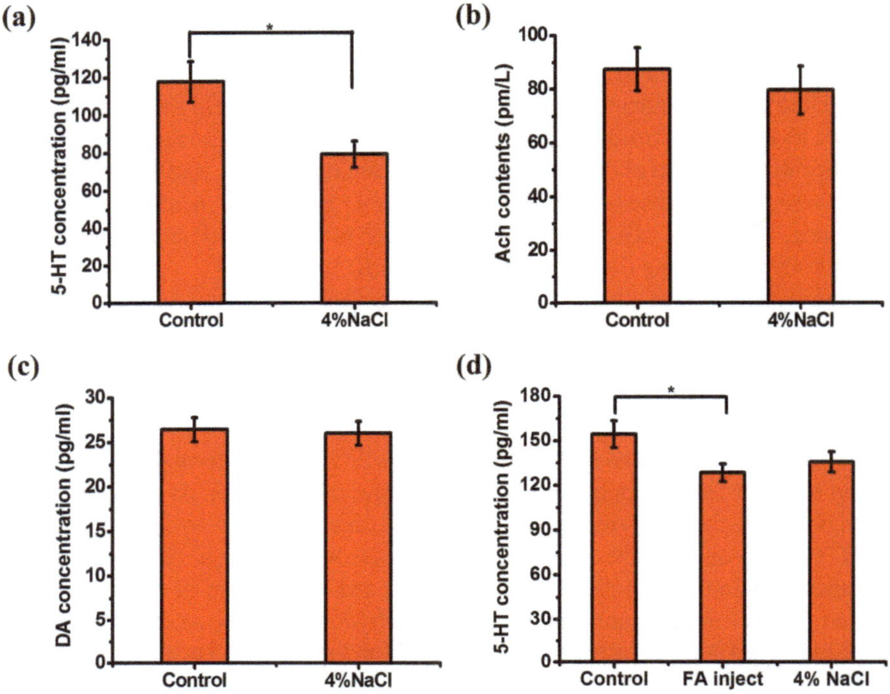

Fig. 4.4 Level changes of 5-HT, Ach, and DA in the mouse brain. The levels of 5-hydroxltryptamine (5-HT, **a**), acetylcholine (Ach, **b**), and dopamine (DA, **c**) in the brain of 4% NaCl-treated mice were detected by ELISA kits for three-month group. Meanwhile, the 5-HT concentration of mice injected intraperitoneally with formaldehyde or 4% NaCl was also detected (**d**). The data are shown as the mean SE; $*P < 0.05$; $**P < 0.01$. The usage of this figure was authorized by the authors and Prog Biochem Biophys

(Li et al. 2016b). All these results suggest a low possibility that long-term 4% NaCl-fed mice suffer from anxiety-like moods. It appears "shuttle box" learning paradigm processes "a punishment with electric shock", which makes mice painful and fearful. 5-HT plays a role in animal behaviors for fear, according to Bocchio and colleagues (Bocchio et al. 2016).

7 Dynamic Changes in Levels of Formaldehyde Daily

To follow changes in concentrations of endogenous formaldehyde daily, Li and her collaborators recruited 20 young participants (9 male, 11 female, 25–35 years old) to take part in "Investigation of the relation between urine formaldehyde and water intake." The participants' demographic characteristics and daily actions were described (Li et al. 2016b). The levels of endogenous formaldehyde reach the highest in the morning before breakfast, remarkably decreases around noon, and then increases again at night. The concentration of urine formaldehyde at noon was noticeably elevated while the participants were not allowed to drink water or any other beverages (water intake deprivation) from 8:30 to 12:30 (Li et al. 2016b). Furthermore, Li and her colleagues gave the participants "fried chicken" without water deprivation for three meals daily. After 2 days, an increase in endogenous formaldehyde of the participants was observed (Li and her colleagues' unpublished data). These data have demonstrated that water intake is important for each individual to exclude endogenous formaldehyde because of the soluble characteristic of the aldehyde.

8 Water Intake Is Beneficial to Exclude Endogenous Formaldehyde

We would like to emphasize that "water intake in the morning" is more effective because of the high levels of formaldehyde occurring at that time of the day (Li et al. 2012). "Water intake at night" is also recommended, but someone should have to get up to urinate at night which may interfere with sleeping. Protection of our brain should be carried out at early stage when neural cells are actively living. To make an analogy (Fig. 4.5), crops that undergo a short-term drought (Fig. 4.5a) can reversibly recover after being watered (Fig. 4.5b). However, they will die of a long-term drought even though you can water them fully (Fig. 4.5c). Albert and colleagues found a positive correlation between the quantity of water intake and MMSE score (Albert et al. 1994). A 21-year investigation performed by Eskelinen and collaborators showed that drinking coffee or tea can decrease the risk of onset of age-related cognitive impairment (Eskelinen et al. 2009). In this laboratory, we observe

Fig. 4.5 Diagram for an analogy to early intervene age-related cognitive impairment in the preclinical stage. We should protect our brain in early stage when neural cells are still actively functioning. To make an analogy, crops undergoing a short-term drought (panel **a**) can recover after being watered (panel **b**). However, they will die of a long-term drought even though they are fully watered (panel **c**)

the protective effects of resveratrol on neural cells in the presence of formaldehyde (Miao et al. 2013; He et al. 2016) Thus, we would like to recommend tea, coffee, or some other beverages containing resveratrol. They are beneficial to clear excess formaldehyde in our body. Hooper and colleagues also proposed the strategies to increase fluid intake in residential care homes (Hooper et al. 2014). They considered that elderly individuals suffer from physical inability to reach drinks, reduce social drinking, and diminish drinking pleasure. Therefore, recognizing and overcoming the barriers impeding water intake for the elderly should be paid attention (Hooper et al. 2014).

According to the guideline released by the China Nutrition Association (CNA) (Chen et al. 2013), water intake for male and female adults is 1.7 L and 1.5 L daily in addition to water contained in food, respectively. As mentioned above, levels of endogenous formaldehyde in the human body are highest in the morning. It is recommended that healthy adults should drink 400–500 ml of water (male) and 350–400 ml of water (female) after getting up before breakfast (Li et al. 2012). Water could be warm boiled water, and mineral water. tea, coffee, or some other beverage, as you like, can be used (Panza et al. 2015). According to Eskelinen and colleagues, coffee drinking at midlife is associated with a decreased risk of dementia/AD later in life (Eskelinen et al. 2009). The green tea polyphenols, especially epigallocatechin gallate, were found to be associated with anti-Alzheimer, and anti-aging properties, possibly preventing system level structural damage (Afzal et al. 2015). Formaldehyde is easy to dissolve in water. Water aids elimination of excess formaldehyde through kidneys. Individuals can consume the remaining water (1000–1200 ml daily) divided in the morning, at noon, in the afternoon, and in the evening, according to one's own habits (Li et al. 2016b). Water intake eliminates not only excess formaldehyde but also other harmful metabolic products.

9 Conclusion

Diminished perception of thirst or/and memory loss may be a cause to chronic dehydration in elderly people with cognitive impairment. Chronic dehydration diminishes the elderly individual's ability to eliminate metabolic products, leading to the accumulation of cytotoxic compound, including formaldehyde. Dehydration and formaldehyde accumulation not only interfere with water intake behavior but also with cognitive ability. Age-related cognitive impairment diminishes water intake sensation and worsens dehydration, leading to formaldehyde accumulation. Consequently, the accumulated brain formaldehyde could impair cognitive ability, for instance, forgetting water intake and worsening the chronic dehydration. That creates and undergoes in a vicious cycle as an elderly individual's aging. The highest concentration of endogenous formaldehyde is found in the morning, followed by a decrease around noon, and an increase in the evening. Thus, regular water intake including tea or other beverages containing resveratrol is recommended to intervene and reduce the accumulation of formaldehyde, especially when it is taken in the morning.

Acknowledgment This project was supported by grants from the Beijing Municipal Science and Technology Project (Z161100000217141; Z161100000216137), the National Key Research and Development Program of China (2016YFC1306300), the National Basic Research Program of China (973 Program) (2012CB911004), the Natural Scientific Foundation of China (NSFC 31270868), the Foundation of Chinese Academy of Sciences CAS-20140909, and the Queensland-Chinese Academy of Sciences Biotechnology Fund (GJHZ201302).

Competing Financial Interests The authors declare no competing financial interests.

References

Afzal M, Safer AM, Menon M (2015) Green tea polyphenols and their potential role in health and disease. Inflammopharmacol 23(4):151–161

Albert SG, Nakra BR, Grossberg GT, Caminal ER (1989) Vasopressin response to dehydration in Alzheimer's disease. J Am Geriatr Soc 37(9):843–847

Albert SG, Nakra BR, Grossberg GT, Caminal ER (1994) Drinking behavior and vasopressin responses to hyperosmolality in Alzheimer's disease. Int Psychogeriatr 6:79–86

Altman PL, Katz DD (1961) Blood and other body fluids. Fed Am Soc Exp Biol 178(1):92

Archer SJ, Barna AE, Holper RM, Forsyth DM, Ellenbecker SM, Smith LK, Clobes JE, Meiers SJ, Malone C, Lapid MI (2015) Associations among fluid intake, cognitive function, and length of stay in psychogeriatric inpatients. Perspect Psychiatr Care 51(1):52–56

Aristotle (350 B.C.) On longevity and Shortness of life (trans: Ross GRT). Internet Classics Archive. http://classics.mit.edu/Aristotle/longev_short.html

Begg DP (2017) Disturbances of thirst and fluid balance associated with aging. Physiol Behav 2017:4. doi:10.1016/j.physbeh.2017.03.003. [Epub ahead of print]

Bocchio M, McHugh SB, Bannerman DM, Sharp T, Capogna M (2016) Serotonin, amygdala and fear: assembling the puzzle. Front Neural Circuits 5(10):24

Buffa R, Mereu RM, Putzu PF, Floris G, Marini E (2010) Bioelectrical impedance vector analysis detects low body cell mass and dehydration in the patients with Alzheimer's disease. JNHA 14(10):823–874

Calka J, Wolf G, Brosz M (1994) Ultrastructural demonstration of NADPH-diaphorase histochemical activity in the supraoptic nucleus of normal and dehydrated rats. Brain Res Bull 34:301–308

Chen YY, Yang YX, Yang XG, Di FY, Guo JS, Su YX, et al (2013) Adequate water intake for Chinese resident. Chinese DRIs handbook by China Nutrition Association. Standards Press of China, Beijing, p 35

Chen J, Sun M, Wang X, Lu J, Wei Y, Tan Y, Liu Y, Götz J, He R, Hua Q (2014) The herbal compound geniposide rescues formaldehyde-induced apoptosis in N2a neuroblastoma cells. Sci China Life Sci 57(4):412–421

Chi YN, Zhang X, Cai J, Liu FY, Xing GG, Wan Y (2012) Formaldehyde increases intracellular calcium concentration in primary cultured hippocampal neurons partly through NMDA receptors and T-type calcium channels. Neurosci Bull 28(6):715–722

Constans T (2005) Diabetic hyperosmolarity: a consequence of loss of autonomy. Diabet Metab 2(6):5S62–5S66

Eskelinen MH, Ngandu T, Tuomilehto J, Soininen H, Kivipelto M (2009) Midlife coffee and tea drinking and the risk of late-life dementia: a population-based CAIDE study. J Alzheim Dis 16:85–91

Friis-Hansen BJ, Holiday M, Stapleton T, Wallace WM (1951) Total body water in children. Pediatrics 7(3):321

Gharibzadeh, S, Hoseini SS (2007) Chronic dehydration may be a preventable risk factor for Alzheimer's disease. Med Hypoth 68(3):718

Gillette GS, Abellan VKG, Alix E, Andrieu S, Belmin J, Berrut G, Bonnefoy M, Brocker P, Constans T, Ferry M, Ghisolfi-Marque A, Girard L, Gonthier R, Guerin O, Hervy MP, Jouanny P, Laurain MC, Lechowski L, Nourhashemi F, Raynaud-Simon A, Ritz P, Roche J, Rolland Y, Salva T, Vellas B (2007) Weight loss and Alzheimer's disease. J Nutr Health Aging 11(1):38–48

Greenleaf JE (1998) Stress, aging and thirst. NTRS, pp 47–58

He X, Li Z, Rizak JD, Wu S, Wang Z, He R, Qin D, Hu X (2016) Resveratrol attenuates formaldehyde induced hyperphosphorylation of tau protein and cytotoxicity in N2a cells. Front Neurosci 10:598

Hooper L, Bunn D, Jimoh FO, Fairweathertait SJ (2014) Water-loss dehydration and aging. Mech Ageing Dev 136-137(1):50–58

Hou JL (2009) To understand the aging theory of Chinese traditional medicine. Jiangsu J Chinese Trad Med 41(9):4–5

Kenney WL, Chiu P (2001) Influence of age on thirst and fluid intake. Med Sci Sports Exerc 33(9):1524–1532

Koch CA, Fulop T (2017) Clinical aspects of changes in water and sodium homeostasis in the elderly. Rev Endocr Metab Disord 18(1):49–66

Koopmans RT, Ekkerink JL, Sirre LG, Verkuylen MM, van den Hoogen HJ, van Weel C (1992) Multi-infarct dementia in nursing home patients; more comorbidity and shorter life expectancy than in Alzheimer's disease. Ned Tijdschr Geneeskd 136(45):2223–2227

Leitl MD, Potter DN, Cheng K, Rice KC, Carlezon WA Jr, Negus SS (2014) Sustained pain-related depression of behavior: effects of intraplantar formalin and complete freund's adjuvant on intracranial self-stimulation (ICSS) and endogenous kappa opioid biomarkers in rats. Mol Pain 10:62

Li T, Qiang M, He RQ (2012) Chronic dehydration and regularly drinking water to mitigate age-related cognitive impairment. Acta Neuropharmacologica 2(3):43–51

Li Y, Song Z, Ding Y, Xin Y, Wu T, Su T, He R, Tai F, Lian Z (2016a) Effects of formaldehyde exposure on anxiety-like and depression-like behavior, cognition, central levels of glucocorticoid receptor and tyrosine hydroxylase in mice. Chemosphere 144:2004–2012

Li T, Su T, He YG, He RQ (2016b) Chronic-dehydrated dysmetabolism of formaldehyde in mouse brain and decline of learning in the shuttle box. Prog Biochem Biophys 43(4):429–438

Li T, Su T, He YG, Lu JH, Mo WC, Wei Y, He RQ (2016c) Brain formaldehyde is related to water intake behavior. Aging Dis 7(5):561–584

Liu Y, Ye Z, Luo H, Sun M, Li M, Fan D, Chui D (2009) Inhalative formaldehyde exposure enhances aggressive behavior and disturbs monoamines in frontal cortex synaptosome of male rats. Neurosci Lett 464(2):113–116

Liu KL, He YG, Yu LX, Chen Y, He RQ (2017) Elevated formaldehyde in the cecum of APP/PS1mouse. Microbiol China. doi:10.13344/j.microbiol.china.176008

Lu J, Miao JY, Su T, Liu Y, He RQ (2013) Formaldehyde induces hyperphosphorylation and aggregation of tau protein both in vitro and in vivo. Biochim Biophys Acta-GS 1830(8):4102–4116

Luckey AE, Parsa CJ (2003) Fluid and electrolytes in the aged. Arch Surg 138(10):1055–1060

Mei Y, Jiang C, Wan Y, , Lv J, Jia J, Wang X, Yang X, Tong Z (2015). Aging-associated formaldehyde-induced norepinephrine deficiency contributes to age-related memory decline. Aging Cell, 14(4):659–668

Miao JY, Lu J, Zhang ZJ, Tone ZQ, He RQ. (2013). The effect of formaldehyde on cell cycle is in a concentration-dependent manner. Prog Biochem Biophys 40(7):641–651

Miller DK, Perry HM, Morley JE (1998) Relationship of dehydration and chronic renal insufficiency with function and cognitive status in older US blacks. J Investig Med 46(7):282A–282A

Morley JE (2002) Nutrition in the elderly. Curr Opin Gastroenterol 18(2):240–245

Mueller KD, Boisen AM (1989) Keeping your patient's water level up. RN 52(7):65–68

Nie CL, Wang XS, Liu Y, Perrett S, He RQ (2007a) Amyloid-like aggregates of neuronal tau induced by formaldehyde promote apoptosis of neuronal cells. BMC Neurosci 8:9

Nie CL, Wei Y, Chen X, Liu YY, Dui W, Liu Y, Davies MC, Tendler SJ, He RG (2007b) Formaldehyde at low concentration induces protein tau into globular amyloid-like aggregates in vitro and in vivo. PLoS One 2(7):e629

Olde Rikkert MG, Deurenberg P, Jansen RW, Van't Hof MA, Hoefnagels WH (1997) Validation of multi-frequency bioelectrical impedance analysis in detecting changes in fluid balance of geriatric patients. J Am Geriatr Soc 45:1345–1351

Panza F, Solfrizzi V, Barulli MR, Bonfiglio C, Guerra V, Osella A, Seripa D, Sabbà C, Pilotto A, Logroscino G (2015) Coffee, tea, and caffeine consumption and prevention of late-life cognitive decline and dementia: a systematic review. J Nutr Health Aging 19(3):313–328

Purdy M (2006) Accelerating weight loss may signal development of Alzheimer's disease. Arch Neurol 63:1312–1317

Qiang M, Xiao R, Su T, Wu BB, Tong ZQ, Liu Y, He RQ (2014) A novel mechanism for endogenous formaldehyde elevation in SAMP8 mouse. J Alzheim Dis 40(4):1039–1053

Saikia M, Jobava R, Parisien M, Putnam A, Krokowski D, Gao XH, Guan BJ, Yuan Y, Jankowsky E, Feng Z, Hu GF, Pusztai-Carey M, Gorla M, Sepuri NB, Pan T, Hatzoglou M (2014) Angiogenin-cleaved tRNA halves interact with cytochrome c, protecting cells from apoptosis during osmotic stress. Mol Cell Biol 34(13):2450–2463

Salemme E, Diano S, Maharajan P, Maharajan V (1996) Nitric oxide, a neuronal messenger. Its role in hippocampal plasticity. Riv Biol 89:87–107

Stoothoff WH, Johnson GV (2001) Hyperosmotic stress-induced apoptosis and tau phosphorylation in human neuroblastoma cells. J Neurosci Res 65(6):573–582

Stout NR, Kenny RA, Baylis PH (1999) A review of water balance in ageing in health and disease. Gerontology 45(2):61–66

Su T, Monte WC, Hu X, He Y, He RQ (2016) Formaldehyde as a trigger for protein aggregation and potential target for mitigation of age-related, progressive cognitive impairment. Curr Top Med Chem 16(5):472–484

Tanii H, Ankarcrona M, Flood F, Nilsberth C, Mehta ND, Perez-Tur J, Winblad B, Benedikz E, Cowburn RF (2000) Alzheimer's disease presenilin-1 exon 9 deletion and L250S mutations sensitize SH-SY5Y neuroblastoma cells to hyperosmotic stress-induced apoptosis. Neuroscience 95(2):593–601

Thomson G (1959) Pervye filosofy 2:145–158

Tong ZQ, Zhang JL, Luo W, Wang W, Li F, Li H, Luo H, Lu J, Zhou J, Wan Y, He RQ (2011) Urine formaldehyde level is inversely correlated to mini mental state examination scores in senile dementia. Neurobiol Aging 32(1):31–41

Tong ZQ, Wang WS, Luo WH, Lv J, Li H, Luo H, Jia J, He RQ (2017) Urine formaldehyde predicts cognitive impairment in post-stroke dementia and Alzheimer's disease. J Alzheimers Dis 55(3):1031–1038

Wang J, Zhou J, Mo WC, He YG, Wei Y, He RQ (2017) Accumulation of simulated pathological level of formaldehyde decreases cell viability and adhesive morphology in neuronal cells. Prog Biochem Biophys 44(7):601–614

Warren JL, Bacon WE, Harris T, McBean AM, Foley DJ, Phillips C (1994) The burden and outcomes associated with dehydration among US elderly. Am J Pub Health 84:1265–1269

Wilson MM, Morley JE (2003) Impaired cognitive function and mental performance in mild dehydration. Eur J Clin Nutr 57:S24

Yang MF, Lu J, Miao JY, Rizak J, Yang J, Zhai R, Zhou J, Qu J, Wang J, Yang S, Ma Y, Hu X, He RQ (2014a) Alzheimer's disease and methanol toxicity (part 1): chronic methanol feeding led to memory impairments and tau hyperphosphorylation in mice. J Alzheimers Dis 41(4):1117–1129

Yang MF, Miao JY, Rizak J, Zhai R, Wang Z, Huma T, Li T, Zheng N, Wu S, Zheng Y, Fan X, Yang J, Wang J, Yang S, Ma Y, Lü L, He RQ, Hu XT (2014b) Alzheimer's disease and methanol toxicity (part 2): lessons from four rhesus macaques (Macacamulatta) chronically fed methanol. J Alzheimers Dis 41(4):1131–1147

Yu J, Su T, Zhou T, He YG, Lu J, Li J, He RQ (2014) Uric formaldehyde negatively correlates with cognitive abilities in healthy older adults. Neurosci Bull 30(2):172–184

Chapter 5
The Role of Formaldehyde in Cell Proliferation and Death

Weichuan Mo and Rongqiao He

Abstract Accumulating public attentions have been paid to the neurotoxicity of formaldehyde during aging and carcinogenesis induced by environmental and occupational formaldehyde exposure. The cytotoxicity and carcinogenesis by formaldehyde manifestly reveal two distinct cellular functions of formaldehyde: inducing cell death and promoting cell proliferation. The degree of the disturbance of the formaldehyde baseline determines the effect of formaldehyde on cells. This chapter classifies the dose ranges of formaldehyde based on its role in cell proliferation and death and discusses the mechanisms underlying the nonlinear dose- and time-dependent effects of formaldehyde, which would give light to our knowledge on the physiological and pathological effects of formaldehyde. Short-term small interruption on the formaldehyde baseline, *i.e.*, formaldehyde elimination by formaldehyde capturers and low concentration of formaldehyde (<0.1 mM), provides a pro-proliferative environment for cancer cell lines by activating extracellular-regulated protein kinase (ERK) pathways and raising genomic instability, which contributes to the carcinogenesis of formaldehyde. The moderate concentration of formaldehyde (0.1–1 mM) further increases the formaldehyde-induced DNA damage (DNA adducts and DNA–protein cross-links), which overwhelms the cellular DNA repair capacity and triggers programmed cell death, especially apoptosis. In addition, moderate concentration of formaldehyde induces oxidative stress, mitochondrial dysfunction, endoplasmic reticulum stress, proteotoxicity, and excitotoxicity delivered by prolonged Ca^{2+} releasing in neurons, which subsequently lead to apoptosis. Formaldehyde plays a role in autophagy and cell senescence, as well. High concentration of formaldehyde (> 1 mM) induces necrosis. Particularly, neuronal cells are more sensitive to the toxicity of formaldehyde. The necrosis-inducing dose of formaldehyde also depends on cell density.

W. Mo
State Key Laboratory of Brain and Cognitive Science, Institute of Biophysics, Chinese Academy of Sciences, Beijing 100101, China

R. He (✉)
State Key Laboratory of Brain and Cognitive Science, Institute of Biophysics, Chinese Academy of Sciences, Beijing 100101, China

University of Chinese Academy of Sciences, Beijing 100049, China
e-mail: rongqiaohe@163.com

R. He, *Formaldehyde and Cognition*, DOI 10.1007/978-94-024-1177-5_5

Keywords Formaldehyde • Carcinogenesis • Neurotoxicity • Apoptosis • Cell proliferation • DNA adduct • DNA–protein cross-link

1 Introduction

High dose of formaldehyde is definitely lethal to living cells. Direct contact with formalin solution causes inflammatory alterations in the skin. Inhalation of formaldehyde gas in the air is irritating to the mucous membranes of the respiratory tract. Necrosis and apoptosis were observed in cells treated with formaldehyde at high concentrations. Aging-related endogenous formaldehyde accumulation leads to apoptosis in neuronal cells and subsequently results in the brain and cognitive damages (Miao et al. 2013; Tang et al. 2013; Tong et al. 2013b). Moreover, the cytotoxicity of formaldehyde has been applied to develop formaldehyde-releasing antitumor drugs (Cutts et al. 2005; Dhareshwar and Stella 2008; Nudelman et al. 2005), which not only provides toxicity to tumor cells (Barthel et al. 2016; Cutts et al. 2005) but also increases the sensitivity of cancer cells to the drug (Kato et al. 2000, 2001). The cytotoxicity of formaldehyde is closely related with DNA damage, oxidative stress, mitochondrial dysfunction, and proteotoxicity.

Meanwhile, accumulating evidence has shown that formaldehyde exposure raises carcinogenic risk (Costa et al. 2011, 2013; Mahboubi et al. 2013; Swenberg et al. 2013). The International Agency for Research on Cancer classified formaldehyde as a human carcinogen in 2004 (IARC 2006). The evidence that formaldehyde is a carcinogen is strong for nasopharyngeal cancer, as the nasopharyngeal airway is the initial site to contact inhaled formaldehyde (McGregor et al. 2006). The threat of widespread low-level formaldehyde exposure and the occupational contact to the environmental formaldehyde have raised public health concerns (Binazzi et al. 2015; Joob and Wiwanitkit 2016). The pro-proliferative effect of low-dose formaldehyde exposure is believed to cause carcinogenesis, which is attributed to the genotoxicity and mutagenesis by formaldehyde.

The cytotoxicity and carcinogenesis by formaldehyde manifestly reveal two distinct cellular functions of formaldehyde: inducing cell death and promoting cell proliferation. Since the cellular activity was conducted under the context of endogenous formaldehyde, the degrees of the disturbance of the endogenous formaldehyde baseline determine the action of formaldehyde on cells. This chapter will discuss the dose ranges of formaldehyde based on its role in cell proliferation and death, and try to clarify the mechanisms underlying the nonlinear dose- and time-dependent effects of formaldehyde, which may give light to our knowledge on the physiological and pathological effects of formaldehyde.

2 The Formaldehyde Baseline

The endogenous formaldehyde concentration serves as the formaldehyde baseline. Balancing the rapidly metabolic processes of formaldehyde and detoxification pathways of exogenous formaldehyde input *(see Chap. 2 written by Xiao, R. and He, P. Q. for details)* results in a physiological formaldehyde level at approximately 10–80 μM in the serum, urine, and cerebrospinal fluid of healthy individuals (He et al. 2010; Tong et al. 2011, 2013a).

For cell cultures, the formaldehyde baseline is the background concentration of formaldehyde in the medium. The residual formaldehyde level in the serum-free medium was ~5 μM (Chen et al. 2014; Sun et al. 2013). Formaldehyde probing showed that the added formaldehyde in the medium crosses the cell membrane immediately and alters the intracellular formaldehyde level (Brewer and Chang 2015; Xu et al. 2016). Therefore, the endogenous formaldehyde level in cell cultures directly reflects the formaldehyde concentration in the medium under the experimental condition. Intracellular formaldehyde is primarily located in the endoplasmic reticulum (ER) , Golgi apparatus, and lysosome, but very little is present in mitochondria (Chen et al. 2017; Tang et al. 2016), indicating that formaldehyde is a by-product of many metabolic biochemical reactions and concentrated in organelles involved in the synthesis, modification, and degradation of biomacromolecules but toxic to energy metabolism.

3 Pro-proliferative Small Interruption on Formaldehyde Baseline

Small elevation of formaldehyde concentration in the cell culture medium (10–100 μM) above the formaldehyde baseline (~5 μM) enhances cell proliferation, and the pro-proliferative dosage is cell-type dependent. The proliferation of HeLa cell and human malignant melanoma SK-MEL-28 cell is stimulated with the addition of 10 μM formaldehyde (Ke et al. 2014; Rizzi et al. 2016). 10–25 μM formaldehyde enhances the proliferation of both human low-invasive A375P (1-day exposure) and high-invasive SK-MEL-28 cells (3-day exposure) (Rizzi et al. 2014, 2016). 50 μM formaldehyde accelerates the proliferation of human neuroblastoma SH-SY5Y cell (Li et al., unpublished data). 100 μM formaldehyde stimulates the proliferation and reduces apoptosis in human colon carcinoma HT-29 cell and human endothelial HUV-EC-C cell (Tyihak et al. 2001).

Meanwhile, a formaldehyde shortage situation is also beneficial for cell proliferation. Szende and colleagues have shown that resveratrol, a formaldehyde capturer, increases the proliferation of many cancer cell lines at low concentration (0.1 to 1.0 μg/mL) and proposed that resveratrol facilitates the elimination of formaldehyde in the medium by direct reaction with formaldehyde (Szende et al. 2000). Noteworthy, their results also suggested that, when resveratrol was at high concentration, the reaction production of resveratrol and formaldehyde would induce apoptosis, instead. Wu and colleagues showed that aldehyde dehydrogenase 5 (ADH5), a formaldehyde degradation enzyme, is a repressor of neural differentiation. The expression of ADH5 decreases during neurogenesis. The overexpression of ADH5 counteracts the differentiation of human neural stem cells, suggesting that proliferative neural stem cells prefer a low formaldehyde environment and that neural maturation progresses under a high formaldehyde condition (Wu et al. 2014).

In addition, the pro-proliferative effect of low-concentration formaldehyde is largely a short-term effect. Long-term low-dose formaldehyde could also induce apoptosis. The recent work of our laboratory confirmed that, instead of the short-term pro-proliferative effect, long-term exposure to formaldehyde at a low concentration (15 μM) gradually induced damage in cell morphology and eventually led to cell death in murine neuroblastoma N2a cell line and primary hippocampal neurons (Wang et al. 2017). Even a 9-day formaldehyde exposure at the concentrations from 1 pM to 10 nM induced apoptosis in human colorectal cancer SW620 cells (Lee et al. 2016), which further confirmed the cytotoxicity of a long-term low-concentration formaldehyde exposure.

3.1 Mutagenesis and Genotoxicity

The pro-proliferative effect of formaldehyde is believed to cause carcinogenesis, which is mainly attributed to the mutagenesis and genotoxicity by formaldehyde. Formaldehyde directly targets various nucleophiles of DNA, proteins, and amino acids, which induce DNA damages and epigenetic changes by forming adducts with DNA and proteins. Careful physical chemistry-based mode of action studies illustrated that formaldehyde mainly induces two types of DNA damages: DNA adducts and DNA–protein cross-links (DPX).

The interaction of formaldehyde to DNA occurs on the amino groups of base pairs, which is pro-mutagenic (Swenberg et al. 2011). Point mutations in p53 tumor suppressor gene were identified from formaldehyde-induced (> 10 ppm) squamous cell carcinomas in the nasal tissues of rats (Recio et al. 1992; Wolf et al. 1995). DPX are genotoxic as a result of their ability to arrest DNA replication (Szende and Tyihak 2010). The accumulation of formaldehyde-induced DNA adducts and DPX induces the regenerative proliferative response of nasal epithelial cells (Hester et al. 2003). The response is believed to be necessary to replace the damaged epithelium and to protect the nasal airway from further irritation by formaldehyde, which is closely associated with subsequent nasal cancer development (Monticello et al. 1996; Swenberg et al. 2013). The genetic instability, caused by the overload of formaldehyde-induced DNA damages and/or interruptions with the DNA repair signaling pathway, is essential for carcinogenesis (Zhang et al. 2010a). DNA adducts and DPX activate DNA repair pathways. Knockout of ADH5 accumulates endogenous formaldehyde and results in the increase of DNA adducts in mice. FANCD2, a DNA cross-link repair protein, is involved in the repair of this kind of damage.

Double knockout of ADH5 and FANCD2 led to bone marrow failure and the ADH5 (−/−) FANCD2 (−/−) mice eventually developed leukemia and fatal malignancies (Pontel et al. 2015).

3.2 *DNA-Damage-Independent Pro-proliferative Effect of Formaldehyde*

Low-concentration formaldehyde also induces cell proliferation in a DNA-damage-independent way. Aizenshtadt and colleagues stimulated the proliferation of human epidermoid carcinoma A431 cells by exposing the cells to 30 μM formaldehyde for 5 min (Aizenshtadt et al. 2011). Such a short-term exposure is not likely to induce the formation of DNA adducts or DPX. The proliferation capacity of formaldehyde-stimulated A431 maintained higher than that of the control cells for 3 days in formaldehyde-free culturing media, and the percentage of S-phase cells increased significantly 24 h after the formaldehyde treatment. In addition, they showed that short-term 30 μM formaldehyde led to reversible actin cytoskeleton disassembly, formation of epidermal growth factor receptor (EGFR) patches, and increased EGFR phosphorylation. Continuous cultivation for 24 h in the presence of 30 μM formaldehyde also increased the S-phase cell percentage in HEK293 and rat fibroblasts and accelerated the proliferation of the two types of cells. However, the pro-proliferation effect of 5 min formaldehyde stimulation on A431 cell was nonlinearly in a dose-dependent manner. The proliferation of A431 was inhibited when the stimulation concentration of formaldehyde was over 60 μM (Aizenshtadt et al. 2011).

In the case of formaldehyde-induced acceleration of proliferation in A375P (1-day exposure) and SK-MEL-28 cells (3-day exposure), the acute increase in ERK phosphorylation was detected with 1 h formaldehyde treatment in both cell lines (Rizzi et al. 2014). The short-term effect of formaldehyde on the phosphorylation of ERK would be a molecular event that happened earlier than formaldehyde-induced DNA damage. 10 μM formaldehyde could also induce acute increase in ERK phosphorylation in spontaneously immortalized keratinocyte HaCaT cell when cell proliferation was not affected after 1-day formaldehyde treatment (Rizzi et al. 2014). ERK is a converging point of multiple signal transduction pathways involved in cell proliferation and DNA damage response (Lin et al. 2013). Thus, ERK signaling pathway is probably the initial responding pathway to formaldehyde exposure, which might lead to downstream regulation of cell proliferation and DNA damage repair.

3.3 The Production and Accumulation of Formaldehyde in Cancer

Tumor tissues endogenously secrete formaldehyde. Elevation of endogenous formaldehyde concentration was detected in tumor tissues and cancer cell lines. The urine formaldehyde concentration in the patients with prostate and bladder cancer is elevated for two–eightfold compared with the normal participants (Spanel et al. 1999). Concentrations of formaldehyde in lung cancer (0.72 ± 0.06 mM) and breast cancer (0.75 ± 0.12 mM) tissues were significantly higher than that in the normal tissues adjacent to the tumor (0.1 ~ 0.2 mM) (Tong et al. 2010). The formaldehyde level in the expired air from tumor-bearing transgenic mice (1.43–2.98 μM) was two–threefold higher than the control mice (0.77–1.01 μM) (Ebeler et al. 1997). Tong and colleagues observed that concentrations of formaldehyde in the medium of cancer cell lines and in the ascitic fluid from mice inoculated with cancer cells gradually increase over time (Tong et al. 2010). Therefore, formaldehyde released from cancer cells and tumor tissues would provide a pro-proliferative microenvironment and probably stimulate tumor progression. Lysine-specific demethylase 1 contributes to the production of endogenous formaldehyde in cultured MRMT-1 breast cancer cells (Liu et al. 2013). Given that tumor cells are characterized as general DNA demethylation and specific DNA hypermethylation of certain genes (Kisseljova and Kisseljov 2005), Tong and colleagues proposed that the elevated endogenous formaldehyde level in cancer cells could also be sourced from the methyl groups dropped from the chromosome (Tong et al. 2015).

There was a lack of demonstration of the short-term delivery of inhaled formaldehyde to distant sites at present (Heck and Casanova 2004; Lu et al. 2010; Kleinnijenhuis et al. 2013; Salthammer 2015; Zhang et al. 2014). Exogenous formaldehyde-induced DPX (6 ppm, 2 days, 6 h/day; 15 ppm, 1–4 day, 6 h/day) was only detected in rat nasal respiratory tissues but were absent in tissues distant to the site of contact (bone marrow and liver) (Lai et al. 2016).

However, the exogenous formaldehyde-induced accumulation of formaldehyde in the brain is evidential. Acute high-dose formaldehyde injection could increase brain formaldehyde level. The concentration of brain formaldehyde in C57BL/6 mice acutely increased from the baseline (13 μM) to ~15 μM, 2 h after tail vein injection of formaldehyde (3.7 μg per gram of body weight), and the elevation of brain formaldehyde level maintained for at least 10 h (Lu et al. 2013). Li and colleagues showed that the brain formaldehyde concentration significantly increased 7 days after the intraperitoneal injection of formaldehyde (0.2 μg per gram of body weight) in C57BL/6 mice (Li et al. 2016a). 7-day exposure to vapor formaldehyde (2 ppm; 2 h/day) was found to elevate brain formaldehyde level in adult male Kunming mice (Li et al. 2016b). Since the DNA synthesis processes in most brain cells (neurons and glia cells) are largely quiescent, the elevated brain formaldehyde level is not likely to induce cancerous changes in these developed brain cells. However, given the pro-proliferation effect of low-dose formaldehyde, abnormally increased physiological formaldehyde level, e.g., in the brain of senior individual and Alzheimer's disease (AD) patients (80–200 μM) (Tong et al. 2013b), would become a potential pro-proliferation factor for metastatic cancer cells in the brain, which raises the risk of metastatic brain tumor in practice.

3.4 *Neuroprotection*

In addition to the pro-proliferative effect, low-concentration formaldehyde is also proved to have protective effect on neuronal cells. Tsukahara and colleagues found the neuroprotective effect of 12-week extremely low-concentration formaldehyde inhalation (400 ppb 16 h/day; 5 days/week) against apoptosis in ovalbumin-immunized mice hippocampus, indicated by an increase of Bcl-2/Bax expression ratio, accompanied with increase in nerve growth factor production (Fujimaki et al. 2004; Tsukahara et al. 2006). But this effect cannot be detected in mice without ovalbumin immunization. Therefore, the neural protective effects of formaldehyde are probably delivered by its synergistic effect with other protective factors.

Qi and colleagues recorded that the 12-h formaldehyde treatment (0.1–1 mM) attenuated the staurosporine-induced apoptosis in primary hippocampus neurons. They attributed the anti-apoptosis effect of formaldehyde to the inhibitory effect of formaldehyde on the outward potassium currents (Liu et al. 2011). However, such a formaldehyde concentration is sufficient to induce apoptosis based on the data obtained from cell experiments. In addition, the inhibitory of outward potassium currents can only rescue the apoptosis-related cell volume shrinkage but cannot reverse chromatin condensation and internucleosomal DNA fragmentation during late apoptosis (McCarthy and Cotter 1997). Therefore, their conclusion needs to be reassessed by further experiments.

4 Moderate Concentration of Formaldehyde and Programmed Cell Death

Programmed cell death, particularly apoptosis, is triggered when the formaldehyde concentration was elevated >100 μM but lower than the lethal concentration. Such moderate concentrations of formaldehyde induced DNA damage in a faster rate than low concentration of formaldehyde and led to oxidative stress, mitochondrial dysfunction, ER stress, excitotoxicity, and proteotoxicity to the cell, which all contribute to the formaldehyde-triggered programmed cell death.

4.1 Apoptosis

Formaldehyde released from its intracellular carrier (Szende et al. 1998) and sufficient concentration of formaldehyde added into the cell culture medium could enhance apoptosis. The expression of ADH3, a formaldehyde degradation enzyme, in human oral keratinocytes was lowered when cell cultures have reached full confluent (Nilsson et al. 2004), suggesting that cells autonomously increase the endogenous formaldehyde level under weakened proliferation state. Rat lymphocytes exposed to 100 μM formaldehyde for 24 h showed increased population of cells with shrinking morphology and hypodiploid DNA, reflecting the progress of apoptosis (Nakao et al. 2003). A 4-h exposure to 0.35 mM formaldehyde induces apoptosis in Calu-3 and 16HBE cell lines (Kastner et al. 2011). The presence of 0.5 mM formaldehyde or soluble semicarbazide-sensitive amine oxidase, a formaldehyde production enzyme, in the medium, induces apoptosis in rat aorta A7r5 and human aortic smooth muscle cells (Rougeot et al. 2003).

In addition, it is necessary to be noticed that the apoptotic dose of formaldehyde also depends on the incubation time. As we have previously mentioned in Sect. 3, low concentration of formaldehyde exposure around the formaldehyde baseline is able to induce apoptosis after a long period of exposure time. Such phenomenon is also observed in animals and human subjects exposed to low concentrations of formaldehyde, e.g., inhalation and occupational contact. Formaldehyde-induced apoptosis in peripheral blood lymphocytes was detected in personnel under occupational formaldehyde exposure (Jakab et al. 2010). Low-dose formaldehyde inhalation in *cynomolgus macaques*, 2 ppm (6 h/day), induces miRNA transcriptional changes in the nasal epithelial, and the differentially expressed miRNAs were enriched in apoptosis signaling. At high inhalation dose, 6 ppm and above, genes related with cell cycle are upregulated, and DNA repair and apoptosis were enriched in rats (Andersen et al. 2010). The numbers of apoptotic cells and oxidative damage in the lung tissue were significantly increased in young and adult rats after 6 weeks 6 ppm (8 h/day) formaldehyde inhalation (Im et al. 2006; Sandikci et al. 2009; Sul et al. 2007).

Intraperitoneal injection of formaldehyde induces apoptosis in the brain, lung, liver, blood, and reproductive system. Two weeks after formaldehyde injection, the malondialdehyde-related neuronal apoptosis in the prefrontal cortex of rat was recorded (Zararsiz et al. 2007). The formaldehyde-induced (0.1–20 mg/kg body weight) apoptosis in the testis and oocytes of mice is correlated with the Fas signaling pathway (Peng et al. 2010a, b; Ozen et al. 2008; Zhou et al. 2006). The formaldehyde-induced (10 mg/kg body weight) apoptosis in the liver and kidney was delivered by oxidative damage and dysfunction in mitochondria (Bakar et al. 2015a, b). By intracerebroventricular formaldehyde injection, the special memory and novel object recognition of rat are reduced, the formaldehyde-induced apoptosis and lipid peroxidation increase, and H_2S generation in the hippocampus is detected (Tang et al. 2013; Tong et al. 2013b). A five-day nasogastric tube formaldehyde administration (40 mg/kg/day) in rabbit caused increased apoptosis in the brain, indicated by a decrease in Bcl-2 expression and an increase in active caspase-3 and Bax expressions, and the number of GFAP-positive glia cells is also increased (Arici et al. 2014). Subcutaneous formalin injection into the ventral surface of the right hind paw of rat is a common paradigm to induce periphery inflammatory pain (Tabata-Imai et al. 2014). Hu and colleagues confirmed that rats subjected to the inflammatory pain model developed neuronal apoptosis and upregulation of p53 in the hippocampus, suggesting that formaldehyde-induced pain would enhance the brain damage in special memory (Hu et al. 2009). Prolonged anesthesia causes widespread apoptosis in the central nervous system. Nociceptive stimulation with formalin enhances the anesthesia-induced neuroapoptosis and cognitive impairment during the synaptogenic period in the rat (Shu et al. 2012).

4.1.1 DNA Damage Repair

The accumulation of DNA adducts and DPX leads to cell cycle arrest. Excessive DNA repair burden eventually initiates the apoptosis signaling pathway and results in tissue damage, inflammation, and organ dysfunction. Miao and colleagues observed that 0.3 mM formaldehyde treatment (48 h) led to S-phase arrest in SH-SY5Y cell and that early and late apoptosis was markedly observed when cells were exposed to 0.1–0.2 mM formaldehyde (Miao et al. 2013). By using nucleotide excision repair-deficient CHO cell lines, Kumari and colleagues observed delayed DNA double-strand break repair in formaldehyde-treated DNA repair endonuclease XPF-deficient cells, indicating the involvement of XPF in the formaldehyde-induced genomic instability (Kumari et al. 2012). Several proteins in Fanconi anemia pathway, including FANCD2 and FANCD1/BRCA2, are essential to counteract formaldehyde-caused DPX in a chicken B-lymphocyte DT40 cell line (Ridpath et al. 2007). Ren and colleagues showed that FANCD2 mediates the repair of the formaldehyde-induced DPX in human lymphoblastoid cells. A 24-h formaldehyde treatment (50–150 μM) induces more DPX production and apoptosis in FANCD2-deficient cells as compared to cells that ectopically express the FANCD2 protein (Ren et al. 2013).

4.1.2 Oxidative Stress and Mitochondrial Dysfunction

Accumulating evidence has shown that formaldehyde-induced apoptosis is accompanied with oxidative stress. By mimicking occupational formaldehyde exposure, using 3 mg/m^3 formaldehyde in air for 2 weeks, BALB/c mice showed increasing formaldehyde-induced apoptosis in the nucleated spleen and colony-forming unit-granulocyte-macrophage, as well as increasing oxidative damages (Wei et al. 2016; Zhang et al. 2013). Formaldehyde-induced (10 nM; 9 days) apoptosis in human colorectal cancer SW620 cells is accompanied with elevated reactive oxygen species (ROS) (Lee et al. 2016).

Excessive oxidative damage leads to mitochondrial dysfunction. Formaldehyde induces apoptosis and mitochondrial clustering around the nucleus in rabbit corneal epithelial cells, which is attributed to the downregulation of ERK and persistent JNK activation (Lai et al. 2013). Zhang and colleagues indicated that formaldehyde-induced apoptosis in Chinese hamster lung fibroblast (V79-4) cells by elevating intracellular ROS production and inhibiting the mitochondrial function via inhibiting the c-JNK pathway (Zhang et al. 2010b). In sensitive neuronal SK-N-SH cells, oxidative stress was induced after formaldehyde exposure (7.5–27.5 μM) for 2 hours. Formaldehyde (13.5 μM, for 12 h) treatment greatly reduced cellular ATP level and mitochondrial membrane potential, indicating the dysfunction of mitochondria. The treatment further caused a concentration-dependent increase in nuclear fragmentation and activated the apoptosis-initiator caspase-9 and apoptosis-effector caspase-3/caspase-7 (Zerin et al. 2015). Tang and colleagues confirmed that formaldehyde causes apoptosis by mitochondria-related oxidative stress in mouse neuroblastoma PC12 cell, in which the inhibition of paraoxonase-1 expression and activity was involved. They also showed neural protective effect of H_2S against the formaldehyde-induced neuroapoptosis via the BDNF–TrkB pathway (Jiang et al. 2015; Tang et al. 2011, 2012).

4.1.3 ER Stress

Formaldehyde-induced disturbance in redox balance would subsequently induce ER stress, indicated by the elevated expressions of the ER markers, glucose-regulated protein 78 and C/EBP homologous protein, and eventually apoptosis in PC12 cell. Knockout of the modulator of ER stress and inhibitor of apoptosis increased the susceptibility of PC12 cells to formaldehyde-induced neurotoxicity (Luo et al. 2012). Ginsenoside Rgl rescued formaldehyde-induced apoptosis in SH-SY5Y, PC12, and N2a cells and suppresses formaldehyde-induced ER stress by inducing Trx-1 expression in PC12 cells (Chen et al. 2014; Luo et al. 2012; Sun et al. 2013). H_2S inhibits formaldehyde-induced ER stress by upregulation of SIRT-1 in PC12 cell (Bowling et al. 2014; Luo et al. 2012). By using lung epithelial A549 cells, Lim and colleagues demonstrated that formaldehyde induces apoptosis by decreasing Prx-2 expression via p38 MAPK pathway (Lim et al. 2010). They showed that formaldehyde treatment increases the phosphorylation of ER stress proteins, IRE1alpha, PERK, and eIF-2alpha, as well as increases the expressions of pro-apoptotic proteins, such as Bax, C/EPB homologous protein, cleaved PARP, and cleaved caspase-3, but decreases the expression of the anti-apoptotic protein Bcl-2. Overexpression of dimethylarginine dimethylaminohydrolase 1 reverses the formaldehyde-induced apoptosis and ER stress (Lim et al. 2013).

4.1.4 Excitotoxicity

Formaldehyde would probably induce a prolonged over-excitation of the hippocampal neuron, exhibiting the excitotoxicity. This will provide another explanation for the apoptosis induced by formaldehyde (Chi et al. 2012). Chi and colleagues showed that 1 mM formaldehyde increases intracellular calcium concentration in primary cultured hippocampal neurons, which is an indication for neural activation. Stimulation of Ca^{2+} release-activated calcium channels was shown to trigger Ca^{2+} entry into the cells and subsequent Ca^{2+} oscillations in proliferating cells. However, sustained Ca^{2+} entry through Ca^{2+}-permeable cation channels is able to trigger apoptosis (Lang et al. 2005, 2006).

4.1.5 Proteotoxicity

For the pathologic effects on non-proliferative tissues, proteotoxicity is another important factor in cell injury caused by formaldehyde. Formaldehyde could modify the cysteine residues via cysteine thiazolidination, which is anticipated to serve as a biomarker for oxidative stress and formaldehyde exposure (Liu et al. 2016). Recent studies showed that formaldehyde-treated human cells rapidly accumulate large amounts of damaged proteins undergoing K48 polyubiquitination (Ortega-Atienza et al. 2016). Formaldehyde could also induce Tau protein aggregation and hyperphosphorylation in vitro and in vivo, which has been demonstrated to play an important role in formaldehyde-induced cognitive impairment *(see Chap. 7 written by He, R. Q. for details*).

4.2 Autophagy and Cell Senescence

Besides apoptosis, formaldehyde could also induce programmed cell death in the way of autophagy and cell senescence. The evidence for formaldehyde-induced autophagy and cell senescence is limited at present. Formaldehyde inhalation (5–10 mg/m^3; 8 h/day; 4 weeks) could induce autophagy in rat testicular tissue, indicated by decreasing expression of mTOR (Han et al. 2015). Zhan and colleagues found that formaldehyde could accelerate cellular senescence in HT22 cells, in which leptin pathway is possibly involved (Zhan et al. 2016). However, the underlying mechanism remains unclear.

5 The Acute Cytotoxicity of High Concentration of Formaldehyde

The critical concentration for the acute toxicity of formaldehyde is around 1 mM. Most cells cannot survive when incubated with the presence of formaldehyde at concentrations higher than 1 mM after certain incubation period. Such a high formaldehyde concentration disrupts the integrity of membrane and genetic materials, which results in necrosis, indicated by uncontrolled release of cell death products into the extracellular space. After 10 mM formaldehyde treatment, instant high degree of cell damage and practically eradication were recorded in human colon carcinoma HT-29 and human endothelial HUV-EC-C cell cultures (Tyihak et al. 2001). Exposure to 3.5–7.0 mM formaldehyde for 24 h leads to large amount of lactate dehydrogenase leakage in bronchial epithelial Calu-3 and 16HBE cell lines, indicating significant formaldehyde-induced necrosis (Kastner et al. 2011). Primary human NK cells exhibit shrinking in morphology and cell colony disappearance when treated with >0.4 mM formaldehyde for 1 h, indicating the acute toxicity of formaldehyde, and 1 h exposure to >1.6 mM formaldehyde leads to significant increase in necrosis and late apoptosis (Li et al. 2013).

Neuronal cells were more sensitive to formaldehyde. The necrosis-inducing formaldehyde concentration for neuronal cell lines is 0.2~0.5 mM, lower than that for other cancer cells. Exposure to >0.3 mM formaldehyde for 24 h reduces ~50% of the viability of human neuroblastoma SH-SY5Y cell, and the increase in necrosis was detected (Miao et al. 2013; Song et al. 2016). Chen and colleagues reported that 24 h exposure to >0.2 mM formaldehyde in serum-free medium completely killed mouse neuroblastoma N2a (1×10^4 cells/well) (Chen et al. 2014).

The necrosis-inducing formaldehyde dose also depends on cell density. A higher seeding density could partially relieve the stress of formaldehyde on cell survival. Lu and colleagues showed that 24-h 0.5 mM formaldehyde only killed ~40% SH-SY5Y cells when cells were seeded at a high density (1×10^5 cells/well) under the serum-free condition. They observed the acute formaldehyde-induced morphological changes in cells but no sign of necrosis could be detected (Lu et al. 2013). When cells were allowed to grow to confluent (48 h) before formaldehyde treatment, the addition of 1 mM formaldehyde in the medium induced apoptosis in human tumor (HT-29, SW-620, HT-1080) and endothelial (HUV-EC-C) cells instead of inducing necrosis (Tyihak et al. 2001).

6 Conclusion

Accumulating public attentions have been paid to the neurotoxicity of formaldehyde during aging and the carcinogenesis of environmental and occupational formaldehyde exposure. The cytotoxicity and carcinogenesis of formaldehyde manifestly reveal two distinct cellular functions of formaldehyde: inducing cell death and

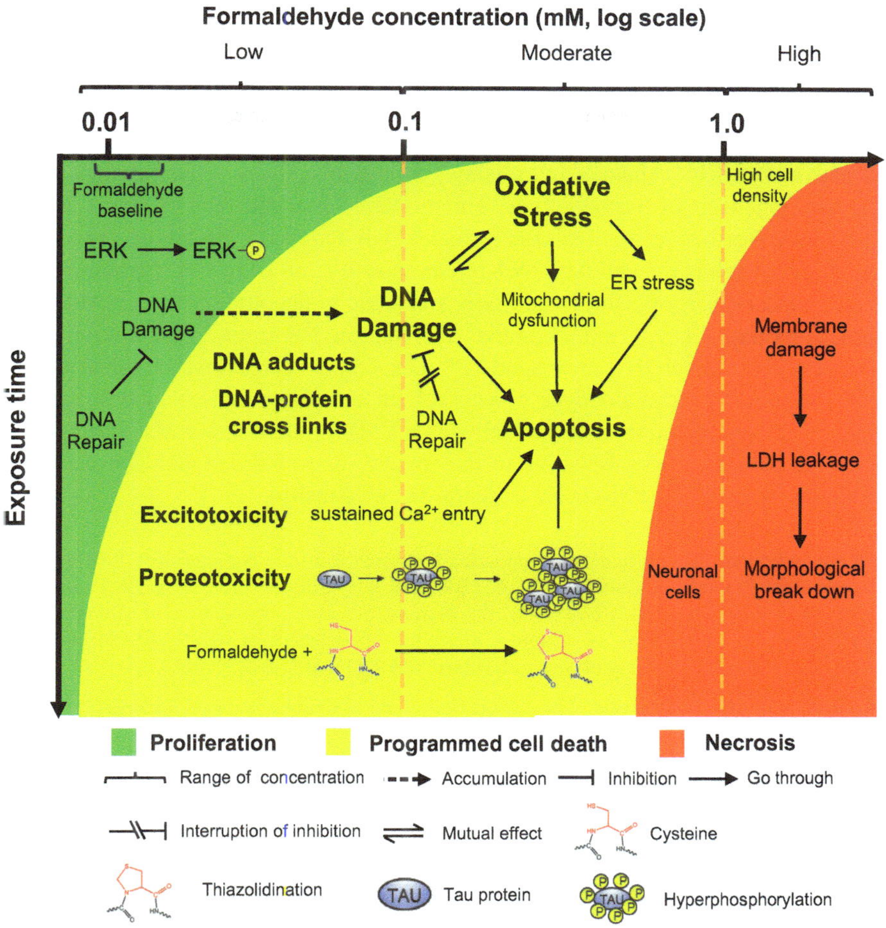

Fig. 5.1 The schematic diagram for effects of formaldehyde at different concentrations on cell proliferation and cell death. Pro-proliferation of cells could be observed in the presence of low concentrations of formaldehyde (<0.1 mM) in a short-term cultivate. Moderate concentrations of formaldehyde (0.1–1 mM) induce apoptosis. High concentrations of formaldehyde (>1 mM) promote necrosis. Low concentrations of formaldehyde treatment could trigger apoptosis, when the exposure time is prolonged enough. Cells that grow in their high density could partially attenuate the pro-necrosis effect of high concentration of formaldehyde. The literatures are referred to the effects of low concentration of formaldehyde on cells: Recio et al. 1992; Wolf et al. 1995; Szende et al. 2000; Hester et al. 2003; He et al. 2010; Zhang et al. 2010a; Tyihak et al. 2001; Szende and Tyihak 2010; Tong et al. 2011; Tong et al. 2013a; Sun et al. 2013; Chen et al. 2014; Ke et al. 2014; Rizzi et al. 2014; Wu et al. 2014; Pontel et al. 2015; Rizzi et al. 2016; Wang et al. 2017; for those of moderate concentrations of formaldehyde: Szende et al. 1998; Nakao et al. 2003; Rougeot et al. 2003; Ridpath et al. 2007; Lim et al. 2010; Kastner et al. 2011; Tang et al. 2011; Chi et al. 2012; Kumari et al. 2012; Luo et al. 2012; Tang et al. 2012; Lim et al. 2013; Miao et al. 2013; Sun et al. 2013; Ren et al. 2013; Zhang et al. 2013; Bowling et al. 2014; Chen et al. 2014; Han et al. 2015; Jiang et al. 2015; Zerin et al. 2015; Lee et al. 2016; Liu et al. 2016; Ortega-Atienza et al. 2016; Wei et al. 2016; Zhan et al. 2016; Zhang et al. 2010b; He et al., Chapter VII; and for those of high concentrations of formaldehyde: Tyihak et al. 2001; Kastner et al. 2011; Li et al. 2013; Lu et al. 2013; Miao et al. 2013; Chen et al. 2014; Song et al. 2016

promoting cell proliferation. The cellular effect of formaldehyde is time- and dose-dependent. The effective dosages and the mode of actions of formaldehyde were summarized in Fig. 5.1. Short-term small interruption on the formaldehyde baseline, i.e., formaldehyde elimination by formaldehyde capturers and low-concentration formaldehyde (<0.1 mM), stimulates the proliferation of cancer cells by activating ERK pathways and raising genomic instability, which contributes to the carcinogenesis of formaldehyde. The moderate concentration of formaldehyde (0.1–1 mM) further accumulates the formaldehyde-induced DNA adducts and DPX, which overwhelms the cellular DNA repair capacity and triggers programmed cell death, especially apoptosis. In addition, moderate concentration of formaldehyde induces oxidative stress, mitochondrial dysfunction, ER stress, proteotoxicity, and excitotoxicity delivered by prolonged Ca^{2+} releasing in neurons, which subsequently lead to apoptosis. Formaldehyde plays a role in autophagy and cell senescence, as well. High concentration of formaldehyde (> 1 mM) induces necrosis. Particularly, neuronal cells are more sensitive to the toxicity of formaldehyde. The necrosis-inducing formaldehyde dose also depends on cell density.

Acknowledgment This project was supported by grants from the National Key Research and Development Program of China (2016YFC1306300), the National Basic Research Program of China (973 Program) (2012CB911004), the Natural Scientific Foundation of China (NSFC 31270868), the Foundation of Chinese Academy of Sciences (CAS-20140909), and the Queensland-Chinese Academy of Sciences Biotechnology Fund (GJHZ201302).

Competing Financial Interests The authors declare no competing financial interests.

References

Aizenshtadt AA, Burova EB, Zenin VV, Bobkov DE, Kropacheva IV, Pinaev GP (2011) Effect of formaldehyde in low concentrations on the proliferation and organization of the cytoskeleton of cultured cells. Tsitologiia 53:978–985

Andersen ME, Clewell HJ 3rd, Bermudez E, Dodd DE, Willson GA, Campbell JL, Thomas RS (2010) Formaldehyde: integrating dosimetry, cytotoxicity, and genomics to understand dose-dependent transitions for an endogenous compound. Toxicol Sci 118:716–731

Arici S, Karaman S, Dogru S, Cayli S, Arici A, Suren M, Karaman T, Kaya Z (2014) Central nervous system toxicity after acute oral formaldehyde exposure in rabbits: an experimental study. Hum Exp Toxicol 33:1141–1149

Bakar E, Ulucam E, Cerkezkayabekir A (2015a) Investigation of the protective effects of proanthocyanidin and vitamin E against the toxic effect caused by formaldehyde on the liver tissue. Environ Toxicol 30:1406–1415

Bakar E, Ulucam E, Cerkezkayabekir A (2015b) Protective effects of proanthocyanidin and vitamin E against toxic effects of formaldehyde in kidney tissue. Biotech Histochem 90:69–78

Barthel BL, Mooz EL, Wiener LE, Koch GG, Koch TH (2016) Correlation of in situ Oxazolidine formation with highly synergistic cytotoxicity and DNA cross-linking in cancer cells from combinations of doxorubicin and formaldehyde. J Med Chem 59:2205–2221

Binazzi A, Ferrante P, Marinaccio A (2015) Occupational exposure and sinonasal cancer: a systematic review and meta-analysis. BMC Cancer 15:49

Bowling AJ, Pence HE, Church JB (2014) Application of a novel and automated branched DNA in situ hybridization method for the rapid and sensitive localization of mRNA molecules in plant tissues. Appl Plant Sci 2(4):apps.1400011
Brewer TF, Chang CJ (2015) An Aza-cope reactivity-based fluorescent probe for imaging formaldehyde in living cells. J Am Chem Soc 137:10886–10889
Chen JY, Sun MR, Wang XH, Lu J, Wei Y, Tan Y, Liu Y, Götz J, He RQ, Hua Q (2014) The herbal compound geniposide rescues formaldehyde induced apoptosis in N2a neuroblastoma cells. Sci China Life Sci 57:412–421
Chen XX, Su T, He YG, He RQ (2017) Effect of oxidative stress on accommodation and transportation of formaldehyde by lysosome. Prog Biochem Biophys 44:486–494
Chi YN, Zhang X, Cai J, Liu FY, Xing GG, Wan Y (2012) Formaldehyde increases intracellular calcium concentration in primary cultured hippocampal neurons partly through NMDA receptors and T-type calcium channels. Neurosci Bull 28:715–722
Costa S, Pina C, Coelho P, Costa C, Silva S, Porto B, Laffon B, Teixeira JP (2011) Occupational exposure to formaldehyde: genotoxic risk evaluation by comet assay and micronucleus test using human peripheral lymphocytes. J Toxicol Environ Health A 74:1040–1051
Costa S, Garcia-Leston J, Coelho M, Coelho P, Costa C, Silva S, Porto B, Laffon B, Teixeira JP (2013) Cytogenetic and immunological effects associated with occupational formaldehyde exposure. J Toxicol Environ Health A 76:217–229
Cutts SM, Nudelman A, Rephaeli A, Phillips DR (2005) The power and potential of doxorubicin-DNA adducts. IUBMB Life 57:73–81
Dhareshwar SS, Stella VJ (2008) Your prodrug releases formaldehyde: should you be concerned? No! J Pharm Sci 97:4184–4193
Ebeler SE, Clifford AJ, Shibamoto T (1997) Quantitative analysis by gas chromatography of volatile carbonyl compounds in expired air from mice and human. J Chromatogr B 702:211–215
Fujimaki H, Kurokawa Y, Kakeyama M, Kunugita N, Fueta Y, Fukuda T, Hori H, Arashidani K (2004) Inhalation of low-level formaldehyde enhances nerve growth factor production in the hippocampus of mice. Neuroimmunomodulation 11:373–375
Han SP, Zhou DX, Lin P, Qin Z, An L, Zheng LR, Lei L (2015) Formaldehyde exposure induces autophagy in testicular tissues of adult male rats. Environ Toxicol 30:323–331
He R, Lu J, Miao J (2010) Formaldehyde stress. Sci China Life Sci 53:1399–1404
Heck HD, Casanova M (2004) The implausibility of leukemia induction by formaldehyde: a critical review of the biological evidence on distant-site toxicity. Regul Toxicol Pharmacol 40:92–106
Hester SD, Benavides GB, Yoon L, Morgan KT, Zou F, Barry W, Wolf DC (2003) Formaldehyde-induced gene expression in F344 rat nasal respiratory epithelium. Toxicology 187:13–24
Hu YY, Li QJ, Li WB, Guo LH, Chu SC (2009) Formalin inflammatory pain induced hippocampal neuronal apoptosis of rats. Zhongguo Ying Yong Sheng Li Xue Za Zhi 25:190–194
IARC Working Group on the Evaluation of Carcinogenic Risks to Humans (2006) Formaldehyde, 2-butoxyethanol and 1-tert-butoxypropan-2-ol. IARC Monogr Eval Carcinog Risks Hum 88:1–478
Im H, Oh E, Mun J, Khim JY, Lee E, Kang HS, Kim E, Kim H, Won NH, Kim YH et al (2006) Evaluation of toxicological monitoring markers using proteomic analysis in rats exposed to formaldehyde. J Proteome Res 5:1354–1366
Jakab MG, Klupp T, Besenyei K, Biro A, Major J, Tompa A (2010) Formaldehyde-induced chromosomal aberrations and apoptosis in peripheral blood lymphocytes of personnel working in pathology departments. Mutat Res 698:11–17
Jiang JM, Zhou CF, Gao SL, Tian Y, Wang CY, Wang L, Gu HF, Tang XQ (2015) BDNF-TrkB pathway mediates neuroprotection of hydrogen sulfide against formaldehyde-induced toxicity to PC12 cells. PLoS One 10:e0119478
Joob B, Wiwanitkit V (2016) Formaldehyde level in anatomy laboratory teaching room and risk for cancer. Workplace Health Saf 64:132
Kastner PE, Casset A, Pons F (2011) Formaldehyde interferes with airway epithelium integrity and functions in a dose- and time-dependent manner. Toxicol Lett 200:109–116

Kato S, Burke PJ, Fenick DJ, Taatjes DJ, Bierbaum VM, Koch TH (2000) Mass spectrometric measurement of formaldehyde generated in breast cancer cells upon treatment with anthracycline antitumor drugs. Chem Res Toxicol 13:509–516
Kato S, Burke PJ, Koch TH, Bierbaum VM (2001) Formaldehyde in human cancer cells: detection by preconcentration-chemical ionization mass spectrometry. Anal Chem 73:2992–2997
Ke YJ, Qin XD, Zhang YC, Li H, Li R, Yuan JL, Yang X, Ding SM (2014) In vitro study on cytotoxicity and intracellular formaldehyde concentration changes after exposure to formaldehyde and its derivatives. Hum Exp Toxicol 33:822–830
Kisseljova NP, Kisseljov FL (2005) DNA demethylation and carcinogenesis. Biochemistry (Mosc) 70:743–752
Kleinnijenhuis AJ, Staal YCM, Duistermaat E, Engel R, Woutersen RA (2013) The determination of exogenous formaldehyde in blood of rats during and after inhalation exposure. Food Chem Toxicol 52:105–112
Kumari A, Lim YX, Newell AH, Olson SB, McCullough AK (2012) Formaldehyde-induced genome instability is suppressed by an XPF-dependent pathway. DNA Repair (Amst) 11:236–246
Lai LJ, Hsu WH, Wu AM, Wu JH (2013) Ocular injury by transient formaldehyde exposure in a rabbit eye model. PLoS One 8:e66649
Lai Y, Yu R, Hartwell HJ, Moeller BC, Bodnar WM, Swenberg JA (2016) Measurement of endogenous versus exogenous formaldehyde-induced DNA-protein crosslinks in animal tissues by stable isotope labeling and ultrasensitive mass spectrometry. Cancer Res 76:2652–2661
Lang F, Foller M, Lang KS, Lang PA, Ritter M, Gulbins E, Vereninov A, Huber SM (2005) Ion channels in cell proliferation and apoptotic cell death. J Membr Biol 205:147–157
Lang F, Shumilina E, Ritter M, Gulbins E, Vereninov A, Huber SM (2006) Ion channels and cell volume in regulation of cell proliferation and apoptotic cell death. Contrib Nephrol 152:142–160
Lee HM, Kim CW, Hwang KA, Choi DW, Choi KC (2016) Three components of cigarette smoke altered the growth and apoptosis of metastatic colon cancer cells via inducing the synthesis of reactive oxygen species and endoplasmic reticulum stress. Environ Toxicol Pharmacol 45:80–89
Li Q, Mei Q, Huyan T, Xie L, Che S, Yang H, Zhang M, Huang Q (2013) Effects of formaldehyde exposure on human NK cells in vitro. Environ Toxicol Pharmacol 36:948–955
Li T, Su T, He YG, He RQ (2016a) Chronic-dehydrated Dysmetabolism of formaldehyde in mouse brain and decline of learning in the shuttle box. Prog Biochem Biophys 43:429–438
Li Y, Song ZY, Ding YJ, Xin Y, Wu T, Su T, He RQ, Tai FD, Lian ZM (2016b) Effects of formaldehyde exposure on anxiety-like and depression-like behavior, cognition, central levels of glucocorticoid receptor and tyrosine hydroxylase in mice. Chemosphere 144:2004–2012
Lim SK, Kim JC, Moon CJ, Kim GY, Han HJ, Park SH (2010) Formaldehyde induces apoptosis through decreased Prx 2 via p38 MAPK in lung epithelial cells. Toxicology 271:100–106
Lim SK, Choi H, Park MJ, Kim DI, Kim JC, Kim GY, Jeong SY, Rodionov RN, Han HJ, Yoon KC et al (2013) The ER stress-mediated decrease in DDAH1 expression is involved in formaldehyde-induced apoptosis in lung epithelial cells. Food Chem Toxicol 62:763–769
Lin X, Yan J, Tang D (2013) ERK kinases modulate the activation of PI3 kinase related kinases (PIKKs) in DNA damage response. Histol Histopathol 28:1547–1554
Liu P, Guo J, Qi Z (2011) Formaldehyde suppresses neuronal apoptosis via inhibition of outward K(+) currents in rat hippocampus. Can J Physiol Pharmacol 89:837–844
Liu J, Liu FY, Tong ZQ, Li ZH, Chen W, Luo WH, Li H, Luo HJ, Tang Y, Tang JM et al (2013) Lysine-specific demethylase 1 in breast cancer cells contributes to the production of endogenous formaldehyde in the metastatic bone cancer pain model of rats. PLoS One 8:e58957
Liu J, Chan KK, Chan W (2016) Identification of protein Thiazolidination as a novel molecular signature for oxidative stress and formaldehyde exposure. Chem Res Toxicol 29(11):1865–1871
Lu K, Collins LB, Ru HY, Bermudez E, Swenberg JA (2010) Distribution of DNA adducts caused by inhaled formaldehyde is consistent with induction of nasal carcinoma but not leukemia. Toxicol Sci 116:441–451

Lu J, Miao JY, Su T, Liu Y, He RQ (2013) Formaldehyde induces hyperphosphorylation and polymerization of tau protein both in vitro and in vivo. BBA-Gen Subjects 1830:4102–4116
Luo FC, Zhou J, Lv T, Qi L, Wang SD, Nakamura H, Yodoi J, Bai J (2012) Induction of endoplasmic reticulum stress and the modulation of thioredoxin-1 in formaldehyde-induced neurotoxicity. Neurotoxicology 33:290–298
Mahboubi A, Koushik A, Siemiatycki J, Lavoue J, Rousseau MC (2013) Assessment of the effect of occupational exposure to formaldehyde on the risk of lung cancer in two Canadian population-based case-control studies. Scand J Work Environ Health 39:401–410
McCarthy JV, Cotter TG (1997) Cell shrinkage and apoptosis: a role for potassium and sodium ion efflux. Cell Death Differ 4:756–770
McGregor D, Bolt H, Cogliano V, Richter-Reichhelm HB (2006) Formaldehyde and glutaraldehyde and nasal cytotoxicity: case study within the context of the 2006 IPCS human framework for the analysis of a cancer mode of action for humans. Crit Rev Toxicol 36:821–835
Miao JY, Lu J, Zhang ZJ, Tong ZQ, He RQ (2013) The effect of formaldehyde on cell cycle is in a concentration-dependent manner. Prog Biochem Biophys 40:641–651
Monticello TM, Swenberg JA, Gross EA, Leininger JR, Kimbell JS, Seilkop S, Starr TB, Gibson JE, Morgan KT (1996) Correlation of regional and nonlinear formaldehyde-induced nasal cancer with proliferating populations of cells. Cancer Res 56:1012–1022
Nakao H, Umebayashi C, Nakata M, Nishizaki Y, Noda K, Okano Y, Oyama Y (2003) Formaldehyde-induced shrinkage of rat thymocytes. J Pharmacol Sci 91:83–86
Nilsson JA, Hedberg JJ, Vondracek M, Staab CA, Hansson A, Hoog JO, Grafstrom RC (2004) Alcohol dehydrogenase 3 transcription associates with proliferation of human oral keratinocytes. Cell Mol Life Sci 61:610–617
Nudelman A, Levovich I, Cutts SM, Phillips DR, Rephaeli A (2005) The role of intracellularly released formaldehyde and butyric acid in the anticancer activity of acyloxyalkyl esters. J Med Chem 48:1042–1054
Ortega-Atienza S, Rubis B, McCarthy C, Zhitkovich A (2016) Formaldehyde is a potent Proteotoxic stressor causing rapid heat shock transcription factor 1 activation and Lys48-linked Polyubiquitination of proteins. Am J Pathol 186:2857–2868
Ozen OA, Kus MA, Kus I, Alkoc OA, Songur A (2008) Protective effects of melatonin against formaldehyde-induced oxidative damage and apoptosis in rat testes: an immunohistochemical and biochemical study. Syst Biol Reprod Med 54:169–176
Peng G, Tian Y, Jin O (2010a) The effects of formaldehyde on ovarian histopathology structure, oocyte ultrastructure and apoptosis of female rats. Xi Bao Yu Fen Zi Mian Yi Xue Za Zhi 26:865–867
Peng G, Zhong C, Zhang Q, Xie Y, Gong F (2010b) Effect of formaldehyde on expressions of Fas apoptosis pathway-related genes of ovary tissues in female rats. Zhong Nan Da Xue Xue Bao Yi Xue Ban 35:341–345
Pontel LB, Rosado IV, Burgos-Barragan G, Garaycoechea JI, Yu R, Arends MJ, Chandrasekaran G, Broecker V, Wei W, Liu L et al (2015) Endogenous formaldehyde is a hematopoietic stem cell Genotoxin and metabolic carcinogen. Mol Cell 60:177–188
Recio L, Sisk S, Pluta L, Bermudez E, Gross EA, Chen Z, Morgan K, Walker C (1992) p53 mutations in formaldehyde-induced nasal squamous cell carcinomas in rats. Cancer Res 52:6113–6116
Ren X, Ji Z, McHale CM, Yuh J, Bersonda J, Tang M, Smith MT, Zhang L (2013) The impact of FANCD2 deficiency on formaldehyde-induced toxicity in human lymphoblastoid cell lines. Arch Toxicol 87:189–196
Ridpath JR, Nakamura A, Tano K, Luke AM, Sonoda E, Arakawa H, Buerstedde JM, Gillespie DA, Sale JE, Yamazoe M et al (2007) Cells deficient in the FANC/BRCA pathway are hypersensitive to plasma levels of formaldehyde. Cancer Res 67:11117–11122
Rizzi M, Cravello B, Reno F (2014) Textile industry manufacturing by-products induce human melanoma cell proliferation via ERK1/2 activation. Cell Prolif 47:578–586

Rizzi M, Cravello B, Tonello S, Reno F (2016) Formaldehyde solutions in simulated sweat increase human melanoma but not normal human keratinocyte cells proliferation. Toxicol In Vitro 37:106–112

Rougeot C, Messaoudi M, Hermitte V, Rigault AG, Blisnick T, Dugave C, Desor D, Rougeon F (2003) Sialorphin, a natural inhibitor of rat membrane-bound neutral endopeptidase that displays analgesic activity. Proc Natl Acad Sci U S A 100:8549–8554

Salthammer T (2015) The formaldehyde dilemma. Int J Hyg Environ Health 218:433–436

Sandikci M, Seyrek K, Aksit H, Kose H (2009) Inhalation of formaldehyde and xylene induces apoptotic cell death in the lung tissue. Toxicol Ind Health 25:455–461

Shu Y, Zhou Z, Wan Y, Sanders RD, Li M, Pac-Soo CK, Maze M, Ma D (2012) Nociceptive stimuli enhance anesthetic-induced neuroapoptosis in the rat developing brain. Neurobiol Dis 45:743–750

Song YX, Miao JY, Qiang M, He RQ, Wang XM, Li WW (2016) Icariin protects SH-SY5Y cells from formaldehyde-induced injury through suppression of tau phosphorylation. Chin J Integr Med 22:430–437

Spanel P, Smith D, Holland TA, Al Singary W, Elder JB (1999) Analysis of formaldehyde in the headspace of urine from bladder and prostate cancer patients using selected ion flow tube mass spectrometry. Rapid Commun Mass Sp 13:1354–1359

Sul D, Kim H, Oh E, Phark S, Cho E, Choi S, Kang HS, Kim EM, Hwang KW, Jung WW (2007) Gene expression profiling in lung tissues from rats exposed to formaldehyde. Arch Toxicol 81:589–597

Sun P, Chen JY, Li J, Sun MR, Mo WC, Liu KL, Meng YY, Liu Y, Wang F, He RQ et al (2013) The protective effect of geniposide on human neuroblastoma cells in the presence of formaldehyde. BMC Complement Altern Med 13:152

Swenberg JA, Lu K, Moeller BC, Gao L, Upton PB, Nakamura J, Starr TB (2011) Endogenous versus exogenous DNA adducts: their role in carcinogenesis, epidemiology, and risk assessment. Toxicol Sci 120(Suppl 1):S130–S145

Swenberg JA, Moeller BC, Lu K, Rager JE, Fry RC, Starr TB (2013) Formaldehyde carcinogenicity research: 30 years and counting for mode of action, epidemiology, and cancer risk assessment. Toxicol Pathol 41:181–189

Szende B, Tyihak E (2010) Effect of formaldehyde on cell proliferation and death. Cell Biol Int 34:1273–1282

Szende B, Tyihak E, Trezl L, Szoke E, Laszlo I, Katay G, Kiraly-Veghely Z (1998) Formaldehyde generators and capturers as influencing factors of mitotic and apoptotic processes. Acta Biol Hung 49:323–329

Szende B, Tyihak E, Kiraly-Veghely Z (2000) Dose-dependent effect of resveratrol on proliferation and apoptosis in endothelial and tumor cell cultures. Exp Mol Med 32:88–92

Tabata-Imai A, Inoue R, Mori H (2014) Increased sensitivity to inflammatory pain induced by subcutaneous formalin injection in serine racemase knock-out mice. PLoS One 9:e105282

Tang XQ, Ren YK, Chen RQ, Zhuang YY, Fang HR, Xu JH, Wang CY, Hu B (2011) Formaldehyde induces neurotoxicity to PC12 cells involving inhibition of paraoxonase-1 expression and activity. Clin Exp Pharmacol Physiol 38:208–214

Tang XQ, Ren YK, Zhou CF, Yang CT, Gu HF, He JQ, Chen RQ, Zhuang YY, Fang HR, Wang CY (2012) Hydrogen sulfide prevents formaldehyde-induced neurotoxicity to PC12 cells by attenuation of mitochondrial dysfunction and pro-apoptotic potential. Neurochem Int 61:16–24

Tang XQ, Zhuang YY, Zhang P, Fang HR, Zhou CF, Gu HF, Zhang H, Wang CY (2013) Formaldehyde impairs learning and memory involving the disturbance of hydrogen sulfide generation in the hippocampus of rats. J Mol Neurosci 49:140–149

Tang Y, Kong X, Liu ZR, Xu A, Lin W (2016) Lysosome-targeted turn-on fluorescent probe for endogenous formaldehyde in living cells. Anal Chem 88:9359–9363

Tong Z, Luo W, Wang Y, Yang F, Han Y, Li H, Luo H, Duan B, Xu T, Maoying Q et al (2010) Tumor tissue-derived formaldehyde and acidic microenvironment synergistically induce bone cancer pain. PLoS One 5:e10234

Tong ZQ, Zhang JL, Luo WH, Wang WS, Li FX, Li H, Luo HJ, Lu J, Zhou JN, Wan Y et al (2011) Urine formaldehyde level is inversely correlated to mini mental state examination scores in senile dementia. Neurobiol Aging 32:31–41

Tong Z, Han C, Luo W, Wang X, Li H, Luo H, Zhou J, Qi J, He R (2013a) Accumulated hippocampal formaldehyde induces age-dependent memory decline. Age 35:583–596

Tong ZQ, Han CS, Luo WH, Li H, Luo HJ, Qiang M, Su T, Wu BB, Liu Y, Yang X et al (2013b) Aging-associated excess formaldehyde leads to spatial memory deficits. Sci Rep-Uk 3:1807

Tong Z, Han C, Qiang M, Wang W, Lv J, Zhang S, Luo W, Li H, Luo H, Zhou J et al (2015) Age-related formaldehyde interferes with DNA methyltransferase function, causing memory loss in Alzheimer's disease. Neurobiol Aging 36:100–110

Tsukahara S, Yamamoto S, Tin Tin Win S, Ahmed S, Kunugita N, Arashidani K, Fujimaki H (2006) Inhalation of low-level formaldehyde increases the Bcl-2/Bax expression ratio in the hippocampus of immunologically sensitized mice. Neuroimmunomodulation 13:63–68

Tyihak E, Bocsi J, Timar F, Racz G, Szende B (2001) Formaldehyde promotes and inhibits the proliferation of cultured tumour and endothelial cells. Cell Prolif 34:135–141

Wang J, Zhou J, Mo WC, He YG, Wei Y, He RQ, Yi FP (2017) Accumulation of simulated pathological level of formaldehyde decreases cell viability and adhesive morphology in neuronal cells. Prog Biochem Biophys 44:601-614

Wei C, Wen H, Yuan L, McHale CM, Li H, Wang K, Yuan J, Yang X, Zhang L (2016) Formaldehyde induces toxicity in mouse bone marrow and hematopoietic stem/progenitor cells and enhances benzene-induced adverse effects. Arch Toxicol 91:921–933

Wolf DC, Gross EA, Lyght O, Bermudez E, Recio L, Morgan KT (1995) Immunohistochemical localization of p53, PCNA, and TGF-alpha proteins in formaldehyde-induced rat nasal squamous cell carcinomas. Toxicol Appl Pharmacol 132:27–35

Wu K, Ren R, Su W, Wen B, Zhang Y, Yi F, Qiao X, Yuan T, Wang J, Liu L et al (2014) A novel suppressive effect of alcohol dehydrogenase 5 in neuronal differentiation. J Biol Chem 289:20193–20199

Xu J, Zhang Y, Zeng L, Liu J, Kinsella JM, Sheng R (2016) A simple naphthalene-based fluorescent probe for high selective detection of formaldehyde in toffees and HeLa cells via aza-cope reaction. Talanta 160:645–652

Zararsiz I, Kus I, Ogeturk M, Akpolat N, Kose E, Meydan S, Sarsilmaz M (2007) Melatonin prevents formaldehyde-induced neurotoxicity in prefrontal cortex of rats: an immunohistochemical and biochemical study. Cell Biochem Funct 25:413–418

Zerin T, Kim JS, Gil HW, Song HY, Hong SY (2015) Effects of formaldehyde on mitochondrial dysfunction and apoptosis in SK-N-SH neuroblastoma cells. Cell Biol Toxicol 31:261–272

Zhan K, Ning M, Wang C, Tang Y, Gu H, Yan C, Tang X (2016) Formaldehyde accelerates cellular senescence in HT22 cells: possible involvement of the leptin pathway. Acta Biochim Biophys Sin Shanghai 48:771–773

Zhang L, Freeman LE, Nakamura J, Hecht SS, Vandenberg JJ, Smith MT, Sonawane BR (2010a) Formaldehyde and leukemia: epidemiology, potential mechanisms, and implications for risk assessment. Environ Mol Mutagen 51:181–191

Zhang R, Lee IK, Kang KA, Piao MJ, Kim KC, Kim BJ, Lee NH, Choi JY, Choi J, Hyun JW (2010b) Cytoprotective effects of triphlorethol-a against formaldehyde-induced oxidative damage and apoptosis: role of mitochondria-mediated caspase-dependent pathway. J Toxicol Environ Health A 73:1477–1489

Zhang Y, Liu X, McHale C, Li R, Zhang L, Wu Y, Ye X, Yang X, Ding S (2013) Bone marrow injury induced via oxidative stress in mice by inhalation exposure to formaldehyde. PLoS One 8:e74974

Zhang J, Sun RL, Chen Y, Tan KH, Wei HY, Yin LH, Pu YP (2014) Small molecule metabolite biomarker candidates in urine from mice exposed to formaldehyde. Int J Mol Sci 15:16458–16468

Zhou DX, Qiu SD, Zhang J, Wang ZY (2006) Reproductive toxicity of formaldehyde to adult male rats and the functional mechanism concerned. Sichuan Da Xue Xue Bao Yi Xue Ban 37:566–569

Chapter 6
Gut Microbiota, Formaldehyde Dysmetabolism, and Cognitive Impairment

Kaili Liu and Rongqiao He

Abstract Formaldehyde has been shown to play important roles in contributing to age-related cognitive impairment including Alzheimer's disease (AD) and vascular dementia (VD). Gut microbiota, also called "the second brain", can modulate brain function and participate in pathogenesis of neurodegeneration, such as AD, Parkinson's disease (PD), and multiple sclerosis (MS). However, whether there is a link among formaldehyde, gut microbiota and cognitive impairment remains unclear. In this review, we briefly summarized the research progress of formaldehyde and gut microbiota and discussed their relationships with age-related cognitive impairment. We propose a probable link between gut microbiota and formaldehyde based on the probable functions of 5-HT and NE during these neurodegenerative diseases. We hypothesize that formaldehyde derived from gut microbiota may be involved in cognitive function and its dysmetabolism may be a risk factor of AD or even lead to cognitive decline. It is also intriguing to predict that, within several specific AD-related gut microorganisms reported to date, the shortage of hexulose-6-phosphate synthase (HPS)/6-phospho-3-hexuloisomerase (PHI) in these probiotics may contribute to the onset of AD and cognitive impairment. Whether the other formaldehyde metabolism-related genes were key factors needs further investigation. In order to clarify that dysmetabolism of formaldehyde in intestinal microbiota is involved in AD, we compared the concentration of intestinal formaldehyde in APP/PS1 double transgenic mice and that of C57BL/6j wild-type mice (as control). There is a significantly elevated level of formaldehyde in the cecum digestion content of the APP/PS1 transgenic mice. Moreover, the formaldehyde level in the small intestinal wall for the APP/PS1 transgenic mice is also higher than that of the control. These data suggest that intestinal microbiota is a main source of formaldehyde, which may have a link to cognitive impairment. To prevent or delay the progress of

---

K. Liu
State Key Laboratory of Brain and Cognitive Science, Institute of Biophysics, Chinese Academy of Sciences, Beijing 100101, China

R. He (✉)
State Key Laboratory of Brain and Cognitive Science, Institute of Biophysics, Chinese Academy of Sciences, Beijing 100101, China

University of Chinese Academy of Sciences, Beijing 100049, China
e-mail: rongqiaohe@163.com

R. He, *Formaldehyde and Cognition*, DOI 10.1007/978-94-024-1177-5_6

neurodegeneration, it is reasonable to use formaldehyde scavengers, such as probiotics, in addition to previously proposed anti-oxidative and anti-inflammatory chemicals to prevent, delay, attenuate, and/or block the progress of age-related cognitive impairment, therefore improving the human health and quality of life.

Keywords Formaldehyde • Gut microbiota • Brain • Cognitive ability • Cognitive impairment • Metabolism • Dysmetabolism • Serotonin • Norepinephrine • Probiotics • APP/PS1 transgenic mouse • Cecum digestion contents • Small intestinal wall

1 Introduction

Formaldehyde is a naturally occurring organic compound and the simplest aldehyde existing in several distinct forms. As a gas, it is colorless with irritating odor. This gas form is able to convert to a number of other forms upon condensation. One of important derivatives is the cyclic trimer metaformaldehyde (MFA) or 1,3,5-trioxane with the formula $(CH_2O)_3$. One of a linear polymer form, called paraformaldehyde (PFA), is commonly used in scientific researches because of its practical value. Formaldehyde is a significant risk for human health due to its chemical activity, toxicity, and global use. Although formaldehyde has been abundantly studied, so far, almost no laboratory reported formaldehyde production in human intestinal microbiota.

It has been generally known that there are approximately 10^{12} human cells while 10^{13}–10^{14} bacteria cells in our body, indicating that human body is actually composed of more bacteria than our own cells (O'Hara and Shanahan 2006; Whitman et al. 1998). These trillions of bacteria and other microorganisms are called the microbiota (Kim et al. 2013; O'Hara and Shanahan 2006). Most of those bacteria locate/live in human's gut, referred to as the gut/intestinal microbiota, and they play numerous roles in human development, physiology, and pathology (Qin et al. 2010). Intestinal microbiota play important roles in physical and psychological health via their own neural network: the enteric nervous system (ENS), which is a comprehensive system composed of ~100 million nerves found in the lining of the gut (Cryan and Dinan 2012). The ENS is also called the "second brain" and it actually arises from the same tissues as the central nervous system (CNS) during embryonic development (Cryan and Dinan 2012). Therefore, it has many structural and chemical parallels and similarities to the brain, which functions via hormones, neurotransmitters, electrical impulses, and immunoreactions (Hyland and Cryan 2016; Minter et al. 2016).

Previous studies indicated that widespread use of gluten and sugar in food and lack of healthy fat intake would cause the systemic inflammation, eventually attack the CNS functions, and affect the brain health (Cryan and Dinan 2012). New evidences showed that the influence of diet on brain health is not because of the diet-induced inflammatory responses but the disruption of gut microbiota (Clark and Mach 2016). Formaldehyde is also endogenously produced and metabolized in animal and human guts (Ohsawa et al. 2015; Wang et al. 2015). Furthermore, formaldehyde is easily dissolved in blood, freely crosses through blood-brain barrier (BBB), and exists in all cells including neurons and glial cells (Shcherbakova et al. 1986). It has been reported that gut microbiota influences human emotion and cognition (Finegold et al. 2010; Naseribafrouei et al. 2014). According to Hu and colleagues, the dysregulation of gut microbiota is correlated with cognitive impairment and neurodegenerative diseases (Hu et al. 2016). Therefore, the relationships between cognitive impairment and gut microbiota ("the second brain") associated with formaldehyde will be discussed in this chapter.

2 Formaldehyde Metabolism and Associated Enzymes in Microbes

2.1 *Formaldehyde Assimilation*

Formaldehyde is a type of simple but highly reactive compound that has a toxic impact on almost all living organisms. As formaldehyde is a key compound in one-carbon metabolisms, most organisms have their own mechanisms for its detoxification (Bruemmer et al. 2017). During the evolution, the formaldehyde detoxification system is highly developed in methylotrophic organisms, which can assimilate C1 compounds. Here, we discuss the main pathways of formaldehyde assimilation.

2.1.1 Ribulose Monophosphate (RuMP) Pathway

Methanotrophic bacteria can produce formaldehyde from the oxidation of methane and methanol. Formaldehyde is then assimilated to form intermediates of the central metabolic pathways which are subsequently used for cellular material biosynthesis (Anthony 1991). RuMP is one of the most important formaldehyde assimilation pathways and contributes significantly to formaldehyde detoxification in methylotrophs. This RuMP pathway was first described by Quayle and his colleagues (Johnson and Quayle 1965; Kemp and Quayle 1967; Strom et al. 1974) and can be interpreted as three main parts – fixation, cleavage, and rearrangement – among which formaldehyde assimilation is present in fixation step (Fig. 6.1).

During fixation procedure, formaldehyde and D-ribulose 5-phosphate (RuMP) are catalyzed by hexulose-6-phosphate synthase (HPS) to generate hexulose 6-phosphate (HuMP or Hu6P), which in turn is converted to β-D-fructofuranose 6-phosphate (FMP or F6P) by 6-phospho-3-hexuloisomerase (HPI). HPS and PHI are two important enzymes unique to this RuMP pathway. HPS and PHI were two important enzymes and initially believed to be unique to the metabolism of one-carbon compounds by methanotrophs and some other methylotrophs that do not utilize methane. However, it was later reported that these proteins are also found in heterotrophs such as *Bacillus subtilis*, where they play a role in formaldehyde detoxification (Yasueda et al. 1999). Therefore, it is accepted to date that the RuMP

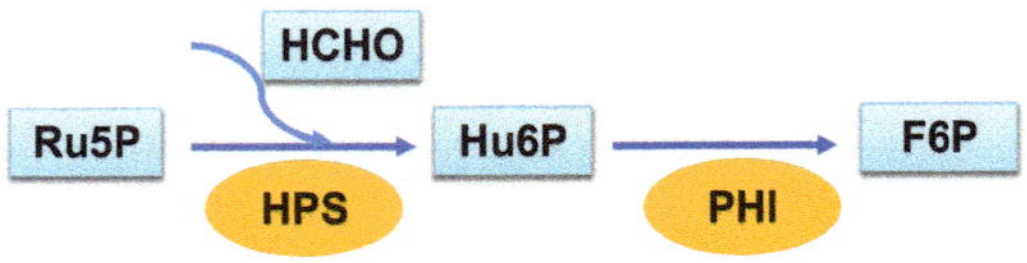

Fig. 6.1 Formaldehyde assimilation: ribulose monophosphate pathway (RuMP). The reaction between formaldehyde and D-ribulose-5-phosphate is catalyzed by 3-hexulose-6-P synthase, forming hexulose-6-phosphate, which is then catalyzed by 6-phospho-3-hexuloisomerase and generates fructofuranose-6-phosphate. Genes coding HPS and PHI were found in AD-related gut probiotics (including *Lactobacillus brevis*, *Bifidobacterium dentium*, and *Lactobacillus fermentum NS9*) while not in other AD-associated non-probiotics and pathogenic bacteria. *HCHO* formaldehyde, *Ru5P* D-ribulose-5-phosphate, *Hu6P* hexulose-6-phosphate, *F6P* fructofuranose-6-phosphate, *HPS* 3-hexulose-6-P synthase, *PHI* 6-phospho-3-hexuloisomerase (Source: Adapted from Kato et al. 2006)

pathway is a prevalent prokaryotic pathway involved in either formaldehyde fixation or detoxification. The HPS and PHI orthologs are also found in many non-methylotrophic organisms (Mitsui et al. 2003).

2.1.2 Serine Pathway

Methanotrophic bacteria are capable to use serine pathway to assimilate formaldehyde, thus forming intermediates for biosynthesis of central metabolic pathways. During the first step of serine pathway, serine is formed from the reaction of formaldehyde with glycine (Fig. 6.2). This reaction is catalyzed by serine hydroxymethyltransferase (SHMT), an enzyme that uses tetrahydropteroyl mono-L-glutamate (THF) as a cofactor. When formaldehyde bounds to it, it forms 5,10-methylenetetrahydropteroyl mono-L-glutamate ($N^{5,10}CH_2 = THF$). During this reaction, the formaldehyde is transferred from 5,10-methylenetetrahydropteroyl mono-L-glutamate to the glycine, forming L-serine. Such enzyme for assimilation of formaldehyde is found in *Methylobacterium extorquens AM1* (Chistoserdova et al. 2003; O'Connor and Hanson 1975).

2.2 *Formaldehyde Dissimilation*

It is of necessity for most life forms to enable the detoxification of the highly toxic formaldehyde. In the case of methylotrophic bacteria, formaldehyde is not just a toxic compound but also a central intermediate. Several different pathways for formaldehyde oxidation/dissimilation are known in bacteria.

Glutathione-dependent formaldehyde dissimilation/oxidation is one of the best characterized pathways (Fig. 6.3a). Glutathione (GSH) and formaldehyde (HCHO) can form the adduct S-(hydroxymethyl)glutathione (HMGSH or $GSCH_2OH$) both spontaneously and under the catalysis by S-(hydroxymethyl)glutathione synthase, which is also named glutathione-dependent formaldehyde-activating enzyme.

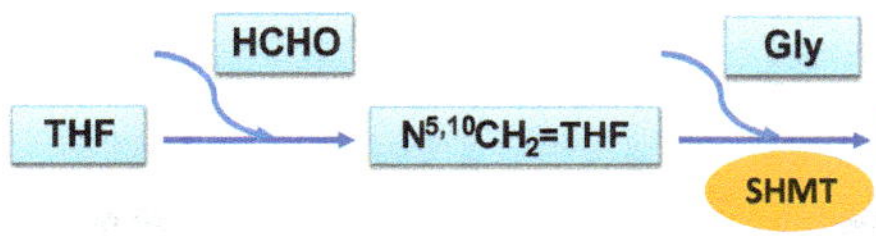

Fig. 6.2 Formaldehyde assimilation: serine pathway. Formaldehyde bounds to tetrahydropteroyl mono-L-glutamate and generates 5,10-methylenetetrahydropteroyl mono-L-glutamate, which then reacts with glycine under the catalyzation of serine hydroxymethyltransferase and forms serine. Genes encoding SHMT were found in all AD-related gut bacteria, showing no preference in between probiotics and pathogens. *HCHO* formaldehyde, *THF* tetrahydropteroyl mono-L-glutamate, $N^{5,10}CH_2 = THF$ 5,10-methylenetetrahydropteroyl mono-L-glutamate, *Gly* glycine, *Ser* serine, *SHMT* serine hydroxymethyltransferase (Source: Adapted from Chistoserdova et al. 2003)

$GSCH_2OH$ is in turn catalyzed by NAD- and glutathione-dependent formaldehyde dehydrogenase (GSH-FDH) as its substrate and forms S-formylglutathione (GSCHO) (Barber et al. 1996; Ras et al. 1995; Sanghani et al. 2002).

Another formaldehyde dissimilation pathway is performed by enzymes that utilize either the cofactor tetrahydropteroyl mono-L-glutamate (H_4F) (Fig. 6.3b) or the cofactor tetrahydromethanopterin (H_4MPT) (Fig. 6.3c). In the methylotroph *Methylobacterium extorquens AM1*, formaldehyde condenses with one of the two pterin cofactors, tetrahydropteroyl mono-L-glutamate or tetrahydromethanopterin, and forms the methylene derivative 5,10-methylenetetrahydropteroyl mono-L-glutamate and 5,10-methylene-tetrahydromethanopterin, respectively. The reaction of formaldehyde with tetrahydropteroyl mono-L-glutamate is spontaneous (Kallen and Jencks 1966; Marx et al. 2003). However, the condensation of formaldehyde with tetrahydromethanopterin is catalyzed by a formaldehyde-activating enzyme (Vorholt et al. 2000).

3 Gut Microbiota-Derived Formaldehyde and Cognition

3.1 AD-Related Gut Bacteria Encode Formaldehyde Metabolism

Since it has been reported that the imbalance of formaldehyde metabolism plays crucial roles during the aging and cognition impairment (Hua and He 2002) and endogenous formaldehyde is related to cognitive impairment including sporadic Alzheimer's disease (Tong et al. 2017), we hypothesize that gut microbiota may affect endogenous formaldehyde metabolism and thus affect neurodegenerative disorder. We would suppose that there are two probable ways. On one hand, probiotic bacteria in AD patients may reduce its capacity in formaldehyde application and detoxification, which includes both formaldehyde assimilation and dissimilation approaches. On the other hand, the harmful microbiota/gut pathogens may increase the endogenous formaldehyde production and accumulation. Both probable situations can theoretically contribute to the genesis of neurodegeneration and cognitive impairment.

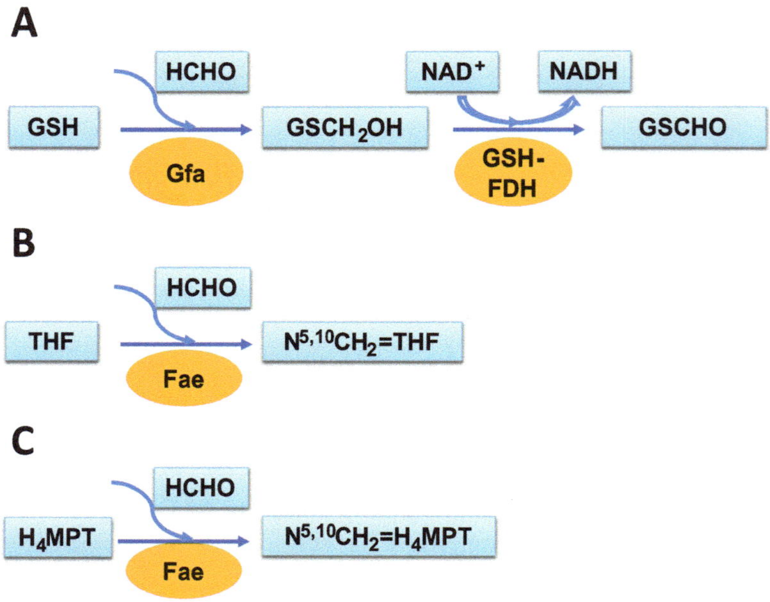

Fig. 6.3 Formaldehyde oxidation/dissimilation pathways. Formaldehyde dissimilation pathways mainly in bacteria contain glutathione pathway (**a**), THF pathway (**b**), and THMPT pathway (**c**). In glutathione pathway (**a**), S-(hydroxymethyl)glutathione synthase, also named glutathione-dependent formaldehyde-activating enzyme, catalyzes the reaction between formaldehyde and glutathione and forms S-(hydroxymethyl)glutathione. In THF pathway (**b**), the reaction between formaldehyde and tetrahydropteroyl mono-L-glutamate is spontaneous and forms 5,10-methylenetetrahydropteroyl mono-L-glutamate. In THMPT pathway (**c**), formaldehyde-activating enzyme catalyzes the reaction between formaldehyde and tetrahydromethanopterin and forms 5,10-methylene-tetrahydromethanopterin. In these three formaldehyde dissimilation/oxidation pathways, no Gfa, GSH-FDH, or Fae can be searched from (https://www.ncbi.nlm.nih.gov/protein) except GSH-FDH in *Citrobacter rodentium* (WP_012904961) and *Salmonella enterica* subsp. *enterica serovar Typhimurium* (OFB34987). *HCHO* formaldehyde, *GSH* glutathione, *$GSCH_2OH$* S-(hydroxymethyl)glutathione, *GSCHO* S-formylglutathione, *Gfa* glutathione-dependent formaldehyde-activating enzyme/S-(hydroxymethyl)glutathione synthase, *GSH-FDH* glutathione-dependent formaldehyde dehydrogenase, *Fae* formaldehyde-activating enzyme, *THF/H_4F* tetrahydropteroyl mono-L-glutamate, $N^{5,10}CH_2 = THF$ 5,10-methylenetetrahydropteroyl mono-L-glutamate, *H_4MPT* tetrahydromethanopterin, $N^{5,10}CH_2 = H_4MPT$ 5,10-methylene-tetrahydromethanopterin (Source: Adapted from Marx et al. 2003 and Vorholt et al. 2000)

In order to prove this hypothesis, we examine whether AD-related gut bacteria mentioned above encode formaldehyde metabolism-related genes (Figs. 6.1, 6.2, and 6.3). The protein sequences of these bacteria were retrieved from "NCBI protein" (https://www.ncbi.nlm.nih.gov/protein). Evolutionary analyses were conducted in MEGA7 (Kumar et al. 2016) by using the maximum likelihood method based on the JTT matrix-based model (Jones et al. 1992). Since *Escherichia coli* (*E.coli*) is known as a common bacterium species existing in human gut (Hudault et al. 2001), we use this bacterium species to find whether these formaldehyde metabolism genes mentioned above can be retrieved from the "NCBI protein."

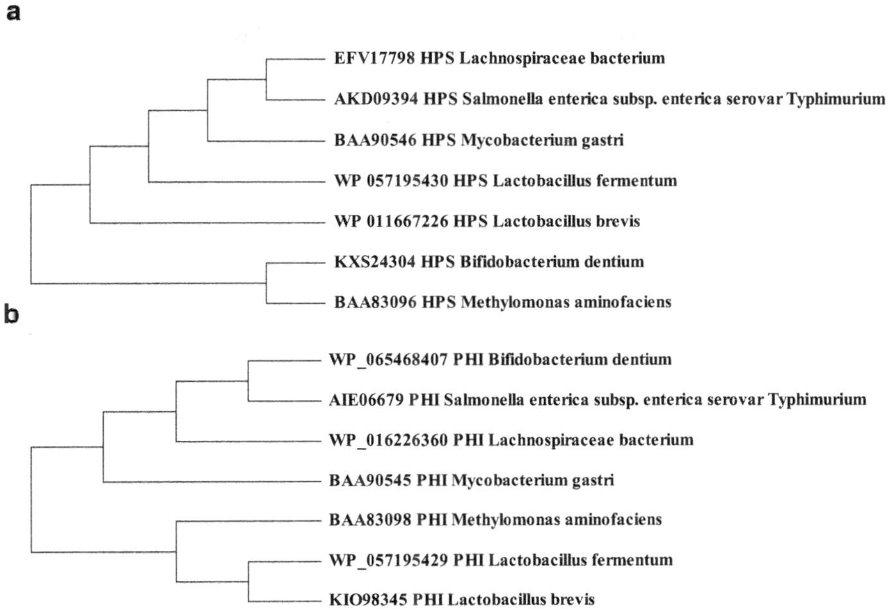

Fig. 6.4 Molecular phylogenetic analysis of HPS and PHI. Molecular phylogenetic analysis of HPS (**a**) and PHI (**b**) is demonstrated by using by maximum likelihood method. AD-associated FA metabolism enzyme/protein sequences were used for multiple alignments. Protein accession number, protein name, and name of bacterium were shown. *Methylomonas aminofaciens* and *Mycobacterium gastri* were used as positive controls because HPS and PHI genes were firstly discovered in them. *Lactobacillus brevis*, *Bifidobacterium dentium*, and *Lactobacillus fermentum* are reported probiotics and contain HPS (**a**) and PHI (**b**). *Salmonella enterica* subsp. *enterica serovar Typhimurium* is probably a gut pathogenic bacterium, while the role of *Lachnospiraceae bacterium* remains to be identified. Other AD neither HPS or PHI genes in other AD-related gut microbiota were found by searching NCBI protein and gene database till the present time (Source: sequences referred to https://www.ncbi.nlm.nih.gov/protein)

As a result, we found genes encoding HPS (KXG93554), PHI (WP_075700053), SHMT (WP_074522665), GSH-FDH (WP_077166504), Gfa (WP_032210276), Fae (KDZ90508), and methanol dehydrogenase (KYO72479) in *E.coli* but not methylamine dehydrogenase. These findings demonstrate that the "NCBI protein" can be effectively employed for searching formaldehyde metabolism-related genes in bacterium and *E.coli* is used as a positive control in searching these genes.

We would like to discuss the involvement of gut microbiota-formaldehyde-brain cassette/axis in age-related cognitive impairment. The observations which support this viewpoint are as follows. First, the protein sequences of HPS and PHI from RuMP pathway of formaldehyde assimilation were analyzed. Probiotics such as *Lactobacillus brevis*, *Bifidobacterium dentium*, and *Lactobacillus fermentum NS9* encode the genes for HPS/PHI, while other AD-related bacteria mentioned above were not found to have HPS (Fig. 6.4a) or PHI (Fig. 6.4b). It is interesting to predict that shortage of HPS/PHI in these AD-related probiotics may contribute to the onset

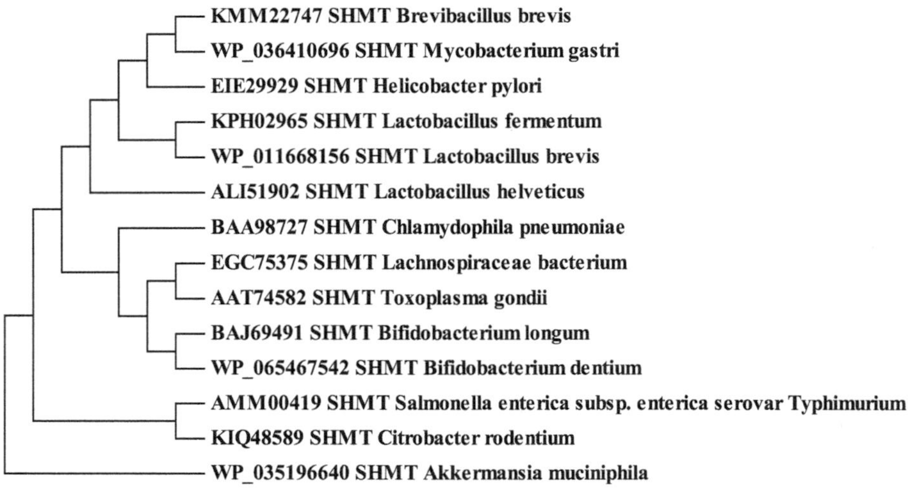

Fig. 6.5 Molecular phylogenetic analysis of SHMT. Molecular phylogenetic analysis of SHMT was shown by using maximum likelihood method. FA metabolism enzyme/protein sequences which are related with AD-associated gut microbiota were used for multiple alignments. Protein accession number, protein name, and name of bacterium were shown in the *alignment diagram.* SHMT proteins were ubiquitously found in all AD-associated gut bacteria (Source: sequences referred to https://www.ncbi.nlm.nih.gov/protein)

of AD and cognitive impairment. Second, genes coding serine hydroxymethyltransferase (SHMT) in formaldehyde serine pathway were found in all AD-related candidate bacteria mentioned above (Fig. 6.5), suggesting that there is no preference for gut probiotics and pathogens in formaldehyde assimilation. Third, formaldehyde dissimilation/oxidation enzymes such as S-(hydroxymethyl)glutathione synthase and Gfa sequences were also searched in AD-related probiotics and pathogens. No gene encoding S-(hydroxymethyl)glutathione synthase or Gfa was found except *Salmonella enterica* subsp. *enterica serovar Typhimurium*. Taken these three aspects together, it suggests that it is more probable for formaldehyde assimilation rather than dissimilation/oxidation to be a candidate factor in aging-related cognitive impairment.

3.2 *The Gut Microbiota in Alleviation of the Risk for Alzheimer's Disease*

Gut bacteria can communicate with the host through the microbiota-gut-brain axis dependent on the immune, neuroendocrine, and neural pathways to influence the brain and behavior (Liang et al. 2015). Accumulating evidence suggests that microbiota-gut-brain axis plays a vital role in maintaining the health of the host and dysregulation of gut microbiota is closely associated with neurodegenerative

Table 6.1 Candidates for AD-related FA-generating gut microbiota

Gut microbiota	Classification	Relation with AD	References
Lactobacillus fermentum strain NS9	Probiotic	Ameliorate cognitive impairment	Park et al. (2012) and Wang et al. (2015)
Lactobacillus helveticus	Probiotic	Ameliorate cognitive impairment	Luo et al. (2014) and Ohsawa et al. (2015)
Bifidobacteria longum	Probiotic	Ameliorate cognitive impairment	Del et al. (2000) and Savignac et al. (2015)
Chlamydia pneumoniae	Pathogen	Associated with AD	Little et al. (2004) and Gerard et al. (2006)
Helicobacter pylori	Pathogen	Associated with AD	Roubaud-Baudron et al. (2012) and Kountouras et al. (2012)
Toxoplasma gondii	Pathogen	Associated with AD	Prandota (2014)

disease and cognitive impairment. Probiotics are the type of commensal bacteria that offer potential health benefit and are able to decrease anxiety, depression, etc. However, intestinal pathogens are harmful and contribute to the risk of neurodegenerative disorder. Specific bacteria of intestinal microbiota are involved in progress of AD and cognitive impairment (Table 6.1).

Alzheimer's disease (AD) is characterized by disruptions in gamma-aminobutyric acid (GABA) in AD patients (Lanctot et al. 2004). Meanwhile, recent report shows that *Lactobacillus brevis* and *Bifidobacterium dentium*, which are regarded as the probiotic gut bacteria, are able to produce GABA (Barrett et al. 2012). It suggests that *Lactobacillus brevis* and *Bifidobacterium dentium* may play a certain role in regulating AD via GABA neurotransmitter. *Lactobacillus fermentum strain NS9*, the probiotic bacteria, normalized the composition of gut microbiota, alleviated the ampicillin-induced inflammation in the gut, and rescued ampicillin-induced memory impairment (Wang et al. 2015). This research indicates that the probiotic gut *Lactobacillus fermentum strain NS9* maintains brain function and prevents mental disorder. Similarly, *Lactobacillus helveticus* has been shown to be useful for the prevention of Alzheimer's disease, enhancing learning and memory in human (Luo et al. 2014) and mice (Ohsawa et al. 2015). *Bifidobacterium longum* can improve learning and memory and has a positive influence on cognition (Savignac et al. 2015).

3.3 *The Gut Microbiota in Aggravation of the Risk for Alzheimer's Disease*

Many gut pathogens display harmful impacts in gut inflammation production and brain function and may induce neurodegenerative disorder. These AD-related candidate gut microbiota include *Chlamydophila pneumoniae* (Gerard et al. 2006; Little et al. 2004), *Helicobacter pylori* (Roubaud-Baudron et al. 2012), *Toxoplasma gondii* (Prandota 2014), *Salmonella enterica* subsp. *enterica serovar Typhimurium* (Sekirov et al. 2008), and *Lachnospiraceae* (Minter et al. 2016). A case-control study showed that AD patients infected with *Helicobacter pylori* displayed lower scores in MMSE and more serious cognitive impairment (Roubaud-Baudron et al. 2012). Additionally, the eradication of *H. pylori* in AD patients can prolong AD patient lifetime, suggesting that *H. pylori* infection may participate in the pathogenesis of AD (Kountouras et al. 2012).

For another example, amyloid deposits resembling plaques in AD brains were formed in the brains of mice following intranasal infection with *Chlamydia (Chlamydophila) pneumoniae* (Little et al. 2004). *Chlamydophila pneumonia* was found in astrocytes, microglia, and neurons of AD patients as host cells. These infected cells showed in close proximity to both neuritic senile plaques and neurofibrillary tangles in the AD brain (Gerard et al. 2006). These findings suggest that the pathogen from gut microbiota might play certain role in the neuropathogenesis of AD.

Toxoplasma gondii is an intracellular protozoan infecting 30% to 50% of global human population. Recently, it has been reported that chronic latent neuroinflammation caused by the parasite may be responsible for the development of several neurodegenerative diseases, including AD. Studies found that serum anti-*T. gondii* immunoglobulin G antibody levels significantly increased in AD patients, suggesting a link between AD and *T. gondii* infection (Prandota 2014). A recent work investigated the function of the host microbiome in regulating amyloidosis in a mouse model of AD and found that in ABX-treated $APP_{SWE}/PS1_{\Delta E9}$ mice, a dramatic expansion of the genus *Akkermansia* and family *Lachnospiraceae* was observed, suggesting a link between these gut bacteria and AD, although the mechanism remains unknown (Minter et al. 2016).

4 Formaldehyde and Age-Related Cognitive Impairment

4.1 Dysmetabolism of Formaldehyde Related to Amyloid β Deposition and Tau Hyperphosphorylation

Oral administration of methanol (the metabolic precursor of FA) in a long term triggers the formation of senile plaques (SPs) and tau hyperphosphorylation in the brains of monkeys with memory decline (Yang et al. 2014). Intraperitoneal injection of formaldehyde results in hyperphosphorylation of tau in wild-type mouse brains and N2a cells via activating glycogen synthase kinase-3β (GSK-3β) (Lu et al. 2013a). Furthermore, formaldehyde at low concentrations can directly induce tau aggregation and amyloid β (Aβ) peptide deposits in vitro (Lu et al. 2013b). Formaldehyde-induced tau aggregation is implicated in cytotoxicity and neural cell apoptosis. These previous data showed that dysmetabolism of endogenous formaldehyde is closely related to age-related cognitive impairment.

In order to provide direct evidence to explain the relationship among formaldehyde, gut microbiota, and AD, we use the APP/PS1 double transgenic mice to detect the formaldehyde concentration/level in the mice gut. Mice guts from APP/PS1 and C57 wild type were segmented, and the duodenum, small intestine, large intestine, and cecum were collected. Feces from the small intestine, large intestine, and cecum were also sampled, respectively. Formaldehyde level determination was conducted according to the method described by Su et al. (2011) as follows.

4.2 *Markedly Elevated Formaldehyde in the Cecum of APP/PS1 Mouse*

Liu and his collaborators compared the concentration of intestinal formaldehyde between APP/PS1 transgenic mouse and C57BL/6j wild-type mouse. Their study was on the basis of these observations:

1. Intestinal microbiota affects the function of central nervous system through the "microbiota-gut-brain axis" (Cryan and Dinan 2012).
2. Alzheimer's disease is related with the microbiota in the intestines.
3. The concentration of endogenous formaldehyde is positively correlated to the cognitive impairment of AD patients (Tong et al. 2011).
4. Formaldehyde can quickly enter humans including crossing BBB (Shcherbakova et al. 1986).
5. APP/PS1 transgenic mouse is widely employed in the study of age-related cognitive impairments, in particular of AD (Bilkei-Gorzo 2014).

The duodenum, small intestine, cecum, and colon were taken and sectioned from APP/PS1 transgenic mice and C57BL/6j wild-type mice, respectively. The digestion contents and intestinal walls were sampled, and the concentrations of formaldehyde were measured with HPLC coupled with 2,4-dinitrophenylhydrazine (DNPH) absorbance. As shown in Fig. 6.6, the levels of the cecum formaldehyde in the digestion contents of APP/PS1 transgenic mice were significantly ($P = 0.036$) higher than those of C57BL/6j wild-type mice. No significant ($P > 0.05$) difference could be observed in the small intestine and colon.

To demonstrate formaldehyde produced by intestinal microbiota is absorbed into intestines, Liu and his collaborators analyzed the aldehyde concentration in intestinal walls. As shown in Fig. 6.7, except for the small intestine, formaldehyde in the duodenum, cecum, and colon walls was not significantly different. In other words, formaldehyde was observably ($P = 0.052$) elevated in small intestinal wall, suggesting that small intestines may also absorb formaldehyde produced by microbiota (Liu et al. 2017). Intestinal microbiota is one of the important sources producing formaldehyde. The elevated concentrations of formaldehyde in the cecum digestion contents and small intestinal wall of APP/PS1 transgenic AD mice indicate that dysmetabolism of formaldehyde in the intestinal microbiota may be involved in age-related cognitive impairment.

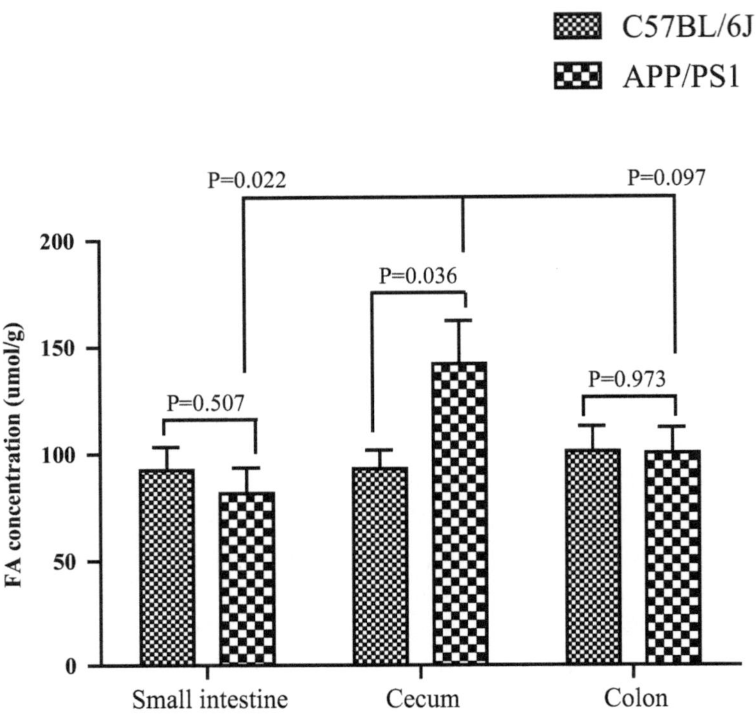

Fig. 6.6 Comparison of concentrations of formaldehyde in the intestinal digestion contents between APP/PS1 transgenic mice and C57BL/6J mice. Concentrations of formaldehyde in the small intestine, cecum, and colon of APP/PS1 and C57BL/6J mice were determined with HPLC coupled with DNPH absorption measurements. The levels of the cecum formaldehyde of APP/PS1 mice ($n = 8$) were significantly ($P < 0.05$) higher than C57BL/6J wild-type mice ($n = 9$) used as control. However, levels of formaldehyde in the small intestine and colon between both APP/PS1 and C57BL/6J mice were not markedly different ($P > 0.05$) (Source: Liu et al. 2017)

4.3 Age-Related Cognitive Impairment May Be Related to Formaldehyde Dysmetabolism of Intestinal Microbiota

Now, there may be a potential link between formaldehyde and gut microbiota, as well as cognitive impairment. First, formaldehyde is a gaseous and water-soluble molecule that can be produced in eukaryotic cells and bacteria (Marx et al. 2003; Vorholt et al. 2000). Second, gut microbiota has been reported to modulate brain function and dysfunction (Mayer et al. 2014). Third, a series of work from this lab show that formaldehyde dysmetabolism, i.e., the accumulation of formaldehyde along with aging, is closely related with the pathogenesis of Alzheimer's disease (Tong et al., 2013). Clinical evidence shows that the degree of cognitive impairment of AD patients is positively correlated with formaldehyde levels in human urine and in the brain (Tong et al. 2011). These data showed that the relation between imbalance of formaldehyde metabolism and age-related cognitive impairment may result from dysmetabolism of intestinal microbiota.

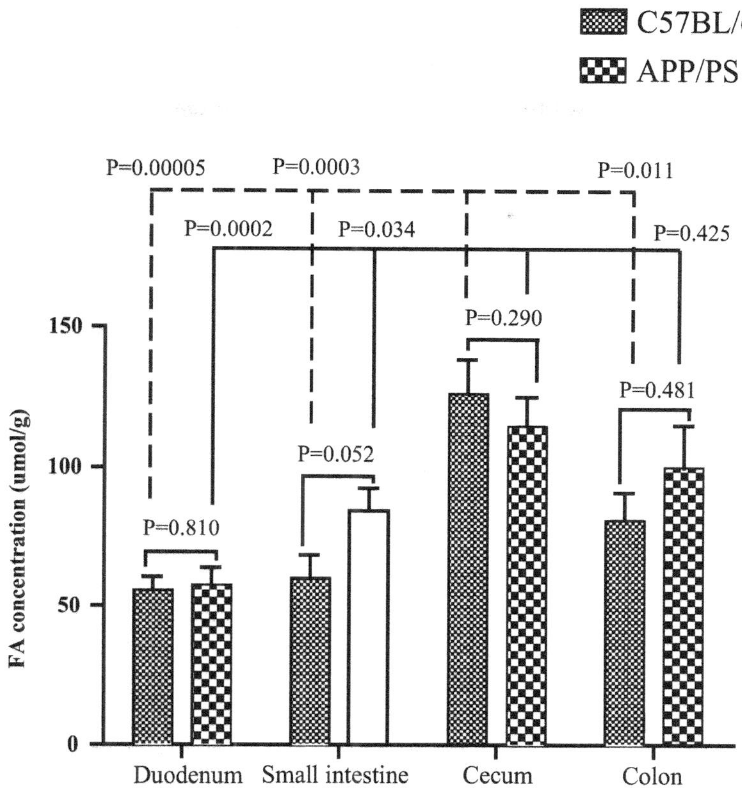

Fig. 6.7 Concentrations of formaldehyde in the intestinal wall of APP/PS1 transgenic mice. Conditions were as for Materials and Methods, except that concentrations of formaldehyde in the intestinal wall were determined. Levels of formaldehyde in small intestinal walls of APP/PS1 and C57BL/6J mice were observably different ($P < 0.052$). However, levels of formaldehyde in the duodenum, cecum, and colon walls were not significantly different ($P > 0.05$) (Source: Liu et al. 2017)

4.4 *Administration of NS Lactobacillus as a Try*

As described by Pérez Martínez and colleagues, gut microbiota in elderly's health will enable intervention through probiotics (Pérez Martínez et al. 2014). The pleiotropic actions of resveratrol may be resulted from its altering gut microbiota and influences stem cell proliferation and differentiation (Diaz-Gerevini et al. 2016). As we know, acetaldehyde dehydrogenase (EC 1.2.1.10) catalyzes both acetaldehydes and formaldehyde. To test whether NS lactobacillus can metabolite formaldehyde, I (the corresponding author) tried to drink alcohol instead because elevation of metabolism of ethanol implies an increase in metabolism of formaldehyde.

Fortunately, Jin and his group (Institute of Psychology, Chinese Academy of Sciences in Beijing) studied the lactobacillus as food nutrition for oral intake. They believed that lactobacillus is beneficial to elderly's brain function through regulation of gut microbiota (Hu et al. 2016). Jin gave me a bag of "NS.8 + 9" lactobacillus. In order to make a sense, I have tried to take his NS lactobacillus (1 capsule, once daily) for 2 weeks. Before taking the NS capsule, I purchased 42% of commercial alcohol made from highland barley (Laochenfengtan, Huzhu County, Qinghai, China). I drank it in the dinner on each Saturday for 3 weeks. The conditions: I drank just myself, from 18:30 to 19:30 each Saturday, with the same food at home. The requirement: I should not feel the effect of alcohol after 3 hours I had drank.

Afterward, I had taken the NS capsule for 2 weeks. I had not got any marked changes except for a mild diarrhea in the first week. Under the same conditions, I drank the alcohol on each Saturday for 3 weeks. Preliminarily, I found myself having a change that I was able to drink more alcohol from ~60 ml to ~90 ml of the alcohol with a measuring cylinder. It appears that NS lactobacillus could enhance the detoxicity of the aldehyde. It was said that some other people also had got the same effects. However, this was just a try and subjective experience to provide a preliminary record, but not a result from the formally scientific investigation.

Probiotics as an intervention to formaldehyde metabolism could be used to those elderly whose endogenous formaldehyde levels are significantly detected higher than their age-match controls. In fact, more experiments on animals and the long-term investigation of large population should be carried out to clarify whether probiotics could be beneficial to our brain and cognitive ability through gut-microbiota-brain axis. Or more rigorous, detailed, and longtime investigations should be performed to demonstrate whether administration of probiotics including the bacterial clearance of formaldehyde can mitigate the onset and progression of age-related cognitive impairment.

5 Formaldehyde, Serotonin (5-HT), and Norepinephrine (NE)

Alzheimer's disease is associated with chronic dehydration, and one of the significant changes that are known to result in metabolic dysfunction is an increase in the endogenous formaldehyde level. The brain formaldehyde levels of mice increased with age, and these increases were followed by decreases in their drinking frequency and water intake (Li et al. 2016b). Additionally, the decrease of serotonin (5-HT), but not other neural transmitters such as acetylcholine or dopamine, in brain of mice was observed, compared with control mice (Li et al. 2016a). Moreover, injection of formaldehyde results in serotonin decline and learning and memory defect indicated by shuttle-box avoidance test, suggesting the specificity of this serotonin in AD pathogenesis.

Hippocampal norepinephrine (NE) exists as a neurotransmitter that is required for learning and memory and affects long-term potentiation (LTP) in the hippocampus (Gray and Johnston 1987; Izumi and Zorumski 1999). Recently, Mei et al. reported that aging-associated formaldehyde-induced norepinephrine deficiency leads to age-related memory decline and cognitive impairment, suggesting that NE is one of the factors that link formaldehyde and Alzheimer's disease during aging (Mei et al. 2015). Additionally, formaldehyde exposure promotes anxiety- and depression-like behaviors, such as open field test, elevated plus-maze test, and forced swimming test (Li et al. 2016c). In addition, gaseous formaldehyde inhibits tyrosine hydroxylase (TH) (Li et al. 2016c), a rate-limiting enzyme for L-DOPA, the precursor of dopamine, which in turn gives rise to NE. Although it shows that formaldehyde can interact directly with NE with a yet unknown mechanism (Mei et al. 2015), it is clearly believed, at least, that formaldehyde may affect cognition via inhibiting TH-DA-NE cassette. This is to say, 5-HT and NE are two specific markers in formaldehyde-derived AD pathogenesis, and formaldehyde increment-related behavior and cognitive impairment may be caused by its inhibitory regulation on 5-HT and NE.

6 Gut Microbiota, Serotonin (5-HT), and Norepinephrine (NE)

Serotonin (5-hydroxytryptamine, 5-HT) is a fundamental and common transmitter in neural physiology and brain activity. More than 95% of the 5-HT are generated in the gut, and thus gut microbiota should be the most important contributor for the synthesis of 5-HT (Spohn and Mawe 2017). According to Yano and colleagues, the content of 5-HT in the blood of germfree (GF) mice was about 60% lower than that of the specific pathogen-free (SPF) mice with a normal gut microbiota, and the concentration was significantly increased when the gut microbiota was reconstructed in GF mice (Yano et al. 2015). Matsumoto and coworkers have analyzed the influence of gut microbiota on cerebral metabolism through assessing the cerebral metabolome of GF mice and Ex-GF mice which were colonized with fecal microbiota obtained from SPF mice (Matsumoto et al. 2013). Results showed that 10 from 38 metabolites with significant changes in concentrations between GF mice and Ex-GF mice were reported to be involved in brain function. The concentrations of tryptophan (Trp), the precursors of 5-HT, were lower in the cerebrum of GF mice than that of Ex-GF mice (Matsumoto et al. 2013). The animal and human clinical trials have showed that selective serotonin reuptake inhibitors (SSRIs) could reduce the Aβ protein production in the brain. The increase extracellular 5-HT levels can effectively reduce Aβ plaque formation and thereby decrease the risk of Alzheimer's disease (AD) (Cirrito et al. 2011). These also indicate that the changed 5-HT biosynthesis caused by gut microbiota disturbance may affect the pathological process of AD.

As described by Oleskin and his colleagues (Oleskin and Shenderov 2016), gut microbiota can produce neurotransmitters and neuropeptides. For example, spore-forming microbes can produce 5-HT (Shishov et al. 2009), and *Escherichia* and *Bacillus* produce norepinephrine (Shishov et al. 2009; Tsavkelova et al. 2000).

Recent report showed that gut microbiota plays a critical role in the production of biologically active catecholamines, which include dopamine (DA) and norepinephrine (NE) (Asano et al. 2012). Results also showed that *L. helveticus NS8* improves chronic restraint stress-induced behavioral (anxiety and depression) and cognitive dysfunction and causes the restoration of hippocampal serotonin (5-HT) and norepinephrine (NE) levels in rats. This gut bacterium has an antidepressant effect, which could be due to the microbiota-gut-brain axis. It is interesting to think of the opposite phenomenon in formaldehyde stress, which reduced the 5-HT and NE levels. Also, it is intriguing to test whether the *L. helveticus NS8* can be used to treat formaldehyde stress-related aging and cognitive impairment.

7 Perspectives

In this review, we summarized the research progress of gut microbiota and formaldehyde in Alzheimer's disease and cognitive impairment and proposed a link between gut microbiota and formaldehyde based on the probable functions of 5-HT and NE during neurodegeneration (Fig. 6.8). Since formaldehyde is increasingly becoming one of the dominant factors that contribute to AD pathogenesis, we hypothesize that gut microbiota-derived formaldehyde may be involved in cognition and its dysmetabolism may be a risk factor of AD or even lead to cognitive decline. Seven-month-old APP/PS1 transgenic mouse has got its intestinal

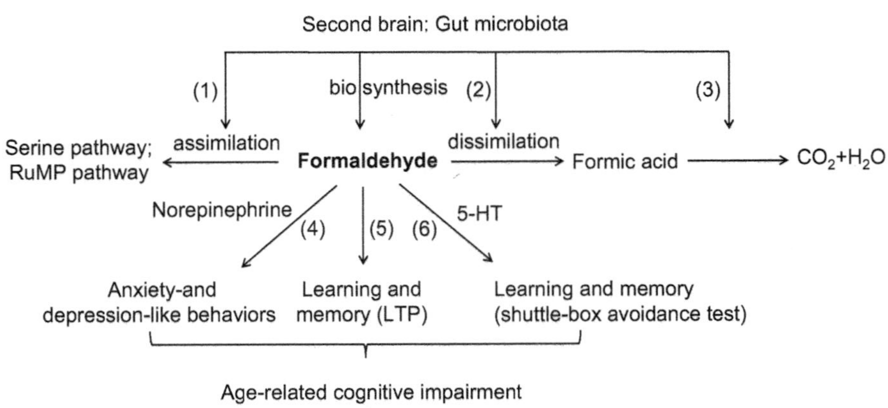

Fig. 6.8 Overview of formaldehyde metabolism and its role in AD pathogenesis. Schematic diagram of gut microbiota-derived formaldehyde metabolism and the regulatory roles in AD pathogenesis. The gut/gastrointestinal tract, which contains an environment of microbes and is also called the second brain, can generate endogenous formaldehyde (Liu et al. 2017). The gut bacteria formaldehyde assimilation and dissimilation pathways may contribute to physiology and pathology of neurodegeneration including AD by regulating neural behaviors through 5-HT and norepinephrine. *(1)*–*(6)* are referred to Kato et al. (2006), Vorholt et al. (2000), Marx et al. (2003), Li et al. (2016b), Mei et al. (2015), and Li et al. (2016a, b, c), respectively

formaldehyde significantly elevated in the cecum digestion contents, compared with the C57BL/6j wild-type mouse as a control. Furthermore, the formaldehyde level in the small intestinal wall for the APP/PS1 transgenic mice is also higher than that of the control. These data suggest intestinal microbiota that is an important source of formaldehyde suffers from dysmetabolism in the AD mouse model. Imbalance of intestinal formaldehyde metabolism by gut microbiota may have a link to cognitive impairment.

By using the NCBI protein sequences to date and analyzing the genes encoding enzymes within formaldehyde assimilation and dissimilation pathways in bacteria, it is interesting to find and predict that within several specific AD-related gut microorganisms, the shortage of HPS/PHI in these AD-related probiotics may contribute to the onset of AD and cognitive impairment, while other formaldehyde metabolism-related genes may not be key factors. Nonetheless, this insight of formaldehyde's contribution to AD derived from gut microbiota has to be validated by experiments in the future. Moreover, to prevent or delay the progress of neurodegeneration, it is reasonable to use formaldehyde scavenger or related anti-oxidative chemicals or anti-inflammatory drugs or probiotics to attenuate and block the reactions that lead to progress and development of the neurodegenerative diseases.

Acknowledgment Many thanks to the "NCBI protein" (https://www.ncbi.nlm.nih.gov/protein) for the information of enzymes involving formaldehyde metabolism. This project was supported by grants from the Beijing Municipal Science and Technology Project (Z161100000217141; Z161100000216137), the National Key Research and Development Program of China (2016YFC1306300), the National Basic Research Program of China (973 Program) (2012CB911004), the Natural Scientific Foundation of China (NSFC 31301880, NSFC 31270868), Foundation of Chinese Academy of Sciences CAS-20140909, the Queensland-Chinese Academy of Sciences Biotechnology Fund (GJHZ201302), Program for Liaoning Excellent Talents in University (LJQ2015057), and Dalian High Level Talent Innovation Support Plan (No. 2015R067).

Competing Financial Interests The authors declare no competing financial interests.

References

Anthony C (1991) Assimilation of carbon by methylotrophs. Biotechnology 18:79–109

Asano Y, Hiramoto T, Nishino R, Aiba Y, Kimura T, Yoshihara K, Koga Y, Sudo N (2012) Critical role of gut microbiota in the production of biologically active, free catecholamines in the gut lumen of mice. Am J Physiol Gastrointest Liver Physiol 303(11):G1288–G1295

Barber RD, Rott MA, Donohue TJ (1996) Characterization of a glutathione-dependent formaldehyde dehydrogenase from Rhodobacter sphaeroides. J Bacteriol 178(5):1386–1393

Barrett E, Ross RP, O'Toole PW, Fitzgerald GF, Stanton C (2012) Gamma-Aminobutyric acid production by culturable bacteria from the human intestine. J Appl Microbiol 113(2):411–417

Bilkei-Gorzo A (2014) Genetic mouse models of brain ageing and Alzheimer's disease. Pharmacol Ther 142(2):244–257

Bruemmer KJ, Brewer TF, Chang CJ (2017) Fluorescent probes for imaging formaldehyde in biological systems. Curr Opin Chem Biol 39:17–23

Chistoserdova L, Chen SW, Lapidus A, Lidstrom ME (2003) Methylotrophy in Methylobacterium extorquens AM1 from a genomic point of view. J Bacteriol 185(10):2980–2987
Cirrito JR, Disabato BM, Restivo JL, Verges DK, Goebel WD, Sathyan A, Hayreh D, D'Angelo G, Benzinger T, Yoon H, Kim J, Morris JC, Mintun MA, Sheline YI (2011) Serotonin signaling is associated with lower amyloid-beta levels and plaques in transgenic mice and humans. Proc Natl Acad Sci U S A 108(36):14968–14973
Clark A, Mach N (2016) Exercise-induced stress behavior, gut-microbiota-brain axis and diet: a systematic review for athletes. J Int Soc Sports Nutr 13:43
Cryan JF, Dinan TG (2012) Mind-altering microorganisms: the impact of the gut microbiota on brain and behaviour. Nat Rev Neurosci 13(10):701–712
Diaz-Gerevini GT, Repossi G, Dain A, Tarres MC, Das UN, Eynard AR (2016) Feb; beneficial action of resveratrol: how and why? Nutrition 32(2):174–178
Del Re B, Sgorbati B, Miglioli M, Palenzona D (2000) Adhesion, autoaggregation and hydrophobicity of 13 strains of Bifidobacterium longum. Lett Appl Microbiol 31(6):438–442
Finegold SM, Dowd SE, Gontcharova V, Liu C, Henley KE, Wolcott RD, Youn E, Summanen PH, Granpeesheh D, Dixon D, Liu M, Molitoris DR, Green JA 3rd (2010) Pyrosequencing study of fecal microflora of autistic and control children. Anaerobe 16(4):444–453
Gérard HC, Dreses-Werringloer U, Wildt KS, Deka S, Oszust C, Balin BJ, Frey WH 2nd, Bordayo EZ, Whittum-Hudson JA, Hudson AP (2006) Chlamydophila (chlamydia) pneumoniae in the Alzheimer's brain. FEMS Immunol Med Microbiol 48(3):355–366
Gray R, Johnston D (1987) Noradrenaline and beta-adrenoceptor agonists increase activity of voltage-dependent calcium channels in hippocampal neurons. Nature 327(6123):620–622
Hu X, Wang T, Jin F (2016) Alzheimer's disease and gut microbiota. Sci China Life Sci 59(10):1006–1023
Hua Q, He RQ (2002) Effect of phosphorylation and aggregation on tau binding to DNA. Protein Pept Lett 9(4):349–357
Hudault S, Guignot J, Servin AL (2001) Escherichia coli Strains colonising the gastrointestinal tract protect germfree mice against salmonella typhimurium infection. Gut 49(1):47–55
Hyland NP, Cryan JF (2016) Microbe-host interactions: influence of the gut microbiota on the enteric nervous system. Dev Biol 417(2):182–187
Izumi Y, Zorumski CF (1999) Norepinephrine promotes long-term potentiation in the adult rat hippocampus in vitro. Synapse 31(3):196–202
Johnson PA, Quayle JR (1965) Microbial growth on C1 compounds. Synthesis of cell constituents by methane- and methanol-grown Pseudomonas methanica. Biochem J 95:859–867
Jones DT, Taylor WR, Thornton JM (1992) The rapid generation of mutation data matrices from protein sequences. Comput Appl Biosci 8(3):275–282
Kallen RG, Jencks WP (1966) The mechanism of the condensation of formaldehyde with tetrahydrofolic acid. J Biol Chem 241(24):5851–5863
Kato N, Yurimoto H, Thauer RK (2006) The physiological role of the ribulose monophosphate pathway in bacteria and archaea. Biosci Biotechnol Biochem 70(1):10–21
Kemp MB, Quayle JR (1967) Microbial growth on C1 compounds. Uptake of [14C]formaldehyde and [14C]formate by methane-grown Pseudomonas methanica and determination of the hexose labelling pattern after brief incubation with [14C]methanol. Biochem J 102(1):94–102
Kim BS, Jeon YS, Chun J (2013) Current status and future promise of the human microbiome. Pediatr Gastroenterol Hepatol Nutr 16(2):71–79
Kountouras J, Boziki M, Zavos C, Gavalas E, Giartza-Taxidou E, Venizelos I, Deretzi G, Grigoriadis N, Tsiaousi E, Vardaka E (2012) A potential impact of chronic helicobacter pylori infection on Alzheimer's disease pathobiology and course. Neurobiol Aging 33(7):e3–e4
Kumar S, Stecher G, Tamura K (2016) MEGA7: molecular evolutionary genetics analysis version 7.0 for bigger datasets. Mol Biol Evol 33(7):1870–1874
Lanctôt KL, Herrmann N, Mazzotta P, Khan LR, Ingber N (2004) GABAergic function in Alzheimer's disease: evidence for dysfunction and potential as a therapeutic target for the treatment of behavioural and psychological symptoms of dementia. Can J Psychiatr 49(7):439–453

Li T, Su T, He YG, He RQ (2016a) Chronic-dehydrated Dysmetabolism of formaldehyde in mouse brain and decline of learning in the shuttle box. Prog Biochem Biophys 43(4):429–438

Li T, Su T, He Y, Lu J, Mo W, Wei Y, He R (2016b) Brain formaldehyde is related to water intake behavior. Aging Dis 7(5):561–584

Li Y, Song Z, Ding Y, Xin Y, Wu T, Su T, He R, Tai F, Lian Z (2016c) Effects of formaldehyde exposure on anxiety-like and depression-like behavior, cognition, central levels of glucocorticoid receptor and tyrosine hydroxylase in mice. Chemosphere 144:2004–2012

Liang S, Wang T, Hu X, Luo J, Li W, Wu X, Duan Y, Jin F (2015) Administration of Lactobacillus helveticus NS8 improves behavioral, cognitive, and biochemical aberrations caused by chronic restraint stress. Neuroscience 310:561–577

Little CS, Hammond CJ, MacIntyre A, Balin BJ, Appelt DM (2004) Chlamydia pneumoniae induces Alzheimer-like amyloid plaques in brains of BALB/c mice. Neurobiol Aging 25(4):419–429

Liu KL, He YG, Yu LX, He RQ (2017) Elevated formaldehyde in the cecum of APP/PS1 mouse. Microbiol China 44(8):1761–1766

Lu J, Miao J, Su T, Liu Y, He R (2013a) Formaldehyde induces hyperphosphorylation and polymerization of tau protein both in vitro and in vivo. Biochim Biophys Acta 1830(8):4102–4116

Lu Y, He HJ, Zhou J, Miao JY, Lu J, He YG, Pan R, Wei Y, Liu Y, He RQ (2013b) Hyperphosphorylation results in tau dysfunction in DNA folding and protection. J Alzheimers Dis 37(3):551–563

Luo J, Wang T, Liang S, Hu X, Li W, Jin F (2014) Ingestion of lactobacillus strain reduces anxiety and improves cognitive function in the hyperammonemia rat. Sci China Life Sci 57(3):327–335

Marx CJ, Chistoserdova L, Lidstrom ME (2003) Formaldehyde-detoxifying role of the tetrahydromethanopterin-linked pathway in Methylobacterium extorquens AM1. J Bacteriol 185(24):7160–7168

Matsumoto M, Kibe R, Ooga T, Aiba Y, Sawaki E, Koga Y, Benno Y (2013) Cerebral low-molecular metabolites influenced by intestinal microbiota: a pilot study. Front Syst Neurosci 7:9

Mayer EA, Savidge T, Shulman RJ (2014) Brain-gut microbiome interactions and functional bowel disorders. Gastroenterology 146(6):1500–1512

Mei Y, Jiang C, Wan Y, Lv J, Jia J, Wang X, Yang X, Tong Z (2015) Aging-associated formaldehyde-induced norepinephrine deficiency contributes to age-related memory decline. Aging Cell 14(4):659–668

Minter MR, Zhang C, Leone V, Ringus DL, Zhang X, Oyler-Castrillo P, Musch MW, Liao F, Ward JF, Holtzman DM, Chang EB, Tanzi RE, Sisodia SS (2016) Antibiotic-induced perturbations in gut microbial diversity influences neuro-inflammation and amyloidosis in a murine model of Alzheimer's disease. Sci Rep 6:30028

Mitsui R, Kusano Y, Yurimoto H, Sakai Y, Kato N, Tanaka M (2003) Formaldehyde fixation contributes to detoxification for growth of a nonmethylotroph, Burkholderia cepacia TM1, on vanillic acid. Appl Environ Microbiol 69(10):6128–6132

Naseribafrouei A, Hestad K, Avershina E, Sekelja M, LinlØkken A, Wilson R, Rudi K (2014) Correlation between the human fecal microbiota and depression. Neurogastroenterol Motil 26(8):1155–1162

O'Connor ML, Hanson RS (1975) Serine transhydroxymethylase isoenzymes from a facultative methylotroph. J Bacteriol 124(2):985–996

O'Hara AM, Shanahan F (2006) The gut flora as a forgotten organ. EMBO Rep 7(7):688–693

Ohsawa K, Uchida N, Ohki K, Nakamura Y, Yokogoshi H (2015) Lactobacillus helveticus-fermented milk improves learning and memory in mice. Nutr Neurosci 18(5):232–240

Oleskin AV, Shenderov BA (2016) Neuromodulatory effects and targets of the SCFAs and gasotransmitters produced by the human symbiotic microbiota. Microb Ecol Health Dis 27:30971

Park JH, Kim Y, Kim SH (2014) Green tea extract (*Camellia sinensis*) fermented by attenuates alcohol-induced liver damage. Biosci Biotechnol Biochem 76(12):2294–2300

Pérez Martínez G, Bäuerl C, Collado MC (2014) Understanding gut microbiota in elderly's health will enable intervention through probiotics. Benef Microbes 5(3):235–246

Prandota J (2014) Possible link between toxoplasma gondii and the anosmia associated with neurodegenerative diseases. Am J Alzheimers Dis Other Demen 29(3):205–214

Qin J, Li R, Raes J, Arumugam M, Burgdorf KS, Manichanh C, Nielsen T, Pons N, Levenez F, Yamada T, Mende DR, Li J, Xu J, Li S, Li D, Cao J, Wang B, Liang H, Zheng H, Xie Y, Tap J, Lepage P, Bertalan M, Batto JM, Hansen T, Le Paslier D, Linneberg A, Nielsen HB, Pelletier E, Renault P, Sicheritz-Ponten T, Turner K, Zhu H, Yu C, Li S, Jian M, Zhou Y, Li Y, Zhang X, Li S, Qin N, Yang H, Wang J, Brunak S, Dore J, Guarner F, Kristiansen K, Pedersen O, Parkhill J, Weissenbach J, Meta HITC, Bork P, Ehrlich SD, Wang J (2010) A human gut microbial gene catalogue established by metagenomic sequencing. Nature 464(7285):59–65

Ras J, Van Ophem PW, Reijnders WN, Van Spanning RJ, Duine JA, Stouthamer AH, Harms N (1995) Isolation, sequencing, and mutagenesis of the gene encoding NAD- and glutathione-dependent formaldehyde dehydrogenase (GD-FALDH) from Paracoccus denitrificans, in which GD-FALDH is essential for methylotrophic growth. J Bacteriol 177(1):247–251

Roubaud-Baudron C, Krolak-Salmon P, Quadrio I, Mégraud F, Salles N. (2012) Impact of chronic *Helicobacter pylori* infection on Alzheimer's disease: preliminary results. Neurobiol Aging 33(5):1009. e11-9

Sanghani PC, Robinson H, Bosron WF, Hurley TD (2002) Human glutathione-dependent formaldehyde dehydrogenase. Structures of apo, binary, and inhibitory ternary complexes. Biochemistry 41(35):10778–10786

Savignac HM, Tramullas M, Kiely B, Dinan TG, Cryan JF (2015) Bifidobacteria modulate cognitive processes in an anxious mouse strain. Behav Brain Res 287:59–72

Sekirov I, Tam NM, Jogova M, Robertson ML, Li Y, Lupp C, Finlay BB (2008) Antibiotic-induced perturbations of the intestinal microbiota alter host susceptibility to enteric infection. Infect Immun 76(10):4726–4736

Shcherbakova LN, Tel'pukhov VI, Trenin SO, Bashilov IA, Lapkina TI (1986) Permeability of the blood-brain barrier to intra-arterial formaldehyde. Biull Eksp Biol Med 102(11):573–575

Shishov VA, Kirovskaia TA, Kudrin VS, Oleskin AV (2009) Amine neuromediators, their precursors, and oxidation products in the culture of Escherichia coli K-12. Prikl Biokhim Mikrobiol 45(5):550–554

Spohn SN, Mawe GM (2017) Non-conventional features of peripheral serotonin signalling – the gut and beyond. Nat Rev Gastroenterol Hepatol. doi:10.1038/nrgastro.2017.51

Strom T, Ferenci T, Quayle JR (1974) The carbon assimilation pathways of Methylococcus capsulatus, Pseudomonas methanica and Methylosinus trichosporium (OB3B) during growth on methane. Biochem J 144(3):465–476

Su T, Wei Y, He RQ (2011) Assay of brain endogenous formaldehyde with 2,4-dinitrophenylhydrazine through UV-HPLC. Prog Biochem Biophys 38(12):1171–1177

Tong Z, Wang W, Luo W, Lv J, Li H, Luo H, Jia J, He R (2017) Urine formaldehyde predicts cognitive impairment in post-stroke dementia and Alzheimer's disease. J Alzheimers Dis 55(3):1031–1038

Tong ZQ, Zhang JL, Luo WH, Wang WS, Li FX, Li H, Luo HJ, Lu J, Zhou JN, Wan Y, He RQ (2011) Urine formaldehyde level is inversely correlated to mini mental state examination scores in senile dementia. Neurobiol Aging 32(1):31–41

Tsavkelova EA, Botvinko IV, Kudrin VS, Oleskin AV (2000) Detection of neurotransmitter amines in microorganisms with the use of high-performance liquid chromatography. Dokl Biochem 372(1–6):115–117

Vorholt JA, Marx CJ, Lidstrom ME, Thauer RK (2000) Novel formaldehyde-activating enzyme in Methylobacterium extorquens AM1 required for growth on methanol. J Bacteriol 182(23):6645–6650

Wang T, Hu X, Liang S, Li W, Wu X, Wang L, Jin F (2015) Lactobacillus fermentum NS9 restores the antibiotic induced physiological and psychological abnormalities in rats. Benef Microbes 6(5):707–717

Whitman WB, Coleman DC, Wiebe WJ (1998) Prokaryotes: the unseen majority. Proc Natl Acad Sci U S A 95(12):6578–6583

Yang M, Lu J, Miao J, Rizak J, Yang J, Zhai R, Zhou J, Qu J, Wang J, Yang S, Ma Y, Hu X, He R (2014) Alzheimer's disease and methanol toxicity (part 1): chronic methanol feeding led to memory impairments and tau hyperphosphorylation in mice. J Alzheimers Dis 41(4):1117–1129

Yano JM, Yu K, Donaldson GP, Shastri GG, Ann P, Ma L, Nagler CR, Ismagilov RF, Mazmanian SK, Hsiao EY (2015) Indigenous bacteria from the gut microbiota regulate host serotonin biosynthesis. Cell 161(2):264–276

Yasueda H, Kawahara Y, Sugimoto S (1999) Bacillus Subtilis yckG and yckF encode two key enzymes of the ribulose monophosphate pathway used by methylotrophs, and yckH is required for their expression. J Bacteriol 181(23):7154–7160

Chapter 7
Effects of Formaldehyde on Protein (Tau) Aggregation and Cytotoxicity

Rongqiao He

Abstract The human body, including the brain, is producing and processing formaldehyde all the time. Formaldehyde is active in the reaction with biomacromolecules, especially with peptides and proteins. So neuronal proteins have chances to react directly with formaldehyde. Among the proteins, neuronal Tau is extremely prone to react with formaldehyde because Tau features in "wormlike" conformation, and its α-/ε-amino groups are exposed to the protein exterior. Tau is a multifunctional protein which is able to bind to microtubule, actin, DNA and RNA, with its proline-rich domain and microtubule-binding domain. Tau promotes the melting temperature of DNA double strands and accelerates refolding of denatured DNA. Interaction between Tau and DNA forms a complex called "DNA-Tauosome", which may be the structure in resistance to the attack of free radicals, for example, reactive oxygen species (ROS). Treatment with formaldehyde inactivates Tau protein in the interaction with microtubule as well as DNA in which the formation of DNA-Tauosome is inhibited. Formaldehyde induces Tau aggregation which features in globular-like deposits stained with Congo red and probed by the fluorescence of thioflavin T (ThT). The cytotoxicity of globular-like aggregate leads to the impairment of cell viability and eventually to cell death. Lysosomes are classically considered as non-specific systems in degradation of protein aggregation. Endogenous formaldehyde is mainly localized in lysosome. Abnormal lysosomes increase as aging and so does endogenous formaldehyde. Dysfunction of lysosome and formaldehyde metabolism could be the major risk factors to impede the cellular degradation and scavenging of protein aggregation. Since formaldehyde is actively and directly reacted with the side chains of peptides and proteins, we mainly discuss the effect of chemical modification with formaldehyde on morphology and function of neuronal Tau in this chapter, except for phosphorylation, glycosylation, and other modifications.

Keywords Formaldehyde • Tau • Protein aggregation • Aggregate cytotoxicity • DNA-Tauosome • Actin • Microtubule • Abnormal lysosome

---

R. He (✉)
State Key Laboratory of Brain and Cognitive Science, Institute of Biophysics, Chinese Academy of Sciences, Beijing 100101, China

University of Chinese Academy of Sciences, Beijing 100049, China
e-mail: rongqiaohe@163.com

R. He, *Formaldehyde and Cognition*, DOI 10.1007/978-94-024-1177-5_7

1 Introduction

Human genetic information inside a cell encodes not only the specific structures and functions of proteins but also the way these structures are attained through the process known as protein folding (Stefani and Dobson 2003). Only in this way the nascent peptides fold into the specific conformational proteins and possess their natural biological functions. However, protein misfolding and deposition could occur, especially under endoplasmic reticulum stress and by abnormal modifications (Evans et al. 2016). The misfolded proteins are often prone to aggregation and/or interact with other cellular components causing impairment of cell viability and eventually cell dysfunction and death (Cohen 1999).

Protein aggregation plays a central role in pathological progression of neurodegenerative diseases. β-Amyloid (Aβ) deposition and Tau aggregation are major pathological hallmarks for Alzheimer's disease (AD) (Hardy and Selkoe 2002). β-Amyloid deposition in the brain occurs in the early stage of preclinical phase, followed by neuronal Tau hyperphosphorylation and paired helical filament formation (Jack et al. 2010). Tau protein is a member of microtubule-associated proteins, promoting microtubule assembly and stabilizing microtubules (Goedert et al. 1989a, b, 1992; Goedert and Jakes 1990). This protein is a major constituent of paired helical filaments (PHF) found in the brain of Alzheimer's patients (Wischik et al. 1988). Native Tau is prone to hydrolysis by proteases because it lacks secondary structures and its proteolytic sites exist on the exterior surface of protein molecule in solution (Schweer et al. 1994). So far, however, what triggers protein abnormal modification and aggregation during neurodegeneration or what occurs ahead of Aβ deposition and Tau hyperphosphorylation acting as an etiological factor is still unclear, though some polyanions such as heparin and polyglutamate were found to induce the Tau filament formation years ago (Friedhoff et al. 1998). To clarify this major issue is beneficial and imperative to understand the early onset and progression of age-related cognitive impairment.

Since it was first introduced as a fixative reagent in 1893, formaldehyde has still been employed for tissue preservation, including fresh-retaining purpose (Fox et al. 1985; Puchtler and Meloan 1985). Formaldehyde is widely employed in paraffin-embedded cells and tissues and also active in the reaction with proteins as a chemical cross-linker (Carpenter 1946; Hopwood et al. 1988). Theis and Lams investigated different reactive conditions to enhance the efficacy of formaldehyde to react with collagen from wool (Theis 1944). Recently, Magdeldin and Yamamoto showed their useful viewpoints to decipher proteomes of formalin-fixed paraffin-embedded (FFPE) tissues (Magdeldin and Yamamoto 2012). However, instead of tissue fixation and preservation, the effect of formaldehyde at low concentrations on protein aggregation and neurotoxicity needs to be investigated.

In the chemical modification of protein with formaldehyde, the α-/ε-amino groups of protein reacted with formaldehyde (Kitamoto and Maeda 1980). In addition to the reaction of the amino groups, the side chain groups of amino acids are also known to be involved in the reaction with formaldehyde (French and Edsall

1945). After reacting with a small amount of excess formaldehyde, His, Trp, Asn, Tyr, Phe, and Cys can form stable products, and the product structures are proposed based on classical methods (French and Edsall 1945; Kitamoto and Maeda 1980). Under the physiological conditions, however, α-/ε-amino group of protein is most sensitive to formaldehyde reaction. The reaction of formaldehyde with Tau protein may be considered to block the positive charges of Lys residues, inducing conformational changes of Tau. Lu and collaborators have studied the structural characterization of formaldehyde-induced cross-links between amino acids and deoxynucleosides and their oligomers. The clarified structures will provide a basis for investigation of the characteristics and properties of DNA-protein cross-links (DPCs) formed in vivo and will be helpful in identifying biomarkers for the evaluation of formaldehyde exposure both at the site of contact and at distant sites (Lu et al. 2010). In this chapter, we discuss Tau function and dysfunction as well as its aggregation and cytotoxicity in the presence of formaldehyde.

2 Multiple Functions of Tau Protein

2.1 One of the Microtubule-Associated Members

In 1975, Weingarten and colleagues found a heat-stable protein essential for microtubule assembly and designated it Tau (Weingarten et al. 1975). Neuronal Tau acts in promoting and stabilizing the microtubule system involved in cellular transport and neuronal morphogenesis (Cleveland et al. 1977; Goedert et al. 1989a, b). The structure of Tau molecule can be subdivided into an amino-terminal proline-rich domain (PRD) that projects from the microtubule surface and a carboxy-terminal microtubule-binding domain (MBD). Incubation of bacterially expressed human Tau with sulfated glycosaminoglycans leads to bulk assembly of Tau filaments (Goedert et al. 1996), making it possible to obtain structural information (Berriman et al. 2003). Tau lacks secondary structures and is considered to be in a "wormlike" conformation with a high flexibility in solution (Schweer et al. 1994). Therefore, the side chains of amino acid residues such as Lys, Cys, Thr, and Ser are mostly exposed and vulnerable to chemical modifications. This is why Tau protein is sensitive to phosphorylation and dephosphorylation (See Chap. 9) as well as modification by formaldehyde.

2.2 As a Role of Chaperone

Chen and his collaborators have observed the interaction between human neuronal Tau-40 (hTau-40) and rabbit muscle D-glyceraldehyde-3-phosphate dehydrogenase (GAPDH, EC 1.2.1.12). First, they found that Tau protein did not influence the activity of GAPDH at room temperature or in solution in the absence of denaturant guanidine hydrochloride (GdnHCl) (Chen et al. 2000). Inactivation of GAPDH incubated with Tau was more distinguishably detected than that of control GAPDH during thermal and GdnHCl denaturation. Their work revealed that Tau binds to the denatured GAPDH but not the native enzyme under the experimental conditions. However, Tau suppressed refolding and reactivation of GAPDH when this enzyme was reactivated by dilution of GdnHCl solution. Tau protein could enhance the aggregation of the nonnative GAPDH in solutions, suggesting an anti-chaperone-like manner of Tau toward GAPDH in vitro, according to the definition for a protein to be regarded as a chaperone by Ellis (1990). For an anti-chaperone, just one characterization is different from the chaperone: it functions to obstruct the correct non-covalent assembly of polypeptide. Tau did not influence the activity of the native GAPDH but bound to the denatured enzyme and obstructed its reactivation. The delayed addition of Tau affecting the reactivation of the denatured GAPDH indicates that Tau does not interact with native GAPDH as a component of the assembled structure of the enzyme. This is why they regard Tau as a new anti-chaperone (Chen et al. 2000). However, Tau was inactivated in its anti-chaperone-like function when it was aggregated or phosphorylated by neuronal cdc2-like protein kinase (NCLK).

In contrast, neuronal Tau showed chaperone-like function to lactate dehydrogenase (LDH, EC1.1.1.27) as described by Tian and her collaborators. Furthermore, Tau protein remarkably enhances reactivation of glutamic dehydrogenase (GDH, EC 1.4.1.3), another carbohydrate metabolic enzyme, and also shows a chaperone-like manner (Tian et al. 2004). Similar results were described by Fukuhara and coworkers. They showed chaperone-like and anti-chaperone-like activities of amyloid fibrils of peptides from αA-crystallin (Fukuhara et al. 2012). These data indicated that either chaperone like- or anti-chaperone like-function of Tau depends on the types of substrate protein.

2.3 *Binding to Actin*

Tau protein interacts with actin in addition to microtubules (Sobue et al. 1985; Correas et al. 1990) and is involved in the organization of the cytoskeletal network (Henriquez et al. 1995). Actin monomers (G-actin) were found to form gel-like bundles in the presence of Tau (Yamauchi and Purich 1993). According to Farias and colleagues (Farias et al. 2002), actin monomers inhibit the association of Tau with tubulin immobilized on a solid-phase support system (Moraga et al. 1993). A stronger inhibition can be observed with preassembled actin filaments. The microtubule-associated protein Tau is able to interact with actin and serves as a cross-linker between the microtubule and actin networks (Kotani et al. 1985; Knowles et al. 1994).

The cross talk between microtubules and actin is essential for cellular functions. The MTBD of Tau protein is involved in its interaction with actin. To address the function of other domains of Tau in this interaction, Hai-Jin He and his collaborators

Table 7.1 The abilities of Tau mutants in binding to actin

		Microtubule	Actin	DNA/RNA
tauN	N-terminal region	–	–	–
tauPRD	Proline-rich domain	+	+	+
tauMTBD	Microtubule-binding domain	++	++	++
tauΔMTBD	Microtubule-binding domain deletion mutant	+	+	+
tauC	C-terminal region	–	–	–
Native Tau	Proline-rich domain/microtubule-binding domain	+++	+++	+++

"+" and "–" represent binding and non-binding to microtubule and actin (He et al. 2009).

constructed several *tau* truncation and deletion mutants (Table 7.1). The proline-rich domain truncation mutant (*tau*PRD), the microtubule-binding domain (*tau*MTBD), and the microtubule-binding domain deletion mutant (*tau*ΔMTBD) are able to promote the formation of actin filaments (F-actin). Native Tau was used as positive controls (He et al. 2009). However, actin assembly could not be observed in the presence of the N-terminal or C-terminal truncation mutant. In other words, the PRD is involved in the association of Tau with G-actin. Further studies with co-sedimentation and electron microscopy demonstrated that the PRD is also capable of binding to F-actin and inducing the formation of F-actin bundles. Using solid-phase assays to analyze apparent dissociation constants for the interaction of Tau or its mutants with F-actin resulted in a sequence of affinity for F-actin: native Tau>MTBD>PRD. Based on these data, we can see that Tau protein binds to and induces bundling F-actin through its PRD and MTBD.

In order to clarify the mechanisms underlying the microtubule-actin organization by cross-linkers, Elie and colleagues exhibited that neuronal Tau binds to actin and microtubules simultaneously, promoting in vitro co-organization and coupled growth of both networks. Tau protein is able to induce the guided polymerization of actin filaments along microtubule tracks and growth of single microtubule along actin filament bundles (Elie et al. 2015). They revealed that at least two of the four Tau-repeated motifs, primarily identified as tubulin-binding sites, are required to associate the two cytoskeleton microtubules and actin. Tau, acting as a linker, enables a coordination of the microtubule and actin networks that might be essential in various neuronal contexts for cellular skeletons.

2.4 Binding to DNA

Tau has a higher affinity for DNA than for microtubules (Corces et al. 1980) and is found in the nuclei of several cell lines such as human neuroblastoma cells (SH-SY5Y), mouse neuro-2a (N2a) cell line, human macrophages, monkey kidney cells, and PC12 cells (Loomis et al. 1990; Brady et al. 1995; Thurston et al. 1996; Cross et al. 2000; Wang et al. 1993). All six isoforms of Tau have been observed to bind to nucleolar organizer regions of acrocentric chromosomes and the fibrillar region of the nucleoli (the site for rRNA transcription) in some non-neuronal cells, for instance, lymphocyte (Turston et al. 1996). Greenwood and colleagues isolated nuclei and observed that Tau was covalently cross-linked to DNA, representing the binding of Tau to DNA in the nucleus (Greenwood and Johnson 1995). Sjoberg and coworkers proposed a functional role for nuclear Tau in relation to nucleolar organization and heterochromatinization of a portion of RNA genes (Sjoberg et al. 2006).

2.4.1 Protecting DNA Double Strands

In 2003, Hua and He proposed a hypothesis: *Tau could protect DNA double helix structure* (Hua and He 2003). The hypothesis is based on the following observations: (1) Tau enhances the melting temperature of calf thymus DNA, showing that Tau is able to stabilize the DNA double helix during thermal conditions (Hua and He 2000). (2) Tau will accelerate the refold of thermally denatured DNA when the temperature decreases (Hua et al. 2003). (3) Ma and coworkers employed the phen-copper complex as a model to study the effect of Tau on the oxidative DNA damage (Ma et al. 1998). Protein Tau could not directly scavenge the reactive oxygen species (ROS) but prevent DNA from ROS attacks in vitro, because the background light emission was observably caused by ROS. (4) Sultan and collaborators proposed that nuclear Tau is a key player in DNA protection. They showed Tau-mediated DNA protection in postmitotic neurons and proposed a scheme illustrating the nuclear Tau protection of DNA integrity in heat-stressed neurons (Sultan et al. 2011). These data demonstrate the protective effect of Tau protein on DNA. In 2014, Lu and her collaborators demonstrated Tau isomers also localized in the nucleus of neurons by immunohistochemical staining and gene knockout (Lu et al. 2014).

As described by Krylova and colleagues, Tau protein is capable of binding to single-stranded DNA with sequence specificity. They tried to demonstrate their viewpoint using an in vitro method with non-equilibrium capillary electrophoresis of equilibrium mixtures. Tau protein bound to one of DNA double strands in a sequence-specific fashion, which could induce dissociation of strands in dsDNA (Krylova et al. 2005). So far, however, the biological function of Tau protein associated with single strand of DNA is still under investigation.

2.4.2 DNA-Tauosome Complex

The complex of Tau associated with DNA features in a "beads-on-string" structure under electron microscopy. Association of Tau bends DNA double strands into a globular-like complex (Hua et al. 2003). Observing using atomic-force microscope (AFM), Qu and her collaborators showed that the complex of Tau protein associated with DNA double strands looks like histone-DNA complex (Qu et al. 2004).

Interacting with native Tau, DNA converted to a folded globular structure (average diameter, 27.42 ± 0.73 nm) (Lu et al. 2013). However, bovine serum albumin (BSA) as a control did not show the same effect on DNA double strands.

As described by Wei and her colleagues, Tau protein binds to the minor groove of DNA helix structure (Fig. 7.1). This viewpoint is supported by these observations: (1) Tau could be displaced from DNA double strands in the presence of the minor groove binders, for instance, distamycin A (Wei et al. 2008). (2) Tau is able to prevent DNA from degradation by DNase I, which also recognizes the minor groove of the double strands (Wei et al. 2008). (3) Histones associated with the minor groove without nucleotide sequence specificity are also able to replace Tau from DNA helix (Qu et al. 2004; Hua et al. 2003). (4) Methyl green, a major groove binder (Simonsson et al. 1998; Kumar and Muniyappa 1992; Tuite et al. 1997), cannot displace Tau from association with DNA. Finally, (5) Tau protein would associate with DNA in a sequence-specific manner if it is bound to the major groove (Roy 1996; Alberts et al. 2002). All these results point out that Tau protein acts as a minor groove binder to the DNA double strands.

As mentioned above, Tau contains proline-rich domain and microtubule-binding domain besides N-terminal and C-terminal region (Lee et al. 1989; Holt and Koffer 2001; He et al. 2009). In author's opinion, both PRD and MTBD may associate simultaneously with the minor groove of DNA double strands, approximately within the same plane though the calibration of the plane needs to be investigated (Wei et al. 2008). This hypothetical binding model is based on the following data: (1) Tau binds stably to a 13-bp or longer polynucleotides. In the case of a 10-bp polynucleotide, the DNA helix is not long enough to provide its minor groove for both PRD and MTBD to bind in the same orientation plane at the same time. (2) DNA chain could bend when protein binds helix in the same orientation plane (Ukiyama et al. 2001). In this case, the association of Tau definitely bends the DNA double strands. (3) In contrast, the DNA strands would not be bent if the two domains bound to the helix in different orientations. Finally, (4) as result from the binding in the same orientation, DNA chain could surround the surface of Tau protein like DNA helix surrounds the histone octamer complex if several Tau molecules are associated with bent DNA. This may be a putative mechanism for the formation of DNA-Tauosome from Tau and DNA (Fig. 7.1). Of course, whether DNA-Tauosome exists in vivo still needs further investigation.

2.4.3 Suppression of DNA Amplification

What is the putative function of the interactions between Tau and DNA? To clarify the effect of Tau on DNA, Li and her coworkers used both PCR and real-time PCR in the experiments. As described (Li et al. 2005), Tau protein is able to repress DNA amplification. A strong repression could be observed when an in vitro DNA replication assay was performed at the physiological temperature (37 °C). The incorporation of dNTP was markedly decreased to approximately 12 or 15% in the presence of Tau23 or Tau40 compared with BSA as a control group. In the competitive

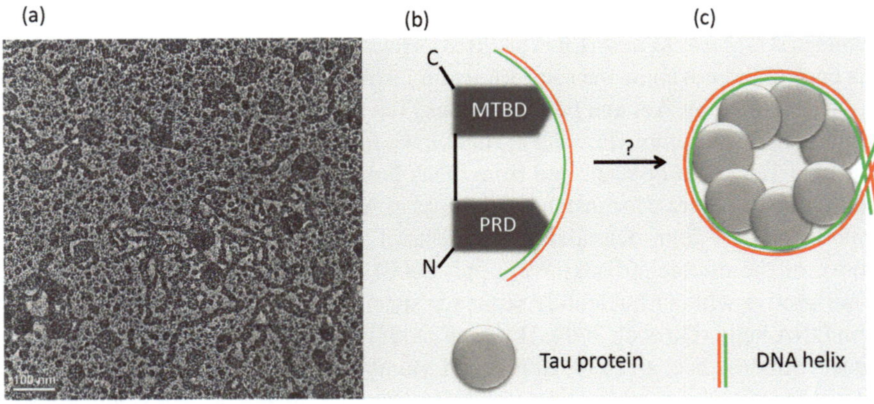

Fig. 7.1 A hypothetic model of DNA-Tauosome. The complexes of DNA chain and Tau protein under electron microscopy (panel **a**) (Hua et al. 2003; Lu et al. 2013). The proline-rich domain and the microtubule-binding domain may bind to the DNA minor groove in the same orientation plane, leading to fold the DNA helix (Wei et al. 2008). A putative morphology of DNA-Tauosome in a painted diagram (panel **c**)

experiments, the PCR product could be restored when the competitor DNA was added (Li et al. 2005). This is to say, the association of Tau with the template gives rise to the repression.

The repression on DNA replication appears that Tau could play a role in gene silence because Tau isomvers localized in both cytoplasm and nucleus (Lu et al. 2014). As we know, the function of Tau is regulated by phosphorylation and dephosphorylation. Tau could dissociate from DNA when it is phosphorylated (Lu et al. 2013). The other modifications also affect Tau's function such as glycosylation (Wang et al. 1996), hydroxymethylation/methylation by formaldehyde (He et al. 2010; Magdeldin and Yamamoto 2012), and acetylation (Cook et al. 2014). Glycosylation appears to be responsible for the maintenance of the PHF structure, although the abnormal phosphorylation might promote aggregation of Tau and inhibition of the assembly of microtubules (Wang et al. 1996). Modification by formaldehyde also impedes Tau binding to DNA, leading to inactivation of Tau protein to suppress DNA replication. Whether Tau plays a role in gene silence, it still remains to be demonstrated in vivo.

2.5 Not Suppressing RNA Transcription Though Binding to RNA

In 1975, the interaction of protein Tau with RNA was investigated by Bryan and colleagues, who proposed that protein Tau bound to RNA and the association inhibited microtubule assembly (Bryan et al. 1975). Later, RNA was found to stimulate aggregation of protein Tau into Alzheimer-like paired helical filaments (Kampers et al. 1996; Hasegawa et al. 1997). Using cytoplasmic acridine orange (AO) histofluorescence, Ginsberg and coworkers abolished with RNase, but not DNase or proteinase K, indicating the relative specificity of AO for RNA species. Their observations demonstrated the presence of RNAs in neurofibrillary tangles (NFTs) and senile plaques (SPs) (Ginsberg et al. 1997). According to Zhang and collaborators, RNA stores Tau protein reversibly in complex coacervates, suggesting that the droplet state can incubate tau and predispose the protein toward the formation of insoluble fibrils (Zhang et al. 2017).

Tau protein contains two major domains (Kanai et al. 1992), a projection domain (Hirokawa et al. 1988) and MTBD (Lee et al. 1988). The projection domain is composed of an acidic and a proline-rich region. As mentioned above, Tau interacts with actin and DNA via the PRD and MTBD. Kampers and coworkers prepared 21 mutants of Tau protein to investigate which domain is involved in the RNA-induced assembly of PHFs (Kampers et al. 1996). The MTBD plays a role in the interaction between Tau and RNA. The three-repeat constructs polymerize most efficiently, and two-repeat constructs are the minimum number required for assembly. Wang and his colleagues investigated the interaction of Tau with RNA via both the MTBD and PRD. Using gel-retardation assay and atomic-force microscopy, they found that Tau binds to tRNA, rRNA, and mRNA in non-sequence specificity. Tau was observed to bind to both the double-stranded RNA (polyA_{15}/U_{15}) and the single-stranded RNA (polyA_{30}) when Tau was incubated with the synthetic polynucleotides (Wang et al. 2006). In the analysis of the net charge distribution of Tau, five mutants were constructed such as the fragments of the N-terminal region, the proline-rich domain, the microtubule-binding domain, and the C-terminal region (Table 7.1). Electrophoretic mobility shift assay showed that PRD and MTBD are involved in the association with RNA, but N-terminal region and the C-terminal region are not. The association of Tau with RNA is observably disturbed in acidic solutions ($pH<5$) or in the presence of NaCl (>0.5 M). In other words, the electrostatic interaction is mainly involved in the binding of Tau to RNA. The results from indirect immunofluorescence assay indicate that Tau protein and RNAs are partially co-localized both in cytoplasm and nucleoplasm in SH-SY5Y cells. These data suggest that Tau probably interacts with RNA in vivo under physiological state.

However, Li and her collaborators investigated whether Tau could affect RNA transcription in vitro. As shown in their results, Tau did not repress the yield of RNA in transcription although the protein binds to DNA in nonspecific manner. It appears that Tau could be replaced or ejected from the template by the elongating T7 RNA polymerase (Li et al. 2005).

3 Formaldehyde Induces Tau Aggregation and Dysfunction

3.1 Tau Aggregation in the Presence of Formaldehyde

In author's laboratory, aggregation of Tau protein was observed in the presence of ethanol with disuccinimidyl suberate (DSS), a chemical cross-linking agent (He et al. 1998). Formation of Tau polymers by fluorescence ($\lambda_{em} = 333$ nm; $\lambda_{ex} = 280$ nm) was observed when the proteins were aggregated in solutions (Luo and He 1999; Luo et al. 2000a). Tau aggregation was also investigated by applying thermal conditions and denaturant of guanidine HCl (Luo et al. 2000b). As described previously, Tau was markedly aggregated when treated with aldehydes including acetaldehyde, glutaraldehyde, and formaldehyde at different concentrations (Luo and He 1999, Hua et al. 2002). The effect of aldehyde on the phosphorylation of Tau protein with NCLK (Chen et al. 1999), which is one of the kinases promoting Tau hyperphosphorylation in the brain of Alzheimer's patients, was also observed (Paudel 1997a).

Yu and his group studied the function of semicarbazide-sensitive amine oxidase (SSAO) and considered that the formation of formaldehyde from adrenaline is a potential risk factor for stress-related angiopathy (Yu 1990; Yu et al. 1997). He proposed a hypothesis that SSAO-associated aldehydes, such as formaldehyde and methylglyoxal, could be crucially important in the pathogenesis of vascular dementia, AD, and diabetic complications (Yu 2001). Yin proposed that the formation of "carbonyl protein" is related with the process of aging (Yin 1993). Hua and RQ He showed that treatment with formaldehyde not only induced Tau aggregation but also inactivated its binding to DNA (Hua and He 2002). Compared with the aggregation of BSA, Tau protein was much more sensitive to the treatment of formaldehyde. Further experiments show that the intensity of light scattering of neuronal Tau-40 solution at 480 nm increased markedly in the presence of low concentrations of formaldehyde. 8-Anilino-1-naphthalenesulfonic acid (ANS) binding assay showed that a putative hydrophobic core is formed in Tau polymers during incubation with formaldehyde (Nie et al. 2005). Native Tau was rapidly hydrolyzed by immobilized earthworm fibrinolytic enzyme II (EFE II), producing a digested fragment (36–37 kDa). However, formaldehyde-treated Tau was not so easy to be digested under the same experimental conditions. The aggregated Tau forms a hydrophobic core and becomes relatively rigid in its polymers. These results suggest that neuronal Tau is one of the target proteins for formaldehyde which may be related to neurodegeneration.

Further study in the summer of 2006 by Nie and his colleagues showed that low concentrations of formaldehyde induced human Tau protein to form neurotoxic aggregation. Their data supported that formaldehyde played a role in the induction of Tauopathies in age-related cognitive impairment (Nie et al. 2007b). In the winter of 2006, Yu and his coworkers incubated Aβ with formaldehyde and showed the effect of aldehydes on Aβ aggregation in vitro. Their results supported the involvement of endogenous aldehydes in amyloid deposition related to AD (Chen et al.

2007). They proposed that increased SSAO-mediated deamination may contribute to protein deposition, formation of plaques, and inflammation and thus may be involved in the pathophysiology of chronic vascular and neurological disorders, such as diabetic complications, atherosclerosis, and AD.

3.2 Tau Aggregation and Inactivation in the Presence of Formaldehyde

Native Tau is prone to misfolding and inactivation in the presence of low concentrations of formaldehyde in vitro though these protein features in a "wormlike" flexible conformation (Nie et al. 2005). As shown in Fig. 7.2, formaldehyde-treated Tau polymerizes into soluble globule-like aggregation with positive staining in both thioflavin T (ThT) and Congo red (Nie et al. 2007a). Under the same conditions, acetaldehyde-treated Tau did not markedly show protein aggregations detected by atomic-force microscopy. To induce Tau aggregation in the presence of acetaldehyde, it needs higher concentrations of the aldehyde and protein. Formaldehyde, unlike acetaldehyde, is not active in volatilization; hence, a precise result can be obtained in the experiment. Definitely, formaldehyde treatment induces inactivation of Tau in its binding to tubulin and stabilizing microtubule.

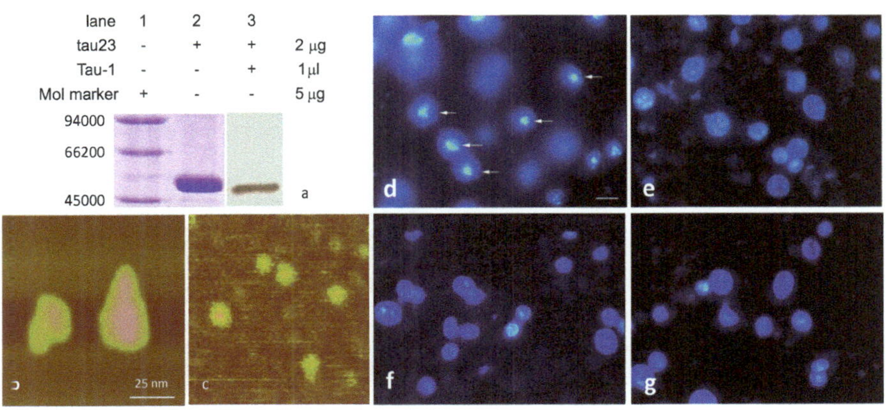

Fig. 7.2 Tau aggregations and cell death. Neuronal Tau was expressed and purified as described (Paudel 1997b). The purified Tau protein was loaded on the SDS-PAGE and analyzed by Western blotting (panel **a**). Tau protein (20 μM) was incubated in 50 mM phosphate buffer (pH 7.2) in the presence of 0.1% formaldehyde at 37 °C overnight (panel **b**). *Bar* = 25 nm for AFM. Tau protein was incubated in the absence of formaldehyde as control (panel **c**). SH-SY5Y cells were cultured in the presence of formaldehyde-treated Tau aggregation (panel **d**), self-aggregated Tau (panel **e**), native Tau (panel **f**), and control culture in DMEM without serum (panel **g**). Cells were stained by Hoechst 22258, and *arrows* designate the apoptotic nuclear profiles. *Bar* = 25 μm for cells. The usage of this figure was authorized by the authors and the journal Prog. Biochem. Biophys

3.3 Formaldehyde Treatment Inactivates Tau in the Association with DNA

As mentioned above, native Tau is active in association with DNA. Thus, the effect of formaldehyde on the interaction of Tau with DNA has been studied by Hua and her colleagues. They incubated Tau with formaldehyde and dialyzed the residual aldehyde. Then they employed electrophoretic mobility shift assay (EMSA) to detect whether formaldehyde-treated Tau is still active in binding to DNA double strands. EMSA depicted a marked inactivation of formaldehyde-treated Tau in interacting with DNA. This is to say, formaldehyde treatment inactivates Tau protein in its association not only with microtubule but also with DNA (Hua and He 2002).

Formaldehyde reacts with α-/ε-amino groups of a protein (Nie et al. 2007a, b), which suppresses Tau protein in the association with DNA. In other words, lysinyl residues are the essential residue for Tau to interact with DNA. The author and collaborators constructed the mutants of tauN and tauC and then incubated them with DNA, respectively. The association of the two mutants with DNA could not be observed on EMSA. This indicates that the Lys residues out of the PRD and MTBD may not be the main contributors to the DNA binding (Wei et al. 2008).

The lysinyl residues in the PRD and MTBD are essential to interact with the minor groove of DNA helix. The PRD (7 Lys) and MTBD (12 Lys) contain 19 lysinyl residues in total, and some of them are supposed to be involved in the interaction between Tau and DNA. This viewpoint is based on the following points: (1) the pK of the ε-amino group of lysinyl residue is 10.8 when interacting with DNA bases in electrostatics; (2) the interaction between Tau and DNA is markedly suppressed at the pH higher than 12, showing the interaction is electrostatic dependent as measured by EMSA; and (3) addition of formaldehyde significantly interferes with the association of Tau with DNA (Wei et al. 2008). Note that since formaldehyde induces Tau aggregation, DNA-Tauosome could not be observed after the protein is treated with the aldehyde.

4 Formaldehyde-Treated Tau Becomes Cytotoxic

4.1 Formaldehyde-Treated Tau Features in Globular-Like Aggregation with Cytotoxicity

As described previously (Nie et al. 2007b), unlike the typical globular protein BSA, the natively unfolded structure of human neuronal Tau was induced to misfold and aggregate in the presence of low concentrations of formaldehyde, leading to the formation of globular-like deposits that appeared as densely staining granules under electron and atomic-force microscopy and bound the amyloid-specific dyes thioflavin T and Congo red. This suggests that formaldehyde-treated Tau may feature in amyloid-like aggregation.

Amyloidosis disease is caused by accumulation of proteins in the form of abnormal, insoluble fibers, known as amyloid fibrils (Pepys 2006). As kinetic study shows, Tau can rapidly polymerize in 60 min in 0.1% formaldehyde solution. SDS-PAGE results also exhibit Tau polymerization in the presence of formaldehyde. However, neuronal Tau misfolds and aggregates into globular-like polymers in formaldehyde solutions. No fibril-like polymerization could be observed, while Tau protein was incubated with formaldehyde for 15 days apart from globular-like aggregation observed by atomic-force microscopy (AFM) and electron microscopy (EM). This suggests that formaldehyde polymerization is involved in Tau aggregation. Such aggregation process is probably linked to the Tau's special "wormlike" conformation, which leaves the ε-amino groups of lysinyl residues and thiol groups of Cys residues exposed to the exterior. Such a structure can easily be reacted with formaldehyde in vitro and in vivo.

It is known that formaldehyde molecules are prone to polymerization. Polymerization of formaldehyde itself results in aggregation of Tau protein. Immunocytochemistry and thioflavin S (ThS) staining of both endogenous and exogenous Tau in the presence of formaldehyde at low concentrations in the cell culture and in animal model (Nie et al. 2007a, b) have shown that formaldehyde can induce Tau into globular-like aggregates in vivo during cell death. However, formaldehyde-treated Tau showed a stronger cytotoxicity than formaldehyde-treated BSA. The significant aggregation of Tau induced by formaldehyde and the severe toxicity of the aggregated Tau to neural cells suggested that chronic toxicosis resulted from formaldehyde is related to neurodegeneration and age-related cognitive impairment (Li et al. 2008, 2016).

To demonstrate the effect of formaldehyde on cell viability, the author and collaborators removed the residual formaldehyde and used the globular-like aggregates of Tau to incubate with SH-SY5Y cells and rat hippocampal cells. The formaldehyde-treated Tau is able to induce apoptosis in both neurotypic SH-SY5Y cell line and the rat hippocampus, as observed by Hoechst 33258 staining, assay of caspase-3 activity, and flow cytometry using annexin V and propidium iodide staining. SH-SY5Y cells and rat hippocampal cells incubated with Tau aggregates, formed in the presence of acetaldehyde or in the absence of additives (and which did not show appreciable binding of ThT or Congo red), did not markedly show signs of apoptosis. Note that the concentration of formaldehyde used was not higher than 0.05 mM and depended upon the inoculation cell size. Further experiments showed that Congo red specifically attenuated the caspase-3 activity which is activated by globular-like deposits of Tau. In other words, low concentrations of formaldehyde may play a role in Tauopathies.

Now that formaldehyde-induced Tau features in globular-like soluble deposits, Nie and his coworkers investigated the effect of low concentrations of formaldehyde on protein misfolding and aggregation in cells. They incubated SH-SY5Y cells with low concentrations of formaldehyde and then probed Tau protein with Tau1 (antibody against total Tau) and protein aggregation with ThS staining. ThS-positive protein aggregation can be seen inside the cell overlapped with Tau protein in the presence of formaldehyde (Nie et al. 2007b). To confirm that the intracellular

ThS-positive aggregation was Tau, the authors performed Western blotting with anti-HA-Tau in the cells. Formaldehyde is able to induce Tau into ThS-positive aggregation. That is to say, formaldehyde causes intracellular protein aggregation and deposit in cells (Nie et al. 2007b). However, whether the ThS-positive stained protein aggregation is localized in the cell or on the membrane needs further investigation.

4.2 Formaldehyde Might Induce Tau Form Pore-Like Aggregation

Quist and colleagues (University of California, Santa Barbara) proposed a hypothesis of "aspecific amyloid ion channels" to explain the pathomechanism of cell dysfunction and death during neurodegenerative diseases (Quist et al. 2005). Pre-fibrillar aggregates may interact with reconstituted phospholipid membranes and with cell membrane where they form aspecific channels disrupting cellular homeostasis. The latter possible mechanism of toxicity is similar to that displayed by antimicrobial peptides, pore-forming eukaryotic proteins, bacterial toxins, and newly synthesized cyclic peptide antibiotics (Stefani and Dopson 2003).

As has been mentioned above, formaldehyde-treated Tau features in globular-like aggregation. However, amyloid-like fibrils of Tau could not be observed in the presence of formaldehyde. This is to say, formaldehyde-treated Tau exists in soluble oligomers and polymers. According to our hypothesis, "Dysmetabolism of endogenous formaldehyde may be one of the risk factors involved in sporadic age-related cognitive impairment", the authors incubated Tau protein with formaldehyde and observed globular-like deposits with marked cytotoxicity. Under the experimental conditions, "pore-like" aggregates of Tau protein could be exhibited on the mica surface by AFM after the protein was treated with high concentrations (>0.5%) of formaldehyde. It seems to be a novel approach to study the mechanism of cellular metabolic disturbance, even cell death, which is induced by formaldehyde-treated neuronal Tau. However, the repeatability of this morphology is low in the observations. The authors did not observe the "pore-like aggregation" by electron microscopy. Therefore, whether the pore-like aggregation could be formed in water solution in the presence of formaldehyde still needs to be investigated in vivo and in vitro.

5 Cellular Formaldehyde, Abnormal Lysosome, and Protein Aggregation

Lysosomes are classically considered as systems for protein degradation (Cuervo and Dice 1998). Recent progress in fluorescent probes allows formaldehyde molecules to be detected in living cells (Brewer and Chang 2015; Descamps et al. 2010).

Tang and coworkers incubated HeLa cells with Na-FA (a two-photon fluorescent probe) and organelle indicators such as ER-Tracker™ Red (BODIPY TR Glibenclamide), LysoTracker Red DND-99, MitoTracker Red CMXRos, and Golgi-Tracker Red (Tang et al. 2016b). Their data indicated that the probe Na-FA is located in the endoplasmic reticulum, the Golgi apparatus, and the lysosome (Jian and Zhu 2016), but very little is present in the mitochondria. Later, they emphasized the lysosome-targeted turn-on fluorescent probe for endogenous formaldehyde in living cells (Tang et al. 2016a). Other laboratories have also demonstrated the localization of cellular formaldehyde in lysosomes detected by their own fluorescent probes (Xu et al. 2016; Brewer and Chang 2015).

The author considered that lysosomes play a key role in accommodation and transportation of formaldehyde in cells. This hypothesis is based on these observations: (1) Utilizing formaldehyde fluorescent probe, Chen and her collaborators observed an increase in the intracellular formaldehyde located in the lysosome in the blood endothelial cell line (bEnd.3). (2) To demonstrate this result, they employed neuroblastoma N2a cells (N2a) from the mouse brain and observed the same result under oxidative stress (Chen et al. 2017). (3) LeuLeuOMe is the agent which can induce the permeabilization of lysosome membrane in bEnd.3. Introducing LeuLeuOMe leads to an increase in intracellular formaldehyde but a decrease in extracellular formaldehyde measured by microplate reader and high-performance liquid chromatography (HPLC). This is to say, lysosome cannot only accommodate endogenous formaldehyde inside a cell but also transport the cellular compound out a cell. (4) In order to get further data to support the hypothesis, Chen and her collaborators isolated the lysosomes from the brain of rats which were operated on surgery with the chronic cerebral hypoperfusion because oxidative stress elevated endogenous formaldehyde levels in rats compared with those with SHAM. They found that the brain formaldehyde was significantly ($P < 0.01$) elevated in the lysosome compared to the cytoplasm (Chen et al. 2017). These data suggest that abnormal lysosome function is related with dysmetabolism of formaldehyde.

In fact, abnormal lysosomes increase as aging (Wyss-Coray 2016), and so does the endogenous formaldehyde (Tong et al. 2013). As previously reported (Wyss-Coray 2016), lysosome dysfunction results in protein aggregation and accumulation (Jian and Zhu 2016). It is reasonable to hypothesize that endogenous formaldehyde is involved in protein aggregation as mentioned above. Dysmetabolism of formaldehyde should affect protein synthesis, modification, and degradation (Fig. 7.3). Abnormal lysosomes should dysfunction in cellular formaldehyde metabolism and thus affect the scavenging of the abnormal biomacromolecules in the oxidative stress (Evans et al. 2016), in particular of abnormal proteins and their aggregation accumulation (He 2016).

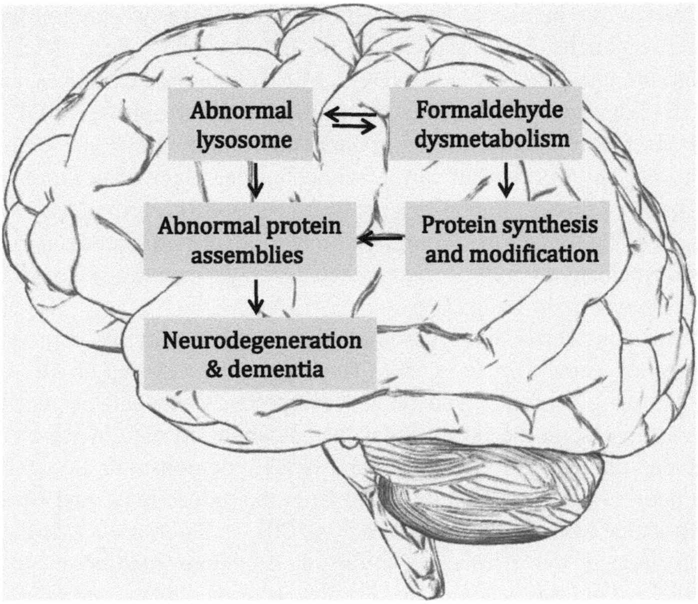

Fig. 7.3 Abnormal lysosome and formaldehyde dysmetabolism. Abnormal lysosome is involved in formaldehyde dysmetabolism, affecting protein homeostasis. This viewpoint is based on these observations: (1) abnormal lysosome increases with aging (Wyss-Coray 2016), and so does endogenous formaldehyde (Tong et al. 2013); (2) administration of formaldehyde induces the formation of senile plaques and promotes Tau hyperphosphorylation accompanied with cognitive impairment for monkeys and mice (Yang et al. 2014a, b); (3) concentrations of endogenous formaldehyde are positively correlated to the severity of cognitive impairment of AD patients (Tong et al. 2011); (4) lysosome is one of the important cellular organelles where formaldehyde is accommodated, metabolized, and transported (Chen et al. 2014; Chen et al. 2017). The usage of this figure was authorized by the authors and the journal Prog. Biochem. Biophys

6 Conclusion

Dysmetabolism of endogenous formaldehyde is regarded as one of the risk factors for the onset and progression of age-related cognitive impairment including Alzheimer's disease and vesicular dementia. Excess extracellular and intracellular formaldehyde induces neuron death, involving in the impairment of cognitive ability. Tau protein is prone to the reaction with formaldehyde, resulting in globular-like protein aggregation with cytotoxicity. Formaldehyde treatment inactivates Tau in its binding to microtubules, actin, and F-actin as well as suppresses Tau's association with DNA and RNA (Fig. 7.4). Tau protein may be modified by formaldehyde in vivo, since the endogenous aldehyde comes from different and multiple sources such as food, oxidative stress, and intestinal microbiota (Hu et al. 2016; Liu et al. 2017). The formaldehyde-treated Tau causes oxidative stress on neural cells and cell apoptosis. In fact, direct intervention of the reaction between formaldehyde and proteins, in particular of neuronal Tau resulting in cytotoxicity, should be addressed

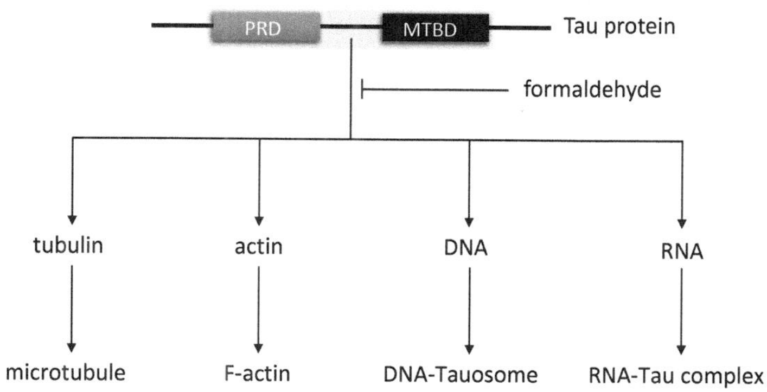

Fig. 7.4 Roles of MTBD and PRD of Tau protein. Tau protein interacts with tubulin, actin, DNA, and RNA through its MTBD and PRD (Wei et al. 2008; He et al. 2009; Wang et al. 2006). The usage of this figure was authorized by the authors and the journal Prog. Biochem. Biophys

at the onset and in the progression of neurodegenerative disease although a great number of researches have been carried out on phosphorylation and glycosylation catalyzed by kinases and enzymes (See Chap. 12). Clarification of the pathomechanism of formaldehyde-modified Tau inducing cell death is beneficial to understand Tauopathies in the age-related cognitive impairment.

Acknowledgments This project was supported by grants from the Beijing Municipal Science and Technology Project (Z161100000217141; Z161100000216137), the National Key Research and Development Program of China (2016YFC1306300), the National Basic Research Program of China (973 Program) (2012CB911004), the National Natural Science Foundation of China (NSFC 31270868), the Foundation of Chinese Academy of Sciences (CAS-20140909), and the Queensland-Chinese Academy of Sciences Biotechnology Fund (GJHZ201302). This project was also supported by grants from the National Natural Science Foundation of China (NSFC 81274093), Shandong Province Natural Science Foundation (ZR2015HL128), and Health Department of Shandong Province (2014WS0478).

Competing Financial Interests The authors declare no competing financial interests.

References

Alberts B, Johnson A, Lewis J, Raff M, Roberts K, Walter P (2002) Molecular biology of the cell, 4th edn. Garland Science Press, New York, pp 381–382

Berriman J, Serpell LC, Oberg KA, Fink AL, Goedert M, Crowther RA (2003) Tau filaments from human brain and from in vitro assembly of recombinant protein show cross-beta structure. Proc Natl Acad Sci U S A 100(15):9034–9038

Brady RM, Zinkowski RP, Binder LI (1995) Presence of tau in isolated nuclei from human brain. Neurobiol Aging 16(3):479–486

Brewer TF, Chang CJ (2015) An Aza-cope reactivity-based fluorescent probe for imaging formaldehyde in living cells. J Am Chem Soc 137(34):10886–10889

Bryan JB, Nagle BW, Doenges KH (1975) Inhibition of tubulin assembly by RNA and other polyanions: evidence for a required protein. Proc Natl Acad Sci U S A 72(9):3570–3574
Carpenter DC (1946) The protein-formaldehyde reaction; the question of methylene bridges and the unreactivity of benzoyl-d (-)-alanine toward formaldehyde. Arch Biochem 9:159–164
Chen YH, Luo JY, Li W, He RQ (1999) Effect of acetaldehyde on phosphorylation of human neuronal tau. J Biochem Mol Biol Biophys 3:197–202
Chen YH, He RQ, Liu Y, Xue ZG (2000) Effect of human neuronal tau on denaturation and reactivation of rabbit muscle D-glyceraldehyde-3-phosphate dehydrogenase. Biochem J 351:233–240
Chen K, Kazachkov M, Yu PH (2007) Effect of aldehydes derived from oxidative deamination and oxidative stress on β-amyloid aggregation: pathological implications to Alzheimer's disease. J Neural Transm 114(6):835–839
Chen J, Sun M, Wang X, Lu J, Wei Y, Tan Y, Liu Y, Götz J, He R, Hua Q (2014) The herbal compound geniposide rescues formaldehyde-induced apoptosis in N2a neuroblastoma cells. Sci China Life Sci 57(4):412–421
Chen XX, Su T, He YG, He RQ (2017) Spatial cognition decline caused by excess formaldehyde in the lysosome. Prog Biochem Biophys 44(6):486–494
Cleveland DW, Hwo SY, Kirschner MW (1977) Purification of tau, a microtubule-associated protein that induces assembly of microtubules from purified tubulin. J Mol Biol 116(2):207–225
Cohen FE (1999) Protein misfolding and prion diseases. J Mol Biol 293(2):313–320
Cook C, Carlomagno Y, Gendron TF, Dunmore J, Scheffel K, Stetler C, Davis M, Dickson D, Jarpe M, DeTure M, Petrucelli L (2014) Acetylation of the KXGS motifs in tau is a critical determinant in modulation of tau aggregation and clearance. Hum Mol Genet 23(1):104–116
Corces VG, Manso R, De La Torre J, Avila J, Nasr A, Wiche G (1980) Effects of DNA on microtubule assembly. Eur J Biochem 105(1):7–16
Correas I, Padilla R, Avila J (1990) The tubulin-binding sequence of brain microtubule-associated proteins, tau and MAP-2, is also involved in actin binding. Biochem J 269(1):61–64
Cross DC, Munoz JP, Hernandez P, Maccioni RB (2000) Nuclear and cytoplasmic tau proteins from human nonneuronal cells share common structural and functional features with brain tau. J Cell Biochem 78(2):305–317
Cuervo AM, Dice JF (1998) Lysosomes, a meeting point of proteins, chaperones, and proteases. J Mol Med (Berl) 76(1):6–12
Descamps MN, Bordy T, Hue J, Mariano S, Nonglaton G, Schultz E, Tran-Thi TH, Vignoud-Despond S (2010) Real-time detection of formaldehyde by a fluorescence-based sensor. Procedia Engineer 5:1009–1012
Elie A, Prezel E, Guérin C, Denarier E, Ramirez-Rios S, Serre L, Andrieux A, Fourest-Lieuvin A, Blanchoin L, Arnal I (2015) Tau co-organizes dynamic microtubule and actin networks. Sci Rep 5:9964
Ellis RJ (1990) The molecular chaperone concept. Semin Cell Biol 1(1):1–9
Evans AM, Fameli N, Ogunbayo OA, Duan J, Navarro-Dorado J (2016) From contraction to gene expression: nanojunctions of the sarco/endoplasmic reticulum deliver site- and function-specific calcium signals. Sci China Life Sci 59(8):749–763
Farias GA, Munoz JP, Garrido J, Maccioni RB (2002) Tubulin, actin, and tau protein interactions and the study of their macromolecular assemblies. J Cell Biochem 85(2):315–324
Fox CH, Johnson FB, Whiting J, Roller PP (1985) Formaldehyde fixation. J Histochem Cytochem 33(8):845–853
French D, Edsall JT (1945) The reactions of formaldehyde with amino acids and proteins. Adv Protein Chem 2:277–335
Friedhoff P, Schneider A, Mandelkow EM, Mandelkow E (1998) Rapid assembly of Alzheimer-like paired helical filaments from microtubule-associated protein tau monitored by fluorescence in solution. Biochemistry 37(28):10223–10230
Fukuhara S, Nishigaki T, Miyata K, Tsuchiya N, Waku T, Tanaka N (2012) Mechanism of the chaperone-like and antichaperone activities of amyloid fibrils of peptides from αA-crystallin. Biochemistry 51(27):5394–5401

Ginsberg SD, Crino PB, Lee VMY, Eberwine JH, Trojanowski JQ (1997) Sequestration of RNA in Alzheimer's disease neurofibrillary tangles and senile plaques. Ann Neurol 41(2):200–209
Goedert M, Jakes R (1990) Expression of separate isoforms of human tau protein: correlation with the tau pattern in brain and effects on tubulin polymerization. EMBO J 9(13):4225–4230
Goedert M, Spillantini MG, Jakes R, Rutherford D, Crowther RA (1989a) Multiple isoforms of human microtubule-associated protein tau: sequences and localization in neurofibrillary tangles of Alzheimer's disease. Neuron 3(4):519–526
Goedert M, Spillantini MG, Potier MC, Ulrich J, Crowther RA (1989b) Cloning and sequencing of the cDNA encoding an isoform of microtubule-associated protein tau containing four tandem repeats: differential expression of tau protein mRNAs in human brain. EMBO J 8(2):393–399
Goedert M, Spillantini MG, Cairns NJ, Crowther RA (1992) Tau proteins of Alzheimer paired helical filaments: abnormal phosphorylation of all six brain isoforms. Neuron 8(1):159–168
Goedert M, Jakes R, Spillantini MG, Hasegawa M, Smith MJ, Crowther RA (1996) Assembly of microtubule-associated protein tau into Alzheimer-like filaments induced by sulphated glycosaminoglycans. Nature 383(6600):550–553
Greenwood JA, Johnson GV (1995) Localization and in situ phosphorylation state of nuclear tau. Exp Cell Res 220(2):332–337
Hardy J, Selkoe DJ (2002) The amyloid hypothesis of Alzheimer's disease: progress and problems on the road to therapeutics. Science 297(5580):353–356
Hasegawa M, Crowther RA, Jakes R, Goedert M (1997) Alzheimer-like changes in microtubule-associated protein tau induced by sulfated glycosaminoglycans. Inhibition of microtubule binding, stimulation of phosphorylation, and filament assembly depend on the degree of sulfation. J Biol Chem 272(52):33118–33124
He RQ, Luo JY, Li W (1998) Effect of ethanol on the aggregation of human neuronal tau protein. Protein Pept Lett 5:279–285
He RQ (2016) Abnormal lysosome, formaldehyde Dysmetabolism and age-related cognitive impairment. Prog Biochem Biophys 43(12):1197
He HJ, Wang XS, Pan R, Wang DL, Liu MN, He RQ (2009) The proline-rich domain of tau plays a role in interactions with actin. BMC Cell Biol 10:81
He RQ, Lu J, Miao JY (2010) Formaldehyde. Stress Sci China Life Sci 53(10):1399–1404
Henriquez JP, Cross D, Vial C, Maccioni RB (1995) Subpopulations of tau interact with microtubules and actin filaments in various cell types. Cell Biochem Funct 13(4):239–250
Hirokawa N, Shiomura Y, Okabe S (1988) Tau proteins: the molecular structure and mode of binding on microtubules. J Cell Biol 107(4):1449–1459
Holt MR, Koffer A (2001) Cell motility: proline-rich proteins promote protrusions. Trends Cell Biol 11(1):3846
Hopwood D, Yeaman G, Milne G (1988) Differentiating the effects of microwave and heat on tissue proteins and their crosslinking by formaldehyde. Histochem J 20(6–7):341–346
Hu X, Wang T, Jin F (2016) Alzheimer's disease and gut microbiota. Sci China Life Sci 59(10):1006–1023
Hua Q, He RQ (2000) Human neuronal tau promoting the melting temperature of DNA. Chin Sci Bull 45(11):999–1002
Hua Q, He RQ (2002) Effect of phosphorylation and aggregation on tau binding to DNA. Prot Pept Lett 9(4):349–357
Hua Q, He RQ (2003) Tau could protect DNA double helix structure. Biochim Biophys Acta 645:205–211
Hua Q, Nie CL, Liu Y, He RQ (2002) Formaldehyde inducing an amyloid-like aggregation of human neuronal tau. Biophys J (Annual Meeting Abstracts) 82(1):504a
Hua Q, He RQ, Haque N, MH Q, del Carmen AA, Grundke-Iqbal I, Iqbal K (2003) Microtubule associated protein tau binds to double-stranded but not single stranded DNA. Cell Mol Life Sci 60(2):413–421
Jack CR Jr, Knopman DS, Jagust WJ, Shaw LM, Aisen PS, Weiner MW, Petersen RC, Trojanowski JQ (2010) Hypothetical model of dynamic biomarkers of the Alzheimer's pathological cascade. Lancet Neurol 9(1):119–128

Jian X, Zhu MX (2016) Regulation of lysosomal ion homeostasis by channels and transporters. Sci China Life Sci 59(8):777–791

Kampers T, Friedhoff P, Biernat J, Mandelkow EM, Mandelkow E (1996) RNA stimulates aggregation of microtubule-associated protein tau into Alzheimer-like paired helical filaments. FEBS Lett 399(3):344–349

Kanai Y, Chen J, Hirokawa N (1992) Microtubule bundling by tau proteins in vivo: analysis of functional domains. EMBO J 11(11):3953–3961

Kitamoto Y, Maeda H (1980) Reevaluation of the reaction of formaldehyde at low concentration with amino acids. J Biochem 87(5):1519–1530

Knowles R, LeClerc N, Kosik KS (1994) Organization of actin and microtubules during process formation in tau-expressing Sf9 cells. Cell Motil Cytoskeleton 28(3):256–264

Kotani S, Nishida E, Kumagai H, Sakai H (1985) Calmodulin inhibits interaction of actin with MAP2 and tau, two major microtubule-associated proteins. J Biol Chem 260(19):10779–10783

Krylova SM, Musheev M, Nutiu R, Li Y, Lee G, Krylov SN (2005) Tau protein binds single-stranded DNA sequence specifically--the proof obtained in vitro with non-equilibrium capillary electrophoresis of equilibrium mixtures. FEBS Lett 6:1371–1375

Kumar KA, Muniyappa K (1992) Use of structure-directed DNA ligands to probe the binding of RecA protein to narrow and wide grooves of DNA and on its ability to promote homologous pairing. J Biol Chem 267(34):24824–24832

Lee G, Neve RL, Kosik KS (1989) The microtubule binding domain of tau protein. Neuron 2(6):1615–1624

Li W, Wang XS, Qu MH, Liu Y, He RQ (2005) Human protein tau represses DNA replication in vitro. Biochim Biophys Acta 1726(3):280–286

Li FX, Lu J, Xu YJ, Tong ZQ, Nie CL, He RQ (2008) Formaldehyde-mediated chronic damage may be related to sporadic neurodegeneration. Prog Biochem Biophys 35(4):393–400

Li T, Su Tao, He YG, He RQ (2016) Chronic-dehydrated dysmetabolism of Formaldehyde in mouse brain and decline of learning in the shuttle box. Prog Biochem Biophys 43(4):429–438

Liu KL, He YG, Yu LX, He RQ (2017) Elevated formaldehyde in the cecum of APP/PS1 mouse. Microbiol China 44(8):1761–1766

Loomis PA, Howard TH, Castleberry RP, Binder LI (1990) Identification of nuclear tau isoforms in human neuroblastoma cells. Proc Natl Acad Sci U S A 87(21):8422–8426

Lu K, Ye W, Zhou L, Collins LB, Chen X, Gold A, Ball LM, Swenberg JA (2010) Structural characterization of formaldehyde-induced cross-links between amino acids and deoxynucleosides and their oligomers. J Am Chem Soc 132(10):3388–3399

Lu Y, He HJ, Zhou J, Miao JY, Lu J, He YG, Pan R, Wei Y, Liu Y, He RQ (2013) Hyperphosphorylation results in tau dysfunction in DNA folding and protection. J Alzheimer's Dis 37(3):551–563

Lu J, Li T, He RQ, Bartlett PF, Götz J (2014) Visualizing the microtubule-associated protein tau in the nucleus. Sci China Life Sci 57(4):422–431

Luo JY, He RQ (1999) Effect of acetaldehyde on aggregation of neuronal tau. Protein Pept Lett 6(2):105–110

Luo JY, Li W, He RQ (2000a) The fluorescent characterization of the aggregating human neuronal tau. Int J Biochem Macromol (UK) 27(4):263–268

Luo JY, Liu Y, Hua Q, He RQ (2000b) Conformational changes of human neuronal tau during thermal and guanidine-Hcl denaturation. Protein Pept Lett 7:133–141

Ma W, Cao EH, Zhang J, Qin JF (1998) Phenanthroline-cu complex-mediated chemiluminescence of DNA and its potential use in antioxidation evaluation. J Photochem Photobiol B 44(1):63–68

Magdeldin S, Yamamoto T (2012) Toward deciphering proteomes of formalin-fixed paraffin-embedded (FFPE) tissues. Proteomics 12(7):1045–1058

Moraga DM, Nunez P, Garrido J, Maccioni RB (1993) A tau fragment containing a repetitive sequence induces bundling of actin filaments. J Neurochem 61(3):979–986

Nie CL, Zhang W, Zhang D, He RQ (2005) Changes in conformation of human neuronal tau during denaturation in formaldehyde solution. Prot Pept Lett 12(1):75–78

Nie CL, Wang XS, Liu Y, Perrett S, He RQ (2007a) Amyloid-like aggregates of neuronal tau are induced by formaldehyde exposure and promote apoptosis of neuronal cells. BMC Neurosci 20(5):A954–A954

Nie CL, Wei Y, Chen XY, Liu YY, Dui W, Liu Y, Davies MC, Tendler SJ, He RQ (2007b) Formaldehyde at low concentration induces protein tau into globular amyloid-like aggregates in vitro and in vivo. PLoS One 2(7):e629

Paudel HK (1997a) Phosphorylation by neuronal cdc2-like protein kinase promotes dimerization of tau protein in vitro. J Biol Chem 272(45):28328–28334

Paudel HK (1997b) The regulatory Ser262 of microtubule-associated protein tau is phosphorylated by phosphorylase kinase. J Biol Chem 272(3):1777–1785

Pepys MB (2006) Amyloidosis. Annu Rev Med 57:223–241

Puchtler H, Meloan SN (1985) On the chemistry of formaldehyde fixation and its effects on immunohistochemical reactions. Histochemistry 82(3):201–204

Qu MH, Li H, Tian R, Nie CL, Liu Y, Han BS, He RQ (2004) Neuronal tau induces DNA conformational changes observed by atomic force microscopy. Neuroreport 15(18):2723–2727

Quist A, Doudevski I, Lin H, Azimova R, Ng D, Frangione B, Kagan B, Ghiso J, Lal R (2005) Amyloid ion channels: a common structural link for protein-misfolding disease. Proc Natl Acad Sci U S A 102(30):10427–10432

Roy KB (1996) DNA recognition and structural specificities. Indian J Biochem Biophys 33(2):83–87

Schweers O, Schonbrunn-Hanebeck E, Marx A, Mandelkow E (1994) Structural studies of tau protein and Alzheimer paired helical filaments show no evidence for beta-structure. J Biol Chem 269(39):24290–24297

Simonsson S, Samuelsson T, Elias P (1998) The herpes simplex virus type 1 origin binding protein. Specific recognition of phosphates and methyl groups defines the interacting surface for a monomeric DNA binding domain in the major groove of DNA. J Biol Chem 273(38):24633–24639

Sjoberg MK, Shestakova E, Mansuroglu Z, Maccioni RB, Bonnefoy E (2006) Tau protein binds to pericentromeric DNA: a putative role for nuclear tau in nucleolar organization. J Cell Sci 119(10):2025–2034

Sobue K, Tanaka T, Ashino N, Kakiuchi S (1985) Ca^{2+} and calmodulin regulate microtubule-associated protein-actin filament interaction in a flip-flop switch. Biochim Biophys Acta 845(3):366–372

Stefani M, Dobson CM (2003) Protein aggregation and aggregate toxicity: new insights into protein folding, misfolding diseases and biological evolution. J Mol Med 81(11):678–699

Sultan A, Nesslany F, Violet M, Bégard S, Loyens A, Talahari S, Mansuroglu Z, Marzin D, Sergeant N, Humez S, Colin M, Bonnefoy E, Buée L, Galas MC (2011) Nuclear tau, a key player in neuronal DNA protection. J Biol Chem 286(6):4566–4575

Tang Y, Kong X, Liu Z, Xu A, Lin W (2016a) Lysosome-targeted turn-on fluorescent probe for endogenous formaldehyde in living cells. Anal Chem 88(19):9359–9363

Tang Y, Kong X, Xu A, Dong B, Lin W (2016b) Development of a two-photon fluorescent probe for imaging of endogenous formaldehyde in living tissues. Angew Chem Int Ed Eng 55(10):3356–3359

Theis ER (1944) The protein-formaldehyde I. Collagen. J. Biol. Chem 154(1):87–97

Thurston VC, Zinkowski RP, Binder LI (1996) Tau as a nucleolar protein in human nonneural cells in vitro and in vivo. Chromosoma 105(1):20–30

Tian R, Nie CL, He RQ (2004) Chaperone-like manner of human neuronal tau towards lactate dehydrogenase. Neurochem Res 29(10):1863–1872

Tong Z, Zhang J, Luo W, Wang W, Li F, Li H, Luo H, Lu J, Zhou J, Wan Y, He R (2011) Urine formaldehyde level is inversely correlated to mini mental state examination scores in senile dementia. Neurobiol Aging 32(1):31–41

Tong Z, Han C, Luo W, Wang X, Li H, Luo H, Zhou J, Qi J, He R (2013) Accumulated hippocampal formaldehyde induces age-dependent memory decline. Age(dordr) 35(3):583–596

Tuite E, Sehlstedt U, Hagmar P, Nordén B, Takahashi M (1997) Effects of minor and major groove-binding drugs and intercalators on the DNA association of minor groove-binding proteins RecA and deoxyribonuclease I detected by flow linear dichroism. Eur J Biochem 243(1–2):482–492

Ukiyama E, Jancso-Radek A, Li B, Milos L, Zhang W, Phillips NB, Morikawa N, King CY, Chan G, Haqq CM, Radek JT, Poulat F, Donahoe PK, Weiss MA (2001) SRYand architectural gene regulation: the kinetic stability of a bent protein-DNA complex can regulate its transcriptional potency. Mol Endocrinol 15(3):363–377

Wang JZ, Grundke-Iqbal I, Iqbal K (1996) Glycosylation of microtubule-associated protein tau: an abnormal posttranslational modification in Alzheimer's disease. Nat Med 2(8):871–875

Wang Y, Loomis PA, Zinkowski RP, Binder LI (1993) A novel tau transcript in cultured human neuroblastoma cells expressing nuclear tau. J Cell Biol 121(2):257–267

Wang XS, Wang DL, Zhao J, Qu MH, Zhou XH, He HJ, He RQ (2006) The proline-rich domain and microtubule-binding domain of protein tau acting as RNA binding domains. Prot Pept Lett 13(7):679–685

Wei Y, Qu MH, Wang XS, Chen L, Wang DL, Liu Y, Hua Q, He RQ (2008) Binding to the minor groove of the double-strand, tau protein prevents DNA from damage by peroxidation. PLoS One 3(7):e2600

Weingarten MD, Lockwood AH, Hwo SY, Kirschner MW (1975) A protein factor essential for microtubule assembly. Proc Natl Acad Sci U S A 72(5):1858–1862

Wischik CM, Novak M, Thogersen HC, Edwards PC, Runswick MJ, Jakes R, Walker JE, Milstein C, Roth M, Klug A (1988) Isolation of a fragment of tau derived from the core of the paired helical filament of Alzheimer disease. Proc Natl Acad Sci U S A 85(12):4506–4510

Wyss-Coray T (2016) Ageing, neurodegeneration and brain rejuvenation. Nature 539(7628):180–186

Xu J, Zhang Y, Zeng L, Liu J, Kinsella JM, Sheng R (2016) A simple naphthalene-based fluorescent probe for high selective detection of formaldehyde in toffees and HeLa cells via aza-cope reaction. Talanta 160:645–652

Yamauchi PS, Purich DL (1993) Microtubule-associated protein interactions with actin filaments: evidence for differential behavior of neuronal MAP-2 and tau in the presence of phosphatidyl-inositol. Biochem Biophys Res Commun 190(3):710–715

Yang MF, Lu J, Miao JY, Rizak J, Yang JZ, Zhai RW, Zhou J, Qu JG, Wang JH, Yang SC, Ma YY, Hu XT, He RQ (2014a) Alzheimer's disease and methanol toxicity (Part 1): chronic methanol feeding led to memory impairments and tau hyperphosphorylation in mice. J Alzheimers Dis 41(4):1117–1129

Yang MF, Miao JY, Rizak J, Zhai RW, Wang ZB, Huma T, Li T, Zheng N, SH W, Zheng YW, Fan XN, Yang JZ, Wang JH, Yang SC, Ma YY, Lu L, He RQ, XT H (2014b) Alzheimer's disease and methanol toxicity (Pt.2): lessons from four rhesus macaques (Macaca mulatta) chronically fed methanol. J Alzheimers Dis 41(4):1131–1147

Yin DZ (1993) Lipofuscin-like fluorophores can result from reactions between oxidized ascorbic acid and glutamine. Carbonyl protein cross-linking may represent a common reaction in oxygen radical and glycosylation-related ageing processes. Mech Ageing Dev 62:35–46

Yu PH (1990) Oxidative deamination of aliphatic amines by rat aorta semicarbazide-sensitive amine oxidase. J Pharm Pharmacol 42(12):882–884

Yu PH (2001) Involvement of cerebrovascular semicarbazide-sensitive amine oxidase in the pathogenesis of Alzheimer's disease and vascular dementia. Med Hypotheses 57(2):175–179

Yu PH, Lai CT, Zuo DM (1997) Formation of formaldehyde from adrenaline in vivo: a potential risk factor for stress-related Angiopathy. Neurochem Res 22(5):615–620

Zhang XM, Lin YX, Eschmann NA, Zhou HG, Rauch JN, Hernandez I, Guzman E, Kosik KS, Han S (2017) RNA stores tau reversibly in complex coacervates. PLoS Biol 6 15(7):e2002183

Chapter 8
Cognitive Ability and Impairment Related to Formaldehyde

Rongqiao He

Abstract Recent studies have shown that dysmetabolism of formaldehyde may be one of the various patho-mechanisms involved in the onset and progression of age-related cognitive impairment including Alzheimer's disease (AD) and vascular dementia (VD). High endogenous formaldehyde levels could be a risk factor for cognitive impairment in the elderly population. Using a double-blind experiment, the correlation between endogenous (urine) formaldehyde and general cognitive abilities was estimated in a community-based elderly population (n = 604). Yu and her colleagues measured the general cognitive abilities of aged people using the Montreal Cognitive Assessment (MoCA) and then correlated them to concentrations of urine formaldehyde. Urine formaldehyde levels were inversely correlated with the MoCA scores of elderly participants. The concentration varied with demographic features: higher odds of a high formaldehyde level occurred among the less educated elderly population. Educational levels are strikingly correlated with the concentration of endogenous formaldehyde for the elderly living in local communities in Beijing. The epidemiological investigation showed that the average concentration of urine formaldehyde of the elderly increases with aging (>75 years old). In clinical study, the concentration of urine formaldehyde of AD patients is inversely correlated with their Mini-Mental State Examination (MMSE) scores. The old patients with the postoperative cognitive dysfunction (POCD) suffered from high levels of endogenous formaldehyde after their surgeries, but those without POCD did not. Monkeys fed with low concentration of methanol suffered from working memory loss. Levels of hippocampal formaldehyde from the patient autopsy are significantly higher than those of age-matched normal controls. Therefore, endogenous formaldehyde could be developed as a noninvasive marker for detection and monitoring of age-related cognitive impairment for both AD and VD, which could be identified with patients' history and further examinations.

Keywords Formaldehyde • Cognitive ability • Cognitive impairment • Metabolism • Dysmetabolism • Educational level • Consciousness

---

R. He (✉)
State Key Laboratory of Brain and Cognitive Science, Institute of Biophysics, Chinese Academy of Sciences, Beijing 100101, China

University of Chinese Academy of Sciences, Beijing 100049, China
e-mail: rongqiaohe@163.com

R. He, *Formaldehyde and Cognition*, DOI 10.1007/978-94-024-1177-5_8

1 Introduction

Alzheimer's disease (AD) is the most common form of senile dementia, affecting daily life of millions of aging people worldwide by deteriorating their working ability. Among many hypothetic pathomechanisms for age-related cognitive impairment, the hypothesis that "dysmetabolism of formaldehyde is involved in senile dementia" is gradually emerging in the progression of age-related cognitive impairment including AD and vascular dementia (VD). Formaldehyde levels in urine of the patients are inversely correlated to their cognitive scores in clinics (Li et al. 2016b; Tong et al. 2017). We hope more researchers are aware of the role of formaldehyde and can repeat and demonstrate the hypothesis with double-blinded assays now and in the future.

Formaldehyde, acting as a basic unit for many molecules in organisms (Pinto et al. 1980), is widely used in manufacturing/building materials and household products. Excessive formaldehyde exposure leads to severe health problems (Tang et al. 2009), including cognitive impairment (Marceaux et al. 2008; Shi and Wang 2013) such as defect in learning and memory (Flyvholm and Menne 1992; Perna et al. 2001). Intentional or unintentional uptake of excess formaldehyde or methanol (a precursor of formaldehyde) revealed marked decline in the neuropsychological integrity of participants (Chen et al. 2007; Maliutina and Taranenko 2014). At the biomolecular level, formaldehyde inactivates neuronal Tau (microtubule-associated protein in neuronal cells) and induces the protein aggregation (Hua and He 2002) to form cytotoxic aggregates (Nie et al. 2007a, b). Excess formaldehyde leads Tau protein to lose protection of DNA and to form stable aggregation as detected by SDS-PAGE (Wei et al. 2008). Further investigation performed on changes in Tau conformation in the presence of formaldehyde demonstrated a "hydrophobic core" inside the stable aggregation (Nie et al. 2005). This is because the aggregation shields the proteolytic sites, so that its cytotoxicity is not easy to be cleared by the proteases. Recently, Tang and colleagues have preliminarily studied the location of cellular formaldehyde with fluorescent probes and found formaldehyde in lysosome, endoplasmic reticulum, and Golgi apparatus (Tang et al. 2016; Jian and Zhu 2016). Chen and her coworkers discovered formaldehyde is accommodated and transported by lysosomes (Chen et al. 2017). Although the toxicity of exogenous formaldehyde to human health has been widely studied (Golden and Valentini 2014), the effects of endogenous formaldehyde on our brain structure and function remain to be mysterious, in particular of the relationship between endogenous formaldehyde and cognitive ability (He 2016a).

In humans, however, endogenous formaldehyde is present in cells such as astrocytes and neurons, as well as cell microenvironment such as cerebrospinal fluid (CSF), blood, and urine (Yu et al. 2003; Li et al. 2008). Human cells generate and degrade endogenous formaldehyde all the time (Shi et al. 2004). The concentration of endogenous formaldehyde tends to maintain equilibrium under physiological conditions. In other words, for normal people, his/her metabolic ability is effective and formaldehyde is quickly degraded to formic acid and then carbon dioxide and water. However, as aging or oxidative stress accumulating, the metabolism of endogenous formaldehyde gradually becomes imbalanced, resulting in chronic cognitive impairment (Tong et al. 2008, 2011a, b). In some sense, AD belongs to a dysmetabolic disease, for instance the glucose dysmetabolism (Crane et al. 2013). Formaldehyde might play a role in cognitive impairment when diabetes progresses and is complicated with dysmetabolism of the aldehyde (He 2016b). Accumulation of endogenous formaldehyde in the brain leads to neurodegenerative diseases (Chen et al. 2006; Miao and He 2012; Tong et al. 2013a; Wang et al. 2012; Chen et al. 2014). Furthermore, clinical investigations have shown that urine formaldehyde concentrations are significantly higher in patients with Alzheimer's disease (AD) than in unaffected aged adults (Tong et al. 2011b; Wang et al. 2010). In this chapter, we would like to discuss the effect of endogenous formaldehyde on the cognitive ability and the cognitive impairment.

2 Endogenous Formaldehyde and Cognitive Ability of Normal Aging Population

2.1 Elevation of Formaldehyde as Aging

So far, very few scientists have paid attention to the relationship between endogenous formaldehyde and cognitive function for normal population, although a considerable literature has been released on the detrimental effects of formaldehyde on DNA, proteins, neurons, respiratory system, brain, and so on. Tong and his coworkers performed an epidemiological investigation and observed changes in the concentration of endogenous formaldehyde as aging (Tong et al. 2013b).

The average concentration of endogenous formaldehyde gradually increases as aging (>75 years old), from either overproduction or declined degradation (Tong et al. 2013a). However, only until the age is over 80, the statistically significant difference in formaldehyde levels was observed in the participants compared with those of young adults. This seemed to be in consistence with the prevalence of AD which is about 2.2% for 65–69-year-old, 3.3% for 70–74-year-old, and 6.5% for 75–79-year-old (see the website: Dementia Guide). The prevalence of a small portion of dementia participants could not make statistically significant difference in concentrations of formaldehyde between aged and young adults. However, as aging, in the group of participants over 80-year-old, the AD prevalence becomes as high as 11.6%, making the correlation of age and concentration of formaldehyde conspicuous. Therefore, one question needs to be answered: whether a correlation between endogenous formaldehyde and cognitive abilities persists in relatively healthy older adult populations.

2.2 Cognitive Ability of Elderly Is Related to Endogenous Formaldehyde

The relationship between the concentration of endogenous formaldehyde and human cognitive ability has been studied in normal aging population (Yu et al. 2014). Since proteins interfere with formaldehyde detection, urine was analyzed instead of blood which is rich in serum protein. Compared with sampling blood, urine formaldehyde measurement has several advantages (Shao et al. 2011), including the noninvasive nature of sample collection and the presence of much fewer proteins. The correlation between urine formaldehyde and general cognitive abilities was estimated in a community-based elderly population in Beijing.

Yu and her colleagues recruited 604 participants and divided them into three groups based on the urine formaldehyde concentrations: low (≤8.00 μmol/L, $n = 192$), medium (8.00–13.20 μmol/L, $n = 200$), and high (≥13.20 μmol/L, $n = 212$) (Yu et al. 2014). To clarify the relationship between endogenous formaldehyde and cognition, they analyzed data from previous work (Yu et al. 2012). Using the MoCA (Montreal Cognitive Assessment) test, they found that the cognitive abilities decreased when formaldehyde levels increased [$F(2, 597) = 11.17$, $P < 0.001$]. Post hoc Newman-Keuls comparisons showed markedly higher MoCA scores in the group with low concentration of formaldehyde than those in the group with medium or high formaldehyde concentrations ($P < 0.05$). That is to say, levels of endogenous formaldehyde are indeed correlated to the cognitive abilities of the aging participants from different communities in Beijing. Besides educational levels, Yu and her colleagues studied the extent and direction in which the correlation to demographic characteristics such as age, gender, and residential region was measured.

2.3 Educational Levels and Endogenous Formaldehyde

Yu and her collaborators analyzed the concentrations of urine formaldehyde with the educational levels (years) for the aging participants in their youth. The endogenous formaldehyde concentrations [$F(2, 601) = 1134.27$, $P < 0.001$] and educational years [$F(2, 597) = 27.05$, $P < 0.001$] were significantly different among the groups (Yu et al. 2014). The group with low concentration of formaldehyde had more educational years than those with the medium or high concentrations of formaldehyde ($P < 0.05$) by using post hoc Newman-Keuls comparisons. Residential regions were different among the three groups ($\chi^2 = 154.84$, $df = 4$, $P < 0.001$). Regarding the residential regions, the group with low concentrations of formaldehyde had the highest proportion of participants living in the new town (91.67%). In other words, those with the medium and high concentrations of formaldehyde were less educated and more likely to live in the old town or rural area, but they had comparable distributions of age and gender with the low group. Note that each group had similar numbers of participants to minimize statistical errors. Sample

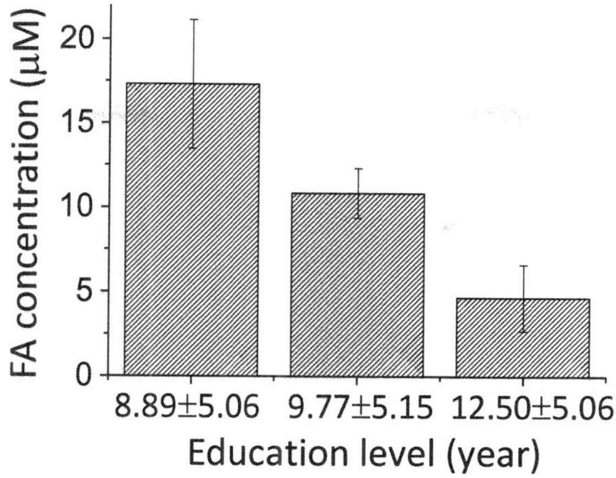

Fig. 8.1 The correlation between urine formaldehyde concentrations and educational levels. The participants were divided into three groups based on the urine formaldehyde concentrations. Their educational years were statistically compared with the endogenous formaldehyde concentrations. The higher the concentrations of formaldehyde is, the less the educational years. *FA* formaldehyde (Source: Adapted from Yu et al. 2014)

characteristics suggest the correlation between the cognitive ability and endogenous formaldehyde of the old participants is related with their educational levels in youth (Yu et al. 2014) (Fig. 8.1).

Wilson and colleagues have studied the educational attainment and cognitive decline in old age. Education is robustly associated with level of cognitive function but not with rate of cognitive decline (Wilson et al. 2009). In Yu and her colleagues' investigation, by controlling age and gender, the MoCA scores were positively correlated ($r = 0.581$, $P < 0.001$) with educational years (Yu et al. 2014). Shpanskaya and colleagues have observed educational attainment and hippocampal atrophy in the AD neuroimaging initiative cohort. A potential protective association between higher education and lower hippocampal atrophy in patients with AD appears consistent with prior epidemiologic data linking higher education levels with lower rates of incident dementia (Shpanskaya et al. 2014). Longitudinal studies are warranted to confirm their findings. Partial correlation analysis also showed the inverse correlation of the urine formaldehyde concentration with the MoCA scores ($r = -0.191$, $P < 0.001$) and educational years ($r = -0.288$, $P < 0.001$) under the same conditions. To clarify whether the demographic variables, education and living region influence the correlation between urine formaldehyde and cognitive ability, Yu and her colleagues performed univariate analysis, but they could not find any significant interactions between these variables in the whole sample. Furthermore, they employed linear regression analysis to assess the direction and strength of the association between these correlates and general cognitive abilities (Yu et al. 2014). The variable urine formaldehyde concentration was no longer significant after education was included in the models, indicating a mediating effect of education on the association between urine formaldehyde concentration and cognitive ability.

As described by Yu and her coworkers, the negative effect of elevated concentrations of formaldehyde could not be apparently observed on the cognitive abilities for the participants with higher educational levels (Yu et al. 2014). This is to say, education is able to counteract the negative effects of endogenous formaldehyde on neural connections or plasticity in subsample analysis. These data are coincident with the "scaffolding theory of aging and cognition" as described by Park and Reuter-Lorenz (2009). In their statement, scaffolding is a normal process involving the use and development of complementary, alternative neural circuits to achieve a particular cognitive goal, and extra learning, cognitive training, and exercise promote or repair the scaffolding during the aging process. Higher educational levels are most probably resulted from more learning, more engagement in intellectual activities, and more participation in cognitive training. More education is able to counteract the decline of cognitive abilities caused by elevated endogenous formaldehyde and to rescue the scaffolding noted here as neuroplasticity. Therefore, education should be one of the protective factors against the abnormal accumulation of formaldehyde and the corresponding cognitive decline. Given the broad spectrum of age-related cognitive decline, it is unlikely that a single biochemical mechanism can fully explain the effect across all individuals (Li et al. 2012). However, the metabolism of endogenous formaldehyde is not only involved in an exclusive pathway (see Chap. 2) but in many biological pathways such as oxidative stress (Kum et al. 2007; Evans et al. 2016), methanol (Li et al. 2008), one-carbon cycle (Goldberg and Mateles 1975), and methylation and demethylation of DNA, RNA, and histones (Su et al. 2015). Dysmetabolism of endogenous formaldehyde could affect a broad spectrum of metabolism in vivo. This is why dysmetabolism of endogenous formaldehyde is related to sporadic age-related cognitive impairment. Furthermore, well-controlled cellular experiments and clinical studies have provided clearer results, suggesting that urine formaldehyde could be employed as a noninvasive marker for age-related cognitive impairment in patients with AD and/or VD (Tong et al. 2017).

3 Endogenous Formaldehyde and Cognitive Impairment

3.1 Formaldehyde Involved in Postoperative Cognitive Dysfunction

More than 10 years ago, my sister-in-law Ms. Yun Feng worked as a nurse supervisor in the operating room in the hospital of Sichuan Province, China. She told me that some of elderly patients could get dementia after a major long-term operation. For example, an old woman has got her femoral head replaced with the artificial one by arthroplasty. Her cognition was impaired after the operation. As described by Moller and colleagues, this is called the postoperative cognitive dysfunction (POCD), a decline in cognitive function following surgical procedures, a matter of concern for anesthesiologists and patients (Moller et al. 1998). POCD was present in 25.8% of patients 1 week after surgery and in 9.9% 3 months after surgery. The cognitive dysfunction is usually exhibited by patients in terms of failure to perform simple cognitive or mental tasks that were previously easily attainable (Hanning, 2015). POCD is commonly seen in the elderly, but the etiology is not clear. A number of perioperative factors have been implicated, but high-confidence evidence for cause, prevention, or treatment is still lacking (Brown and Deiner 2016). Studies on POCD have focused on the impact of increasing age or poor education, as well as perioperative factors including the types of surgery and anesthesia (Monk et al. 2008), hypotension during surgery (Newman et al. 2007), and postoperative complications (Moller et al. 1998). So far, the relationship between Alzheimer's disease (AD) and susceptibility to or exacerbation of POCD is not well understood (Arora et al. 2014), though AD has been thought to be associated with POCD (Xie and Tanzi 2006).

Operation is a striking cause to trigger stress on the patient. Formaldehyde plays a role in the oxidative stress (He et al. 2010). It was a hypothesis that the metabolism of endogenous formaldehyde could be changed during perioperation and postoperation. In addition, no useful extracerebral diagnostic biomarkers have yet been identified as indicators of POCD though aging, duration of anesthesia, postoperative infections, and respiratory complications were regarded as the risk factors for POCD. Therefore, Wang and her collaborators measured the concentrations of endogenous (urine) formaldehyde of the patients before and after operation to find out whether formaldehyde metabolism responded to operation (Wang et al. 2012).

Ninety-five patients (ages 65–80 years) were enrolled and scheduled for major orthopedic or abdominal surgery. Twenty-two young patients (ages 20–40 years) were recruited undergoing the same procedures as controls (Wang et al. 2012). To determine the presence of POCD, all participants received neuropsychological tests 1 day prior and 1 week post procedure, in which the criteria were employed as described (Moller et al. 1998; Monk et al. 2008). As described (Su et al. 2011), midstream of morning urine were sampled 1 day before surgery and on day 1, day 2, and day 7 postoperation. Wang and her colleagues observed that concentrations of urine formaldehyde of all participants either with or without POCD were markedly elevated on the first 2 days after surgery. On day 7, the average concentration of formaldehyde in patients with POCD was still maintained significantly high compared with that in patients without POCD ($P < 0.01$). This suggested that dysmetabolism of endogenous formaldehyde was related to postoperative cognitive dysfunction.

In contrast, the young participants did not show cognitive impairment after surgery, and thus none of them were diagnosed with POCD. As has been mentioned above (Tong et al. 2013a), the average concentration of endogenous formaldehyde was gradually getting higher as aging (>75 years old), showing a decline of the metabolic ability in formaldehyde as aging. Young participants had a high efficiency in metabolism of formaldehyde. Therefore, the concentration of urine formaldehyde of the young participants was significantly lower than that in the elderly ($P < 0.05$), although their formaldehyde levels in perioperative period were similar to the elderly without POCD. In other words, the patients with POCD had high concentrations of urine formaldehyde not only in the perioperative period but also on day 7 after surgery. The concentration of urine formaldehyde is suggested to be used as a biomarker along with neuropsychological assessments to assist the diagnosis of POCD (Wang et al. 2012).

3.2 *Urine Formaldehyde Is Positively Correlated with the Cognitive Impairment of Alzheimer's Patients*

According to the "formaldehyde stress" hypothesis (He et al. 2010), chronic dysmetabolism of endogenous formaldehyde can cause abnormal changes in protein modification and conformation, resulting in reduced long-term potentiation (Tong et al. 2013a), Tau aggregation (Hua and He 2002) and hyperphosphorylation (Lu et al. 2013a, b), amyloid β deposits (Yang et al. 2014a, b), DNA damage (Lu et al. 2013a, b), and even cell death (Szende and Tyihak 2010), followed by associated cognitive disability. The association of endogenous formaldehyde with the severity of cognitive impairment was found in age-related dementia patients (Tong et al. 2011a). The ratio of elevation of urine formaldehyde was approximately ~40% among the patients who were investigated in small samples with either AD or VD.

Li and her collaborators recruited 62 AD participants (81.05 ± 6.36 years old) from the Beijing Geriatric Hospital and 69 normal participants (67.33 ± 6.19 years old) from local community to investigate the relation of urine formaldehyde with cognitive impairment (Li et al. 2016b). The HPLC coupled with DNPH was also used to determine the concentration of formaldehyde (Su et al. 2011). The average concentration of urine formaldehyde (14.27 ± 5.15 μM for male; 13.76 ± 5.12 μM for female) of the AD patients were dramatically higher ($P < 0.001$) than those (10.18 ± 2.17 for male; 9.19 ± 3.28 for female) of the participants with normal cognitive scores. Consequently, 70–80-year-old participants were selected, and the concentrations of urine formaldehyde were compared between the patients and participants with normal cognitive ability. Although the age factor was excluded, the data also showed the same trend (Tong et al. 2013a), namely, urine formaldehyde levels increased not only with aging (>75 years old) but also with senile dementia (Li et al. 2016b).

To further investigate the relationship between endogenous formaldehyde and cognitive impairment, concentration of urine formaldehyde was measured, and cognitive impairment was assessed. To increase the sensitivity of the detection, HPLC coupled with a fluorescent reagent of ampicillin (Luo et al. 2001) was employed to determine urine formaldehyde (Tong et al. 2011b). In the double-blind investigation with the Beijing Geriatric Hospital and Shantou University Medical College, Guangdong, a significant increase of urine formaldehyde levels in patients ($n = 91$) of AD and an increase but not significantly in patients ($n = 50$) with mild cognitive impairment (MCI) were observed. The concentration of urine formaldehyde is inversely correlated with the scores of Mini-Mental State Examination (Rs = −0.441, $P < 0.0001$). This demonstrated urine formaldehyde could be used as a noninvasive biomarker for investigation, diagnosis, and monitoring of the efficacy of medicine for the patients with age-related cognitive impairment (Tong et al. 2011a, 2017).

8.3 Approximately 40% AD Patients with High Levels of Formaldehyde

Let us reconsider the investigation data from previous work. The conditions for recruitment of AD patients and age-matched participants with normal cognition were as described (Li et al. 2016b). The scattering diagram of the relation between MMSE scores and concentrations of urine formaldehyde for normal participants (*panel a'*, Fig. 8.2) shows approximately 9.26% of them are with high concentration of formaldehyde, much less than AD patients (41.38%) (*panel a*, Fig. 8.2). Similar ratios can be observed in analysis of the data by based on the gender of participants (panel b,b',c,c', Fig. 8.2). According to these data, dysmetabolism of endogenous formaldehyde cannot cover all AD patients. This is reasonable because the etiology of AD is so complicated that no single factor could explain the whole pathomechanisms, although formaldehyde hypothesis covers a small part of them. Approximately 9–10% of the age-matched participants as control showed their concentration of urine formaldehyde is higher than the normal. What can we do for the normal participants with high levels of formaldehyde? It is suggested to perform a tracking investigation for them to see whether high levels of formaldehyde in the participants are due to AD or some other causes.

Let us further consider the relation between the endogenous formaldehyde and cognitive ability. The distribution of the data is markedly different between the AD patients and participants with normal cognitive ability. Approximately 40% of the

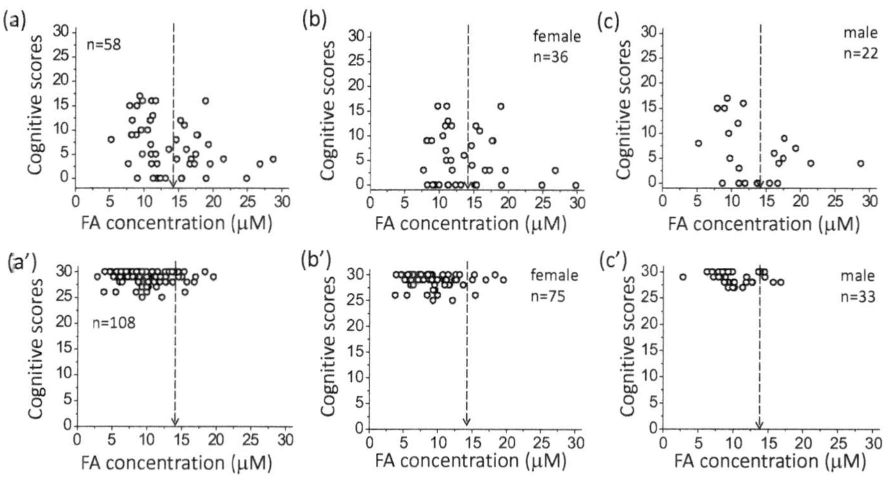

Fig. 8.2 The correlation between urine formaldehyde and cognitive scores. The conditions for recruit of AD patients and normal participants were as described by Li and her colleagues (Li et al. 2016b). The MMSE scores and concentrations of urine formaldehyde were plotted in scattering points from AD patients (panel **a**) and normal participants (panel **a'**). The MMSE scores and concentrations of urine formaldehyde were plotted in scattering points from AD patients of female (panel **b**) and male (panel **c**) as well as normal participants of female (panel **b'**) and male (panel **c'**). The dash line represents the average concentration of formaldehyde for AD patients

total AD patients are with high levels of endogenous formaldehyde, either female (panels b,b', Fig. 8.2) or male (panels c,c′, Fig. 8.2) are. In other words, ~60% AD patients were not related to the dysmetabolism of endogenous formaldehyde. They may suffer from other genetic or environmental factors. In short, a large sample (at least 3000 cases) analysis should be carried out to clarify the relationship between endogenous formaldehyde and cognitive impairment in the future.

4 Other Evidence for Endogenous Formaldehyde Relates to Cognitive Impairment

4.1 Evidence from Primates

Tong and his colleagues analyzed the formaldehyde in the autopsy hippocampus from AD patients. As shown in their results, the level of the hippocampal formaldehyde is significantly ($n = 8$, $P < 0.05$) higher than that of age-matched normal controls (n = 9). This appears that imbalance of formaldehyde metabolism occurs in the brain of AD patients. Note that the autopsy hippocampus tissues were provided by the Netherlands Brain Bank (Tong et al. 2015).

Yang and her coworkers administered low concentrations of methanol in drinking water (an intermediate product from formaldehyde) (Dorokhov et al. 2015). This led monkeys to lose working memory in 2–3 months during the administration. Both senile plaque and Tau hyperphosphorylation could be observed in monkey brain hippocampus and cortex, while the administration was prolonged to 2–3 years (Yang et al. 2014a).

4.2 Evidence from Rodents

4.2.1 Formaldehyde Dysmetabolism Related to Animal Cognitive Impairment

The concentration of brain formaldehyde for 10-month-old mice (C57BL/6J) was significantly higher ($n = 8$, $P < 0.01$) than that for 3-month-old mice (Li et al. 2016b). Determination of hippocampal formaldehyde in SD rats (1-, 6-, 12-, and 24-month-old) showed the concentration was getting higher as the rats aged (Tong et al. 2013a). Formaldehyde in the brain of APP/PS1-transgenic mice elevated at early stage (Tong et al. 2011a, b). Liu and his colleagues found that 7-month-old APP/PS1 mice have got elevated levels of formaldehyde in the cecum microbiota compared with age-matched C57BL/6J wild-type mice (Hu et al. 2016; Liu et al. 2017). All these data demonstrated that dysmetabolism of endogenous formaldehyde is associated with age-related cognitive impairment.

Yang and her collaborators also fed mice (8-week-old male ICR mice from the Kunming Medical College, Yunnan, China) with methanol under the similar conditions. The mice, however, did not show senile plaques but exhibited Tau hyperphosphorylation in hippocampus and memory loss (Yang et al. 2014b). Intraperitoneal injection of formaldehyde induces memory loss and learning retardation. For example, according to the level of formaldehyde in the brain of the transgenic mice, normal mice were injected with formaldehyde (0.5 mM, intraperitoneal administration), and memory loss was observed in Morris water maze test (Yang et al. 2014a). Li and her colleagues administered 4% NaCl in C57BL/6J mice for 3 months and found an elevation of brain formaldehyde accompanied with decline of learning in "shuttle box" assay (Li et al. 2016c).

4.2.2 Dysmetabolism of Formaldehyde in the Aging Mouse

The senescence-accelerated mouse prone 8 (SAMP8) has deficiency in learning and memory and appears to be an excellent model for blood-brain barrier damage in vascular dementia (Ueno et al. 2016). SAMP8 is a model to study age-related cognitive decline with relevance to alterations of the gene expression and protein abnormalities in Alzheimer's disease (Butterfield & Poon, 2005). It is also used as a model to develop therapeutic interventions for AD (Morley et al. 2012). Senescence-accelerated-resistant mice 1 (SAMR1) which did not have distinct cognitive disorders (Butterfield & Poon, 2005) were employed as a control for SAMP8 in cognitive impairment study. To test the metabolism of endogenous formaldehyde in the brain, Qiang and her colleagues measured the levels of formaldehyde in the SAMP8 brain and the enzymes for formaldehyde metabolism. In order to study the relation of endogenous formaldehyde with cognitive impairment for SAMP8, 3-month-old SAMP8 was evaluated of their spatial learning and memory ability by performing the Morris water maze test. The concentrations of formaldehyde in the brain, liver, and kidney were measured, respectively, at the same time. The concentrations of formaldehyde in the brain and liver of SAMP8 were significantly higher than those in age-matched SAMR1. The concentration gradually increased from 1 month up to 10 months of age (Qiang et al. 2014). These data demonstrated that dysmetabolism of formaldehyde may occur with age in mice.

To investigate the dysfunction of formaldehyde metabolism, Qiang and her colleagues analyzed enzymes functioning in synthesis or degradation of formaldehyde in the brain of SAMP8. They measured the activity and expression of an anabolic enzyme (semicarbazide sensitive amine oxidase, SSAO) and a catabolic enzyme (alcohol dehydrogenase III, ADH3). They employed real-time PCR, Western blotting, enzyme assays, and immunohistochemistry techniques in the analysis. In the brain of 3-month-old SAMP8, the expression of SSAO significantly increased, along with decreased in mRNA, protein, and enzyme activity of ADH3, compared with age-matched SAMR1. The imbalance of these metabolic enzymes may represent a causal explanation for the observed formaldehyde elevation in the SAMP8 brain. Such increase could be responsible for the observed Tau hyperphosphorylation and aggregation (see Chap. 9), ultimately leading to cognitive impairment (Lu et al. 2013a, b). These data again indicate that cognitive impairment is related to dysmetabolism of endogenous formaldehyde in aging mice.

Furthermore, the SAMP8 is thought as an excellent model for age-related cardiovascular alterations (Karuppagounder et al. 2017). Microcirculation dysfunction plays an important role in hyperphosphorylation of Tau, Aβ aggregation, and cognitive impairment induced by accumulation of formaldehyde (Chen and Su 2015). Dysfunction of blood vessels is involved in age-related cognitive impairment. For example, the microcirculation disorder is involved in AD, and stroke, blockage of blood vessel, or bleeding in the brain may contribute to VD. Microcirculation dysfunction is regarded as one of the most important pathologies such as senility, degeneration, and immune disorders. Cerebral circulation insufficiency, energetic dysmetabolism, hypoxia-ischemia, and metabolite accumulation in AD and VD have a close relation to microcirculation dysfunction. Taken together, this study gives new insights into the role of metabolic enzymes in age-related accumulation of formaldehyde and thus the establishment of neurodegenerative diseases such as AD and VD.

4.3 Evidence from Drosophila

To study the relation between formaldehyde dehydrogenase (FAD, homolog of mammalian GSNOR) and memory, Hou and colleagues overexpressed FDH in the central nervous system and found defects of the visual pattern memory of *Drosophila* (Hou et al. 2011). The role of FDH in learning and memory was independent upon development. The memory defect induced by FDH overexpression was observed in the fan-shaped body but not in the mushroom body. The visual pattern memory defect could be rescued by co-expression with exogenous cGMP-dependent protein kinase (PKG) (Hou et al. 2011). Their work supports not only NO-cGMP-PKG pathway but also role of formaldehyde metabolism in learning and memory. Mice with ADH3 gene knock-in (overexpression) showed a slowness in its learning and memory in Morris water maze test (our unpublished data). Reducing the levels of formaldehyde in normal *Drosophila* could also disturb the balance of formaldehyde metabolism, leading to memory decline. Further investigation should be carried out to clarify whether overexpression of FDH interferes the metabolism of formaldehyde or affects the NO-cGMP-PKG pathway (Wu et al. 2014).

Furthermore, formaldehyde improves the proliferation of N2a cells when its concentration is lower than ~0.01 mM, but suppresses the proliferation when it is higher than 0.05 mM (Wang et al. 2017). Formaldehyde in excess impaired the processes of N2a cells and neurites of primary cultured rat hippocampal neurons (Lu et al. 2013a, b). However, removal of formaldehyde markedly rescued and regenerated the processes of N2a cells (Yu et al. 2014). In fact, cells produce formaldehyde by themselves and the concentration is about 5 μM in the culture medium (Wang et al. 2017). These results demonstrated a negative correlation between the

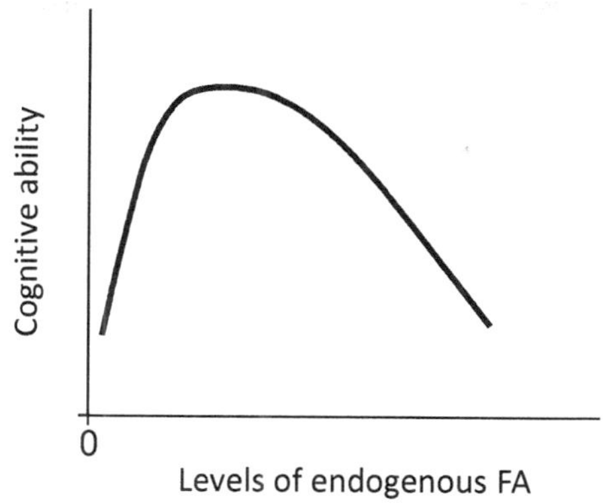

Fig. 8.3 A putative relationship between endogenous formaldehyde and cognitive ability or impairment (diagrammatic drawing). We supposed that excess or lack of formaldehyde resulting from imbalance of metabolism could lead to dysfunction in cognitive ability, for instance, learning and memory. It has been accepted that excess formaldehyde in the brain impairs cognitive ability. However, overexpression of ADH5 acting as scavenger of formaldehyde induces cognitive impairment for *Drosophila* though the formaldehyde dehydrogenase is associated with NO metabolic pathway (Hou et al. 2011; Wu et al. 2014). Similarly, the life span of *Drosophila* also shows in a formaldehyde-dependent manner: a slight increase of formaldehyde extends life span and a marked increase of formaldehyde shortens the life span (see Chap. 11; Li and He 2016)

endogenous formaldehyde and general cognitive abilities. In other words, excess or deficiency of formaldehyde resulted from imbalance of metabolism could lead to decline of cognitive ability as shown in the diagram (Fig. 8.3).

5 Exogenous Formaldehyde Caused Cognitive Impairment

In the winter of 1993, a young chemical engineer and friend of mine happened to be exposed to the steam containing formaldehyde when he and coworkers were working on the synthesis of polyvinyl formal adhesive in a workshop in a far suburb of Beijing. In the pilot plant experiment, he was busy to monitor and control the polymerizing reaction between formaldehyde and polyvinyl alcohol in a 2-ton reactor. Two days after exposure to the steam with formaldehyde, his memory declined. I was deeply concerned about his resultant symptom and health. I attempted to understand why formaldehyde exposure could induce memory decline. After that, I got down to investigate the relationship between formaldehyde and cognitive impairment.

Over 45 years ago, Karapetian reported the cases of nerve system lesions of workers who were exposed to organic chemicals including formaldehyde (Karapetian 1971). Kilburn and coworkers recruited 305 histology technicians and studied them by regression analysis with age, years of cigarette smoking, and hours per day exposure to formaldehyde and other solvents as major independent variables (Kilburn et al. 1987). The results showed that increased daily hours of exposure to formaldehyde were positively correlated with failure in cognitive tests such as reduced performance on visual memory and digit span. More studies of the effects of exogenous formaldehyde on the central nerve system were carried out by the group in the next few years (Kilburn 1994). In 1990, a multispecialty panel of physicians evaluated a case series of 53 composite-materials (including formaldehyde) workers in a large aircraft manufacturing facility who filed workers' compensation claims for illness labeled by the media as the "aerospace syndrome" (Sparks et al. 1990). In the next year, Morton and Feldstein (1991) emphasized the severity of cognitive impairment caused by exposure to formaldehyde.

Epidemiological investigation and animal experiments all show that formaldehyde exposure has a relationship with memory and learning decline. Hawkins and collaborators reported the cases of two subjects persisting memory difficulties following inhalation of formaldehyde-soaked marijuana (Hawkins et al. 1994). Increased daily hours of exposure to formaldehyde were correlated with reduced performance on visual memory, story memory, digit span, pegboard, and sharpened Romberg, as well as with errors on trails (Kilburn et al. 1987). Because of the exposure to low-level neurotoxins mixed with formaldehyde, toluene, and acetone in nail studios, the workers had neurologic complaints as well as perceived problems with cognitive efficiency, memory, and learning (LoSasso et al. 2001). People exposed to indoor air after building renovation or in manufactured homes had neurobehavioral and neuropsychological deficiency though less severe but similar to those with occupational formaldehyde exposures (Kilburn 2000). Formaldehyde exposure in animal experiments showed psychotic disorders accompanied with pathologic and molecular changes of brains (Li et al. 2016a, b, c; Mei et al. 2016). Formaldehyde exposure impaired mice in spatial memory associated with hippocampal neuronal death and decrease of brain melatonin concentrations which is an antidepressant drug in clinic (Mei et al. 2016). Another study has shown that exposure to formaldehyde at 2.6–4.6 ppm inhibited the learning process of rats which were trained in an operant conditioning task, and the data also indicated the possible neurotoxicity of formaldehyde exposures (Pitten et al. 2000). A formaldehyde concentration of 0.5 ppm and above significantly affected the locomotor behavior of adult male and female rats in the open field. After inhaling formaldehyde at 5–6 ppm for 2 h/day and a total of 10 days, the rats used significantly longer time to find the target in the water maze experiment than controls, while the histopathology changes were not observed (Malek et al. 2003). Inhalation of 2 ppm formaldehyde reduced body weight, increased levels of depression-like behavior, and impaired novel object recognition in mice (Li et al. 2016a, b, c). Inhaling 2.44 ppm of gaseous formaldehyde negatively affected learning and memory in mice but lower levels of formaldehyde did not (Lu et al. 2008).

So far, it has been widely known that exogenous formaldehyde exposure induces human cognitive impairment and animal memory loss. Inhalation of exogenous formaldehyde is different from the effect of the endogenous formaldehyde on human health because the inhaled steam goes through our respiratory system and then dissolves in blood. It may damage human respiratory and blood compositions and even induces cancers (Duong et al. 2011). Acute formaldehyde exposure at

5 ppm results in decreased motor activity and increased levels of dopamine (DA) together with its enzymatic metabolites in the hypothalamus of rats (Boja et al. 1985). Furthermore, exposed to the air containing formaldehyde (10.2 ppm, 4 h daily) for 7 days may impair olfactory bulb and hippocampus in rats (Li et al. 2010, 2015, 2016a). Zhang and colleagues reported the impairment of olfactory function and change of SNAP25 proteins in olfactory bulb under repeated formaldehyde inhalation (Zhang et al. 2014; Foveau et al. 2016). In fact, the olfactory impairment (Hu et al. 2017) and odor identification deficit in mild cognitive impairment and AD may be associated with hippocampal and deep gray matter atrophy (Hagemeier et al. 2016). The formaldehyde-induced brain impairment through inhalation should be one of the risk factors related with central nervous system damage and cognitive impairment.

6 A Biomarker for Age-Related Cognitive Impairment Including AD and VD

Tong and his colleagues showed the elevated concentrations of formaldehyde in patients who had got diabetes or hypertension with cognitive impairment. But the formaldehyde elevation could not be observed among the patients without dementia symptoms (Tong et al. 2011b). Chronic inflammation is involved in the progression of both AD (Sawikr et al. 2017) and VD (Rosenberg 2017). SSAO also acts as a factor related to inflammation (our unpublished data). As described by Chen and her collaborators, formaldehyde actually acts as a biomarker for inflammation in the oxidative stress (Chen et al. 2017). Thus, Tong and his coworkers have investigated the relationship between morning urine formaldehyde concentration and cognitive impairment in patients with post-stroke dementia (PSD) or AD in this cross-sectional survey for 7 years (Tong et al. 2017).

They assessed the participants' cognitive abilities (n = 577) by MMSE and divided them into four groups: control (n = 231), stroke (n = 61), PSD (n = 65), and AD (n = 220). Gender- and age-matched participants (n = 42 in each group) were employed from the four groups above. Using an enzyme-linked immunosorbent assay (ELISA) and HPLC, they analyzed the expression levels of SSAO and concentrations of formaldehyde in the blood and urine, respectively. It was demonstrated again that the morning urine formaldehyde levels are inversely correlated with MMSE scores (Tong et al. 2011b). An increase in the concentration of urine formaldehyde could be observed, which may be resulted from upregulation of SSAO in blood. The threshold value (the best cut-off value) of the formaldehyde concentration for predicting cognitive impairment was 0.0418 mM in patients with PSD (sensitivity, 92.3%; specificity, 77.1%) and 0.0449 mM in those with AD (sensitivity, 94.1%; specificity, 81.8%) (Tong et al. 2017). This is to say, the overnight fasting urine formaldehyde can be used as a potentially noninvasive biomarker to evaluate the likelihood of ensuing cognitive impairment for AD and VD. VD could be identified by the patient's history who had suffered from stroke before. Also VD could be distinguished by brain imaging including magnetic resonance imaging (MRI)-based cerebral lesion location and its association with cognitive decline (Quinn and McCleery 2017). The feature of formaldehyde as a biomarker is dependent upon its role in the inflammation and oxidative stress, when SSAO is activated. That is to say, elevation of endogenous formaldehyde could remind us that there occurs inflammation and oxidative stress in the aging brain.

7 Consciousness Impairment and Formaldehyde

7.1 Definition of Consciousness

Consciousness can be described as "a basic brain function that an individual perceives the existence (reality) of oneself and one's environment" (He 2015a). In other words, consciousness works for a person to realize the existence of himself or herself under wakefulness. The perception of the existence is the brain fundamental function for attention, learning, memory, thinking, cognition, etc., because these activities are performed in the presence of the consciousness (He 2015b).

7.2 Consciousness Impairment

My mother (Xiao-Wen He, born on 09/Mar/1917, a tailor) suffered from age-related cognitive impairment. She (aged 91, 26/June/2008) died of heart attack with hypertension. After she got senile dementia, the concentrations of her urine formaldehyde was high (17.12 ± 3.92 μM) analyzed with HPLC coupled DNPH, compared with that of normal control (9.61 ± 2.90 μM) (Li et al. 2016b; Su et al. 2011). This suggested a possible relation between the dysmetabolism of endogenous formaldehyde and her dementia. I lived with her in one apartment as she gradually became dementia. I had observed some of her behaviors that might be related to consciousness impairment in her last life.

She had suffered from loss of memory without dyskinesia symptoms. She could not recognize my eldest brother first, then my second elder brother, and myself at last. I had to employ a family nurse to take care of her. At early stage, she occasionally forgot switching off running water or natural gas in the kitchen. Later, her symptoms became worse. Fortunately, our apartment was not on fire because no one smoked in my family, but she flooded the apartment of our neighbors downstairs with running water. Apart from loss of memory, my mother had got spatial cognitive impairment and usually lost her way home. I had to look for her when she got lost. One day, she was lost again in the street in Zhongguancun in Haidian district (Beijing). I rode a bike, looked for her, and finally found her. On the way home, I asked her, "Where are you going?" She answered, "I am going home." I said, "You may follow a different way home (in Beijing)." She went on, "I know where my

Fig. 8.4 Wrongly perceiving the reality for oneself spatial position in the environment. Consciousness is described, "One of the brain basic functions that an individual perceives the reality (existence) of oneself and one's environment" (He 2015a). The perception of the reality is the brain fundamental function for attention, learning, memory, etc., because these activities are performed in the presence of the consciousness (He 2015b). An age-related cognitively impaired patient who was in the Haidian district of Beijing at the present was looking for the way home at the Goli Gate in Feng Town because she had a wrong perception of the reality of her spatial position and took Haidian district for Feng Town about ~830 kilometers far away in Liaoning Province. Summer Palace in Beijing is shown in the direction indicator

home is. But the way to "Goli Gate" disappeared!" I continued, "This is Beijing." She responded, "It's Feng Town here."

Then I understood why she lost in Beijing and could not find home. Where is the Goli Gate? The gate is in the Feng Town in Liaoning province ~830 kilometers away from Beijing! She in person was in Beijing, but she tried to find the way home in her childhood near the Goli Gate. She could know the way home in the Feng Town in her childhood. She could describe the way to the Goli Gate. In other words, the way was still in her mind. The problem was that she had wrongly perceived the way in Beijing at the present for home of her childhood in the Feng Town. At that moment, in her mind, she herself was living in her childhood in the Feng Town. Furthermore, the Goli Gate in Feng Town is distinctly different from Haidian district. The former is surrounded by villages and cropland, and the latter is modernized scientific and technological city. This could not be a problem of memory loss but a wrong perception for the reality, suggesting her consciousness impairment (Fig. 8.4). Since consciousness is the fundamental function of the brain, consciousness impairment inevitably affects learning and memory.

I am concerned about whether dysmetabolism of formaldehyde affects human's consciousness. At least, we can see, a deeply drunk man with alcohol (its metabolite is acetaldehyde) shows a transient consciousness disorders. My mother had got high levels of endogenous formaldehyde, which was stronger toxic than acetaldehyde. Mice injected with formaldehyde showed spatial disorder that is related with memory loss as mentioned above (Tong et al. 2013b). We did not have any evidence to support this hypothesis yet. Therefore, it is far-fetched to say dysmetabolism of formaldehyde is related to consciousness impairment, which needs to be further investigated and clarified.

7.3 Consciousness Impairment Occurring at Late Stage of AD?

The famous painter Utermohlen's self-portraits showed his perception of himself at different stages of AD (Harrison 2013). Harrison used AD as an exemplar of suffering and described a creative teaching strategy using nine self-portraits that chronicle American-born artist William Utermohlen's deterioration from the disease. This is an extremely great valuable case to describe an AD patient's recognition and perception of himself. Along with the progression of his disease, he gradually lost his recognition and perception of himself, his reality (existence). Eventually, he lost the consciousness on himself.

Auguste Deter was the first patient who had been diagnosed with AD (Alzheimer 1906). Besides memory loss, for instance, she nagged, "I have lost myself," and when asked to write her name, she also suffered from delusion. As described by Maurer and collaborators, "When he released her, she would run out screaming, 'I will not be cut. I do not cut myself' (Maurer et al. 1997)." I am not sure what made her for such a fear. However, as described, she had got delusion or illusion (Heydrich et al. 2010), which could be related to disorder of consciousness. A disorder of consciousness makes a person feel fear. The fear was aroused by wrong perception of the existences such as a beast attacking, the vicious enemy coming, and calamity approaching. Under disorder of consciousness, the patient could wrongly perceive the delusive existence including oneself and one's environment.

7.4 A Hypothesis for MCI Due to AD

Here, I would like to propose a hypothesis that a patient with MCI due to AD may have his/her consciousness mildly impaired because consciousness impairment means disorders of basic function of the brain. Thus, his or her memory loss and learning decline could be resulted from consciousness impairment. It is difficult for us to intervene and ameliorate the basic function impairment. However, a patient with mild impairment of memory alone but not consciousness may not be due to AD.

8 Conclusion

We hypothesized that anabolism and metabolism of endogenous formaldehyde is involved in both normal and abnormal cognitive abilities. This viewpoint is based on these observations as follows:

1. The average concentration of urine formaldehyde of the elderly increases with aging (>75 years old) in a small sample investigation (Tong et al. 2013b).
2. The cognitive ability of the normal elderly is inversely related to the level of endogenous formaldehyde (Yu et al. 2014).
3. Educational levels are strikingly correlated with the concentration of endogenous formaldehyde for the elderly living in local communities in Beijing (Yu et al. 2014).
4. High levels of urine formaldehyde could be observed in the old patients with POCD after 1 week of their surgery, but not in those without POCD (Wang et al. 2012).
5. The concentration of urine formaldehyde is inversely correlated with MMSE scores for AD patients in clinics (Tong et al. 2011b).
6. The concentration of brain formaldehyde of mice and rats increases as they age (Tong et al. 2013a; Li et al. 2016a, b, c).
7. A 7-month-old APP/PS1 mouse has got elevated levels of formaldehyde in the cecum microbiota compared with age-matched C57BL/6J wild-type mouse (Liu et al. 2017).
8. Monkeys fed with water containing a low concentration of methanol suffered from working memory loss in 3 months during the administration (Yang et al. 2014a, b).
9. Levels of hippocampal formaldehyde from the patient autopsy are significantly higher than those of age-matched normal controls (Tong et al. 2015).

Therefore, endogenous formaldehyde can be employed to screen age-related cognitive impairment in epidemiological investigation for a large population, followed by identification of AD from VD in further examination. Actually, it is a procedural requirement to exclude the other diseases in the diagnosis of AD.

Acknowledgment I would like to thank Dr. Jing Lu who works in the Monash Institute of Cognitive and Clinical Neurosciences, School of Psychological Sciences, Monash University, Melbourne, Australia, for her editing language. This project was supported by grants from the Beijing Municipal Science and Technology Project (Z161100000217141; Z161100000216137), the National Key Research and Development Program of China (2016YFC1306300), the National Basic Research Program of China (973 Program) (2012CB911004), the Natural Scientific Foundation of China (NSFC 31301880, NSFC 31270868), the Foundation of Chinese Academy of Sciences CAS-20140909, the Queensland-Chinese Academy of Sciences Biotechnology Fund (GJHZ201302), the Program for Liaoning Excellent Talents in University (LJQ2015057), and the Dalian high level talent innovation support plan (No. 2015R067).

Funding Interests No other funding interest at all.

References

Alzheimer A (1906) Über einen eigenartigen schweren Erkrankungsprozeß der Hirnrinde. Neurologisches Centralblatt 23:1129–1136

Arora SS, Gooch JL, García PS (2014) Postoperative cognitive dysfunction, Alzheimer's disease, and anesthesia. Int J Neurosci 124(4):236–242

Boja JW, Nielsen JA, Foldvary E, Truitt EB Jr (1985) Acute low-level formaldehyde behavioral and neurochemical toxicity in the rat. Prog Neuro Psychopharmacol Biol Psychiatry 9(5–6):671–674

Brown C 4th, Deiner S (2016) Perioperative cognitive protection. Br J Anaesth 117(Suppl 3):iii52–iii61

Chen XX, Su T (2015) Microcirculation dysfunction in age-related cognitive impairment. Prog Biochem Biophys 42(12):1077–1083

Chen XX, Su T (2016) Microcirculation dysfunction in age-related cognitive impairment. Prog Biochem Biophys 42(12):1077–1083

Chen K, Maley J, Yu PH (2006) Potential implications of endogenous aldehydes in β -amyloid misfolding, oligomerization and fibrillogenesis. J Neurochem 99(5):1413–1424

Chen K, Kazachkov M, Yu PH (2007) Effect of aldehydes derived from oxidative deamination and oxidative stress on β-amyloid aggregation; pathological implications to Alzheimer's disease. J Neural Transm 114(6):835–839

Chen JY, Sun MR, Wang XH, Lu J, Wei Y, Tan Y, Götz J, He R, Hua Q (2014) The herbal compound geniposide rescues formaldehyde-induced apoptosis in N2a neuroblastoma cells. Sci China Life Sci 57(4):412–421

Chen XX, Su T, He YG, He RQ (2017) Spatial cognition decline caused by excess formaldehyde in the lysosome. Prog Biochem Biophys 44(6): doi:10.16476/j.pibb.2017.0050

Crane PK, Walker R, Hubbard RA, Li G, Nathan DM, Zheng H, Haneuse S, Craft S, Montine TJ, Kahn SE, McCormick W, McCurry SM, Bowen JD, Larson EB (2013) Glucose levels and risk of dementia. N Engl J Med 369(6):540–548

Dementia Guide. http://www.dementiaguide.com/images/the_prevalence_of_dementia.bmp

Dorokhov YL, Shindyapina AV, Sheshukova EV, Komarova TV (2015) Metabolic methanol: molecular pathways and physiological roles. Physiol Rev 95(2):603–644

Duong A, Steinmaus C, McHale CM, Vaughan CP, Zhang L (2011) Reproductive and developmental toxicity of formaldehyde: a systematic review. Mutat Res 728(3):118–138

Evans AM, Fameli N, Ogunbayo OA, Duan J, Navarro-Dorado J (2016) From contraction to gene expression: nanojunctions of the sarco/endoplasmic reticulum deliver site- and function-specific calcium signals. Sci China Life Sci 59(8):749–763

Flyvholm MA, Menné T (1992) Allergic contact dermatitis from formaldehyde. A case study focusing on sources of formaldehyde exposure. Contact Dermat 27(1):27–36

Foveau B, Albrecht S, Bennett DA, Correa JA, LeBlanc AC (2016) Increased caspase-6 activity in the human anterior olfactory nuclei of the olfactory bulb is associated with cognitive impairment. Acta Neuropathol Commun 4(1):127

Goldberg I, Mateles RI (1975) Growth of pseudomonas C on C1 compounds: enzyme activities in extracts of pseudomonas C cells grown on methanol, formaldehyde, and formate as sole carbon sources. J Bacteriol 122(1):47–53

Golden R, Valentini M (2014) Formaldehyde and methylene glycol equivalence: critical assessment of chemical and toxicological aspects. Regul Toxicol Pharmacol 69(2):178–186

Hagemeier J, Woodward MR, Rafique UA, Amrutkar CV, Bergsland N, Dwyer MG, Benedict R, Zivadinov R, Szigeti K (2016) Odor identification deficit in mild cognitive impairment and Alzheimer's disease is associated with hippocampal and deep gray matter atrophy. Psychiatry Res 255:87–93

Hanning CD (2015) Postoperative cognitive dysfunction. Br J Anaesth 95(1):82–87

Harrison EM (2013) Understanding suffering: Utermohlen's self-portraits and Alzheimer's disease. Nurse Educ 38(1):20–25

Hawkins KA, Schwartz-Thompson J, Kahane AI (1994) Abuse of formaldehyde-laced marijuana may cause dysmnesia. J Neuropsychiatry Clin Neurosci 6(1):67
He RQ, Lu J, Miao JY (2010) Formaldehyde stress. Sci China Life Sci 53(10):1399–1404
He RQ (2015a) Consciousness and anesthesia. Prog Biochem Biophys 42(10):972
He RQ (2015b) Definition of consciousness. ScienceNet 10:21. http://blog.sciencenet.cn/blog-829868-929860.html
He RQ (2016a) Early multidomain intervention to stave off Alzheimer's disease. Chin Sci Bull 61(32):3420–3427
He RQ (2016b) Abnormal lysosome, formaldehyde Dysmetabolism and age-related cognitive impairment. Prog Biochem Biophys 43(12):1197
Heydrich L, Dieguez S, Grunwald T, Seeck M, Blanke O (2010) Illusory own body perceptions: case reports and relevance for bodily self-consciousness. Conscious Cogn 19(3):702–710
Hou Q, Jiang H, Zhang X, Guo C, Huang B, Wang P, Wang T, Wu K, Li J, Gong Z, Du L, Liu Y, Liu L, Chen C (2011) Nitric oxide metabolism controlled by formaldehyde dehydrogenase (fdh, homolog of mammalian GSNOR) plays a crucial role in visual pattern memory in drosophila. Nitric Oxide 24(1):17–24
Hu X, Wang T, Jin F (2016) Alzheimer's disease and gut microbiota. Sci China Life Sci 59(10):1006–1023
Hu B, Geng C, Hou XY (2017) Oligomeric amyloid-β peptide disrupts olfactory information output by impairment of local inhibitory circuits in rat olfactory bulb. Neurobiol Aging 51:113–121
Hua Q, He RQ (2002) Effect of phosphorylation and aggregation on tau binding to DNA. Protein Pept Lett 9(4):349–357
Jian X, Zhu MX (2016) Regulation of lysosomal ion homeostasis by channels and transporters. Sci China Life Sci 59(8):777–791
Karapetian SO (1971) Nerve system lesions in workers at the Echmiadzin plastics factory (Russian). Zh Eksp Klin Med 11(2):100–103
Karuppagounder V, Arumugam S, Babu SS, Palaniyandi SS, Watanabe K, Cooke JP, Thandavarayan RA (2017) The senescence accelerated mouse prone 8 (SAMP8): a novel murine model for cardiac aging. Ageing Res Rev 35:291–296
Kilburn KH (1994) Neurobehavioral impairment and seizures from formaldehyde. Arch Environ Health 49(1):37–44
Kilburn KH (2000) Indoor air effects after building renovation and in manufactured homes. Am J Med Sci 320(4):249–254
Kilburn KH, Warshaw R, Thornton JC (1987) Formaldehyde impairs memory, equilibrium, and dexterity in histology technicians: effects which persist for days after exposure. Arch Environ Health 42(2):117–120
Kum C, Kiral F, Sekkin S, Seyrek K, Boyacioglu M (2007) Effects of xylene and formaldehyde inhalations on oxidative stress in adult and developing rats livers. Exp Anim 56(1):35–42
Li YN, He RQ (2016) The effects of formaldehyde on life span and stress resistance in Drosophila melanogaster. Prog Biochem Biophys 43(4):420–428
Li FX, Lu J, Xu YJ, Tong ZQ, Nie CL, He RQ (2008) Formaldehyde-mediated chronic damage may be related to sporadic neurodegeneration. Prog in Biochem Biophys 35(4):393–400
Li YQ, Chen HH, Yin YF, Han F, Ye XS, Ling SC (2010) Formaldehyde inhalation may damage olfactory bulb and hippocampus in rats. Zhejiang Da Xue Xue Bao Yi Xue Ban 39(3):272–277
Li J, Yu J, Liu YN (2012) Neuropsychological impairment characteristics of MCI and its early detection and intervention: prevent and delay the onset of AD. Prog Biochem Biophys 39(8):804–810
Li G, Yang J, Ling S (2015) Formaldehyde exposure alters miRNA expression profiles in the olfactory bulb. Inhal Toxicol 27(8):387–393
Li Y, Song Z, Ding Y, Xin Y, Wu T, Su T, He R, Tai F, Lian Z (2016a) Effects of formaldehyde exposure on anxiety-like and depression-like behavior, cognition, central levels of glucocorticoid receptor and tyrosine hydroxylase in mice. Chemosphere 144:2004–2012

Li T, Su T, He YG, He RQ (2016b) Chronic-dehydrated Dysmetabolism of formaldehyde in mouse brain and decline of learning in the shuttle box. Prog Biochem Biophys 43(4):429–438

Li T, Su T, He Y, Lu J, Mo W, Wei Y, He R (2016c) Brain formaldehyde is related to water intake behavior. Aging Dis 7(5):561–584

Liu KL, He YG, Yu LX, He RQ (2017) Elevated formaldehyde in the cecum of APP/PS1 mouse. Microbiol China 44(8):1761–1766

LoSasso GL, Rapport LJ, Axelrod BN (2001) Neuropsychological symptoms associated with low-level exposure to solvents and (meth) acrylates among nail technicians. Neuropsychiatry Neuropsychol Behav Neurol 14(3):183–189

Lu Z, Li CM, Qiao Y, Yan Y, Yang X (2008) Effect of inhaled formaldehyde on learning and memory of mice. Indoor Air 18(2):77–83

Lu Y, He HJ, Zhou J, Miao JY, Lu J, He YG, Pan R, Wei Y, Liu Y, He RQ (2013a) Hyperphosphorylation results in tau dysfunction in DNA folding and protection. J Alzheimers Dis 37(3):551–563

Lu J, Miao J, Su T, Liu Y, He R (2013b) Formaldehyde induces hyperphosphorylation and polymerization of tau protein both in vitro and in vivo. Biochim Biophys Acta 1830(8):4102–4116

Luo WH, Li H, Zhang Y, Ang CYW (2001) Determination of formaldehyde in blood plasma by high-performance liquid chromatography with fluorescence detection. J Chromatogr B Biomed Sci Appl 753(2):253–257

Malek FA, Möritz KU, Fanghänel J (2003) A study on the effect of inhalative formaldehyde exposure on water labyrinth test performance in rats. Ann Anat 185(3):277–285

Maliutina NN, Taranenko LA (2014) Cognitive disorders in workers engaged into formaldehyde and methanol production. Med Tr Prom Ekol 5:26–31

Marceaux JC, Dilks LS, Hixson S (2008) Neuropsychological effects of formaldehyde use. J Psychoactive Drugs 40(2):207–210

Maurer K, Volk S, Gerbaldo H (1997) Auguste D and Alzheimer's disease. Lancet 349(9064):1546–1549

Mei Y, Duan C, Li X, Zhao Y, Cao F, Shang S, Ding S, Yue X, Gao G, Yang H, Shen L, Feng X, Jia J, Yang X (2016) Reduction of endogenous melatonin accelerates cognitive decline in mice in a simulated occupational formaldehyde exposure environment. Int J Environ Res Pub Health 13(3).pii E258

Miao J, He R. (2012). Chronic formaldehyde-mediated impairments and age-related dementia. In: Martins LM (ed) Neurodegeneration, 1st edn. InTech, Rijeka, pp 59–76

Moller JT, Cluitmans P, Rasmussen LS, Houx P, Rasmussen H, Canet J, Rabbitt P, Jolles J, Larsen K, Hanning CD, Langeron O, Johnson T, Lauven PM, Kristensen PA, Biedler A, van Beem H, Fraidakis O, Silverstein JH, Beneken JE, Gravenstein JS (1998) Long-term postoperative cognitive dysfunction in the elderly ISPOCD1 study. ISPOCD investigators. International study of post-operative cognitive dysfunction. Lancet 351(9106):857–861

Monk TG, Weldon BC, Garvan CW, Dede DE, van der Aa MT, Heilman KM, Gravenstein JS (2008) Predictors of cognitive dysfunction after major noncardiac surgery. Anesthesiology 108(1):18–30

Morley JE, Farr SA, Kumar VB, Armbrecht HJ (2012) The SAMP8 mouse: a model to develop therapeutic interventions for Alzheimer's disease. Curr Pharm Des 18(8):1123–1130

Morton WE, Feldstein A (1991) Organic cognitive impairment? West J Med 154(2):225–226

Newman S, Stygall J, Hirani S, Shaefi S, Maze M (2007) Postoperative cognitive dysfunction after noncardiac surgery: a systematic review. Anesthesiology 106(3):572–590

Nie CL, Zhang W, Zhang D, He RQ (2005) Changes in conformation of human neuronal tau during denaturation in formaldehyde solution. Protein Pept Lett 12(1):75–78

Nie CL, Wang XS, Liu Y, Perrett S, He RQ (2007a) Amyloid-like aggregates of neuronal tau are induced by formaldehyde exposure and promote apoptosis of neuronal cells. BMC Neurosci 20(5):A954–A954

Nie CL, Wei Y, Chen XY, Liu YY, Dui W, Liu Y, Davies MC, Tendler SJ, He RQ (2007b) Formaldehyde at low concentration induces protein tau into globular amyloid-like aggregates in vitro and in vivo. PLoS One 2(7):e629

Park DC, Reuter-Lorenz P (2009) The adaptive brain: aging and neurocognitive scaffolding. Annu Rev Psychol 60:173–196

Perna RB, Bordini EJ, Deinzer-Lifrak M (2001) A case of claimed persistent neuropsychological sequelae of chronic formaldehyde exposure: clinical, psychometric, and functional findings. Arch Clin Neuropsychol 16(1):33–44

Pinto JP, Gladstone GR, Yung YL (1980) Photochemical production of formaldehyde in Earth's primitive atmosphere. Science 210(4466):183–185

Pitten FA, Kramer A, Herrmann K, Bremer J, Koch S (2000) Formaldehyde neurotoxicity in animal experiments. Pathol Res Pract 196(3):193–198

Qiang M, Xiao R, Su T, Wu B, Tong Z, Liu Y, He R (2014) A novel mechanism for endogenous formaldehyde elevation in SAMP8 mouse. J Alzheimers Dis 40(4):1039–1053

Quinn TJ, McCleery J (2017) Diagnosis in vascular dementia, applying 'Cochrane diagnosis rules' to 'dementia diagnostic tools'. Clin Sci (Lond) 131(8):729–732

Rosenberg GA (2017) Extracellular matrix inflammation in vascular cognitive impairment and dementia. Clin Sci (Lond) 131(6):425–437

Sawikr Y, Yarla NS, Peluso I, Kamal MA, Aliev G, Bishayee A (2017) Neuroinflammation in Alzheimer's disease: the preventive and therapeutic potential of polyphenolic nutraceuticals. Adv Protein Chem Struct Biol 108:33–57

Shao C, Wang Y, Gao Y (2011) Applications of urinary proteomics in biomarker discovery. Sci China Life Sci 54(5):409–417

Shi H, Wang Z (2013) A brief review on studies of Alzheimer's disease in China: its mechanism, imaging and therapy. Sci China Life Sci 56(12):1142–1144

Shi Y, Lan F, Matson C, Mulligan P, Whetstine JR, Cole PA, Shi Y (2004) Histone demethylation mediated by the nuclear amine oxidase homolog LSD1. Cell 119(7):941–953

Shpanskaya KS, Choudhury KR, Hostage C Jr, Murphy KR, Petrella JR, Doraiswamy PM (2014) Educational attainment and hippocampal atrophy in the Alzheimer's disease neuroimaging initiative cohort. J Neuroradiol 41(5):350–357

Sparks PJ, Simon GE, Katon WJ, Altman LC, Ayars GH, Johnson RL (1990) An outbreak of illness among aerospace workers. West J Med 153(1):28–33

Su T, Wei Y, He RQ (2011) Assay of brain endogenous formaldehyde with 2,4-Dinitrophenylhydrazine through UV-HPLC. Prog Biochem Biophys 38(12):1171–1177

Su T, Song D, Li T, Wang XH, He RQ (2015) Nucleic acid methylation/demethylation, endogenous formaldehyde and age-related cognitive impairment. Prog Biochem Biophys 42(3):211–219

Szende B, Tyihák E (2010) Effect of formaldehyde on cell proliferation and death. Cell Biol Int 34(12):1273–1282

Tang X, Bai Y, Duong A, Smith MT, Li L, Zhang L (2009) Formaldehyde in China: production, consumption, exposure levels, and health effects. Environ Int 35(8):1210–1224

Tang Y, Kong X, Liu Z, Xu A, Lin W (2016) Lysosome-targeted turn-on fluorescent probe for endogenous formaldehyde in living cells. Anal Chem 88(19):9359–9363

Tong Z, Wang Y, Luo W, He R (2008) Endogenous formaldehyde and related diseases in human. Prog Nat Sci 18:1202–1210

Tong Z, Han C, Miao J, Lu J, He R (2011a) Excess endogenous formaldehyde induces memory decline. Prog Biochem Biophys 38(6):575–579

Tong Z, Zhang J, Luo W, Wang W, Li F, Li H, Luo H, Lu J, Zhou J, Wan Y, He R (2011b) Urine formaldehyde level is inversely correlated to mini mental state examination scores in senile dementia. Neurobiol Aging 32(1):31–41

Tong Z, Han C, Luo W, Wang X, Li H, Luo H, Zhou J, Qi J, He R (2013a) Accumulated hippocampal formaldehyde induces age dependent memory decline. Age 35(3):583–596

Tong Z, Han C, Luo W, Li H, Luo H, Qiang M, Su T, Wu B, Liu Y, Yang X, Wan Y, Cui D, He R (2013b) Aging associated excess formaldehyde leads to spatial memory deficits. Sci Rep 3:1807

Tong Z, Han C, Qiang M, Wang W, Lv J, Zhang S, Luo W, Li H, Luo H, Zhou J, Wu B, Su T, Yang X, Wang X, Liu Y, He R (2015) Age-related formaldehyde interferes with DNA methyltransferase function on causing memory loss in Alzheimer's disease. Neurobiol Aging 36(1):100–112

Tong Z, Wang W, Luo W, Lv J, Li H, Luo H, Jia J, He R (2017) Urine formaldehyde predicts cognitive impairment in post-stroke dementia and Alzheimer's disease. J Alzheimers Dis 55(3):1031–1038

Ueno M, Chiba Y, Matsumoto K, Murakami R, Fujihara R, Kawauchi M, Miyanaka H, Nakagawa T (2016) Blood-brain barrier damage in vascular dementia. Neuropathology 36(2):115–124

Wang WS, Hao ZH, Zhang L (2010) Research on urine formaldehyde concentration in Alzheimer's disease elderly and normal elderly. Chin J Geriatr Heart Brain Vessel Dis 12:721–722

Wang J, Su T, Liu Y, Yue Y, He R (2012) Postoperative cognitive dysfunction is correlated with urine formaldehyde in elderly noncardiac surgical patients. Neurochem Res 37(10):2125–2134

Wang J, Zhou J, Mo WC, He YG, Wei Y, He RQ, Yi FP (2017) Accumulation of simulated pathological level of formaldehyde decreases cell viability and adhesive morphology in neuronal cells. Prog Biochem Biophys 44(7):601–614

Wei Y, Qu MH, Wang XS, Chen L, Wang DL, Liu Y, Hua Q, He RQ (2008) Binding to the minor groove of the double-strand, tau protein prevents DNA from damage by peroxidation. PLoS One 3(7):e2600

Wilson RS, Hebert LE, Scherr PA, Barnes LL, Mendes de Leon CF, Evans DA (2009) Educational attainment and cognitive decline in old age. Neurology 72(5):460–465

Wu K, Ren R, Su W, Wen B, Zhang Y, Yi F, Qiao X, Yuan T, Wang J, Liu L, Izpisua Belmonte JC, Liu GH, Chen C (2014) A novel suppressive effect of alcohol dehydrogenase 5 in neuronal differentiation. J Biol Chem 289(29):20193–20199

Xie Z, Tanzi RE (2006) Alzheimer's disease and post-operative cognitive dysfunction. Exp Gerontol 41(4):346–359

Yang M, Lu J, Miao J, Rizak J, Yang J, Zhai R, Zhou J, Qu J, Wang J, Yang S, Ma Y, Hu X, He R (2014a) Alzheimer's disease and methanol toxicity (part 1): chronic methanol feeding led to memory impairments and tau hyperphosphorylation in mice. J Alzheimers Dis 41(4):1117–1129

Yang M, Miao J, Rizak J, Zhai R, Wang Z, Huma T, Li T, Zheng N, Wu S, Zheng Y, Fan X, Yang J, Wang J, Yang S, Ma Y, Lü L, He R, Hu X (2014b) Alzheimer's disease and methanol toxicity (Pt.2): lessons from four rhesus macaques (Macaca mulatta) chronically fed methanol. J Alzheimers Dis 41(4):1131–1147

Yu PH, Cauglin C, Wempe KL, Gubisne-Haberle D (2003) A novel sensitive high-performance liquid chromatography/electrochemical procedure for measuring formaldehyde produced from oxidative deamination of methylamine and in biological samples. Anal Biochem 318(2):285–290

Yu J, Li J, Huang X (2012) The Beijing version of the montreal cognitive assessment as a brief screening tool for mild cognitive impairment: a community-based study. BMC Psychiatry 12:156

Yu J, Su T, Zhou T, He Y, Lu J, Li J, He R (2014) Uric formaldehyde negatively correlates with cognitive abilities in healthy older adults. Neurosci Bull 30(2):172–184

Zhang Q, Yan W, Bai Y, Zhu Y, Ma J (2014) Repeated formaldehyde inhalation impaired olfactory function and changed SNAP25 proteins in olfactory bulb. Int J Occup Environ Health 20(4):308–312

Chapter 9
Tau Phosphorylation and Amyloid-β Deposition in the Presence of Formaldehyde

Jing Lu and Rongqiao He

Abstract Alzheimer's disease (AD) is a devastating neurodegenerative disorder with a relentless progression. It is associated with amyloid-β (Aβ) peptide deposition in senile plaques and hyperphosphorylated Tau protein in neurofibrillary tangles in the AD patients' brain. Aβ accumulation occurs in early preclinical stage as a trigger factor for AD pathogenesis followed by silence of synapses and Tau hyperphosphorylation. However, what triggers Aβ deposition and Tau hyperphosphorylation is still under investigation. As an endogenous small-molecule metabolite, formaldehyde is produced by multiple cellular processes, including lipid oxidation, protein denaturation, sugar decomposition, methanol degradation, oxidative stress, DNA demethylation, and semicarbazide-sensitive amine oxidase (SSAO) catalyzation (see Chap. 2). Recent studies demonstrated that formaldehyde plays a pivotal role in the development of age-related neurodegenerative diseases by inducing the cellular component malfunction, such as Tau hyperphosphorylation, Aβ aggregation, and cell apoptosis. However, the underlying molecular mechanisms remain largely elusive. This chapter briefly introduces recent progresses on production and accumulation of formaldehyde with aging and the signaling pathways of formaldehyde, leading to dysfunction of nuclear Tau protein, aggregation of Aβ, as well as dysfunction of ApoE in vitro and in vivo. Formaldehyde acts as an effective trigger of Tau protein hyperphosphorylation in cell lines, primary cultured neurons, mouse brains, as well as monkey brains. During formaldehyde-induced Tau hyperphosphorylation, activation of Tau phosphorylation kinases glycogen synthase kinase 3 beta (GSK-3β) and Ca^{2+}/calmodulin-dependent protein kinase II (CaMKII) as well as suppression of protein phosphatase 2A (PP2A) were shown to be highlighted pathways. Long-term incubation of N2a cells with formaldehyde at the pathological concentration referred to AD patients resulted in formaldehyde accumulation also

J. Lu
Monash Institute of Cognitive and Clinical Neurosciences, Monash University, Melbourne, Australia

R. He (✉)
State Key Laboratory of Brain and Cognitive Science, Institute of Biophysics, Chinese Academy of Sciences, Beijing 100101, China

University of Chinese Academy of Sciences, Beijing 100049, China
e-mail: rongqiaohe@163.com

R. He, *Formaldehyde and Cognition*, DOI 10.1007/978-94-024-1177-5_9

inducing Tau hyperphosphorylation and morphological changes of neuronal processes and neurites. Both Tau hyperphosphorylation and Aβ deposition were found in the hippocampus and cerebral cortex of rhesus monkeys through a long-term feeding with low concentration of methanol in drinking water or cerebral ventricle injection, and those monkeys suffered from decline of cognitive abilities as well. Formaldehyde is shown to be a strong trigger of Aβ deposition and Tau phosphorylation, so that the dysmetabolism of endogenous formaldehyde should occur in the early stage of AD as formaldehyde disturbs nervous system in different pathways and in diverse manners.

Keywords Formaldehyde • Methanol • Tau hyperphosphorylation • Neurofibrillary tangles • Tauopathy • Amyloid-β • Senile plaque • Cognitive impairment • APOE4 • Neurite • Neuronal processes • Cell death • Mouse model • Nonhuman primate • Dephosphorylation • Pathology

1 Introduction

In 1901, German psychiatrist Alois Alzheimer identified the first case of what became known as Alzheimer's disease (AD). The patient called Auguste D was a 50-year-old woman. Alzheimer found senile plaques and neurofibrillary tangles on examining her brain after she died in 1906 (Alzheimer 1907). These became the hallmarks of AD as known by scientists today (Ballard et al. 2011). AD is a progressive, irreversibly neurodegenerative disorder characterized by extracellular deposits of β-amyloid (Aβ) peptides in senile plaques and intracellular aggregates of hyperphosphorylated Tau in neurofibrillary tangles (NFT) in the patient's brain. Accumulation of Aβ has long been considered a leading hypothesis in the early onset of the pathology. Aβ not only disrupts function of neural cells but also triggers hyperphosphorylation of Tau (Oliveira et al. 2015), for instance, leading to a ceramide-Aβ-hyperphosphorylated Tau cascade that ends up with neuronal death (Jembrek et al. 2013). However, it is increasingly evident that the role of Tau hyperphosphorylation playing in destabilization of microtubule assembly, loss of DNA protection, and disturbance of axonal transport is also non-negligibly detrimental in the neurodegenerative process.

The early stages of AD including preclinical AD are pathologically characterized by the formation of mature NFTs in entorhinal cortex and phenotypically characterized by disorientation and confusion when navigating familiar places (Allison et al. 2016; Lithfous et al. 2013). The medial entorhinal cortex contains specialized neurons called grid cells which form part of the spatial navigation system (Fyhn et al. 2004; Hafting et al. 2005). Fu and collaborators have given the evidence that Tau pathology initiated in the entorhinal cortex could lead to deficits in grid cell firing and underlie the deterioration of spatial cognition seen in human AD (Fu et al.

2017). Therefore, abundant work has been carried out to clarify what triggers or induces hyperphosphorylation of Tau (Su et al. 2016).

Etiological factors inducing Tau hyperphosphorylation have been widely investigated, including gene mutation (Kar et al. 2005), changes in epigenetics (Bufill et al. 2013), oxidative stress (Chen et al. 2012a; Evans et al. 2016), inflammation (Barron et al. 2016), tumor or cancer (Serrano et al. 2010), metal ions (Cai et al. 2011; Zhao and Wan 2012), insulin imbalance (Kim et al. 2009; Planel et al. 2007; Qu et al. 2011), starvation (Yanagisawa et al. 1999), some toxicants (Chen et al. 2012b), advanced glycation end products (Li et al. 2012), anesthesia (Run et al. 2009; Whittington et al. 2011), food ingredient (Monte 2012), drug usage (Kim et al. 2012; Rahman et al. 2009), and imbalance of intestinal microbe flora (Hu et al. 2016). Recently, microcirculation dysfunction, which is considered playing an important role in hyperphosphorylation of Tau, Aβ aggregation, and cognitive impairment, was induced by accumulation of formaldehyde (Chen and Su 2015). As we know, microcirculation dysfunction is regarded as one of the important pathologies of senility, degeneration, immune disorder, and many other diseases. Cerebral circulation insufficiency, energetic dysmetabolism, hypoxia-ischemia, and metabolite accumulation in AD have a close relationship with microcirculation dysfunction (Nobutoki and Ihara 2015).

To investigate aldehyde-induced Tau phosphorylation and aggregation in this laboratory, Luo and collaborators employed acetaldehyde and glutaraldehyde on protein Tau aggregation compared to ethanol and methanol (He et al. 1998; Luo and He 1999). We have observed that aldehyde could affect phosphorylation of Tau catalyzed by neuronal Cdc2-like protein kinase (NCLK) (Chen et al. 1999) and also formaldehyde markedly induced Tau protein dysfunction and aggregation (Hua and He 2002). We carried out the study of the relation between dysmetabolism of formaldehyde (He et al. 2016; Lu et al. 2011, 2013a; Su et al. 2016) with cognitive impairment and imbalance in intestinal microbe flora (Liu et al. 2017, submitted) and abnormal lysosome (He 2016; Jian and Zhu 2016).

In Chap. 7, we have described that formaldehyde is able to react with protein directly and modify the side chains of lysine, arginine, tryptophan, and cysteine (Rizak et al. 2014), inducing thioflavin S (ThS)-positive Tau aggregates in human neuroblastoma SH-SY5Y cells and leading to cell death (Nie et al. 2007). In this chapter, among so many potential etiological factors, we would like to discuss how formaldehyde triggers Aβ deposition and Tau hyperphosphorylation due to the following observations: first, formaldehyde is found to be highly produced in vivo and involves in many biological processes; second, levels of formaldehyde are increased with aging in the elderly population (Tong et al. 2013) and correlated with the severity of AD (Tong et al. 2011); and finally, formaldehyde is associated with formation of plaques and neurofibrillary tangles in the brains of animal models (Lu et al. 2013a; Yang et al. 2014a, b).

2 Formaldehyde Triggers Tau Hyperphosphorylation

In 1986, using a combined immunocytochemical and biochemical approach, Grundke-Iqbal and colleagues demonstrated for the first time that the microtubule-associated protein Tau, a normal brain cytoskeletal protein, was a component of the paired helical filaments (PHFs) (Grundke-Iqbal et al. 1986). A major lesion in AD is the presence of numerous neurofibrillary tangles, which are composed of PHFs (Fig. 9.1). The normal Tau usually carries 2–3 mol phosphate per mol protein, while PHF-Tau contents 6–8 mol of phosphate per mol of the protein (Gong et al. 1993; Ksiezak-Reding et al. 1992). The primary function of Tau is promoting the assembly of microtubules (Weingarten et al. 1975) which will be repressed when Tau is hyperphosphorylated (Lindwall and Cole 1984). Thus, in Tau highly phosphorylated AD brain, disassembly of microtubules was associated with atrophied neurons (Ballard et al. 2011).

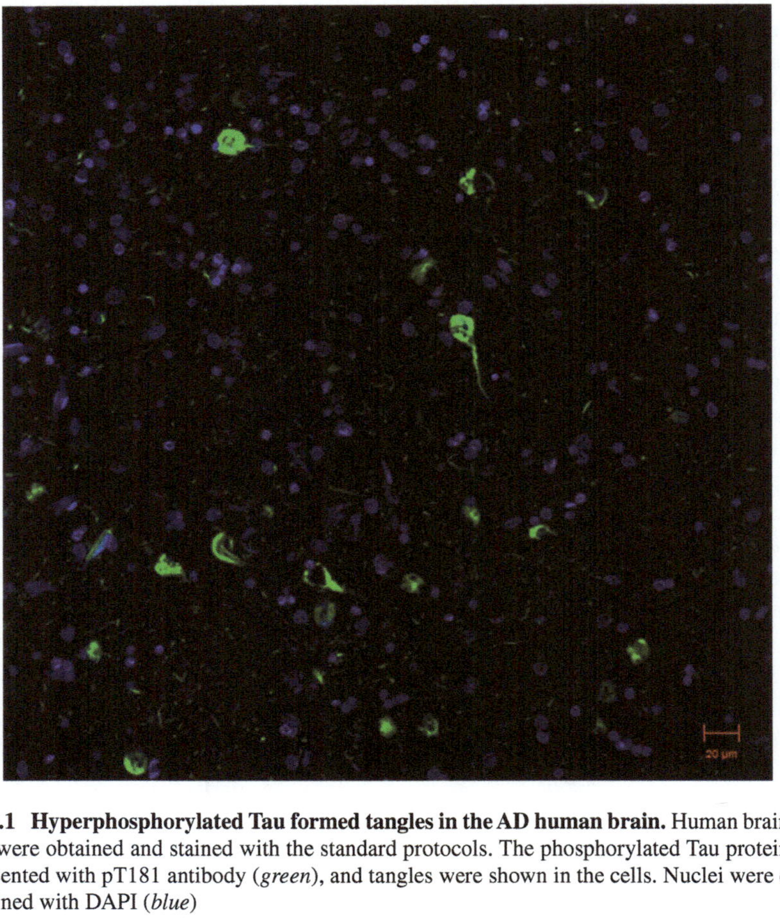

Fig. 9.1 Hyperphosphorylated Tau formed tangles in the AD human brain. Human brain sections were obtained and stained with the standard protocols. The phosphorylated Tau protein was represented with pT181 antibody (*green*), and tangles were shown in the cells. Nuclei were counterstained with DAPI (*blue*)

Phosphorylation and dephosphorylation regulate the function of Tau by decreasing or increasing its affinity of binding to microtubule (Stoothoff and Johnson 2005). Hyperphosphorylation inactivates Tau protein from stabilization of microtubule system and protection of DNA (Lu et al. 2013b). Additionally, hyperphosphorylation leads to Tau misfolding and aggregation and sequesters normal Tau, microtubule-associated proteins 1 and 2, and ubiquitin into PHFs. This insoluble structure damages cytoplasmic structure, interferes with axonal transport, and eventually leads to cell death (Mudher and Lovestone 2002). The hyperphosphorylated Tau-composited PHFs will then form neurofibrillary tangles which deposit in the cytoplasm of neurons (Alzheimer's's 2015; Rhein et al. 2009; Scheuner et al. 1996; Tsuji 2010) and damage the brains of patients suffering from AD (Bird 1993) or other neurodegenerative diseases (Engelborghs and De Deyn 2001; Wu et al. 2012). Therefore, when formaldehyde induced Tau hyperphosphorylation and aggregation in the brains, the above pathological features might occur. Thus, it makes formaldehyde a major contribution to a number of neurodegenerative diseases including AD and primary open-angle glaucoma (one of the neurodegenerative disorders) (Cui et al. 2016).

2.1 *Tau Hyperphosphorylation in N2a Cells in the Presence of Formaldehyde*

As an endogenous metabolite, formaldehyde at physiological concentrations around 0.01 mM was found to promote cell proliferation (Miao et al. 2013). On the other hand, when the formaldehyde concentration increased up to 0.1 mM, cell apoptosis and retardation of cell cycle at S-phase were observed (Miao et al. 2013). Note, it was demonstrated formaldehyde toxicity is correlated to the cell inoculation sizes. The larger the inoculation size, the higher formaldehyde concentrations are needed to decrease the cell viability (Lu et al. 2011). The LD50 of formaldehyde to N2a cells with different inoculation sizes is shown in Table 9.1. Cell process atrophies were observed in the presence of formaldehyde comparing with control (Lu et al. 2011). Similar results were also detected in primary cultured hippocampal neurons incubated with formaldehyde (Yu et al. 2014). The morphological changes of cells indicated that the cellular skeleton is affected by formaldehyde during the treatment.

Table 9.1 LD_{50} of FA on N2a cells at different inoculation densities

LD_{50} (mM)	Inoculum density (cells/mL)		
Time (h)	1×10^5	2×10^5	5×10^5
12	0.43	0.55	1.21
24	0.19	0.41	0.85
48	0.15	0.32	0.65

Since the microtubule system plays a critical role in neuronal activities, Lu and colleagues investigated Tau phosphorylation in N2a cells under formaldehyde treatment (Fig. 9.2) (Lu et al. 2011). Tau hyperphosphorylation was dramatically increased after overloading of formaldehyde (strongest 2–8 h), and the cell processes shrank along with Tau phosphorylation (Lu et al. 2013a). Phosphorylation of Tau protein at sites of Thr181 and Ser396 was detected by Western blotting under the same experimental conditions. The phosphorylation of Tau at Thr181 and Ser396 was also observed with immunocytofluorescence, not only in the cytoplasm but also in the nucleus. Further examination indicated that phosphorylated Tau in the nucleus was not overlapping with DNA stained with Hoechst-33258. Notably, the phosphorylation level of Tau was decreased after 8 h of formaldehyde treatment, accompanying with regrowth of cell processes. These data manifested that overload of formaldehyde led to Tau hyperphosphorylation and subsquent severe process atrophies and possibly DNA damage.

In order to explore whether formaldehyde induces polymerization of hyperphosphorylated Tau, which is a key procedure cause diseases in tauopathies, Lu and coworkers examined polymers inside N2a cells using ThS staining and Western blotting. Based on their results (Lu et al. 2013a), Tau polymer bands with higher apparent molecular masses (~170 kDa for pThr181 and ~130 kDa for pSer396) were observed 2–4 h after the onset of formaldehyde incubation. The polymers

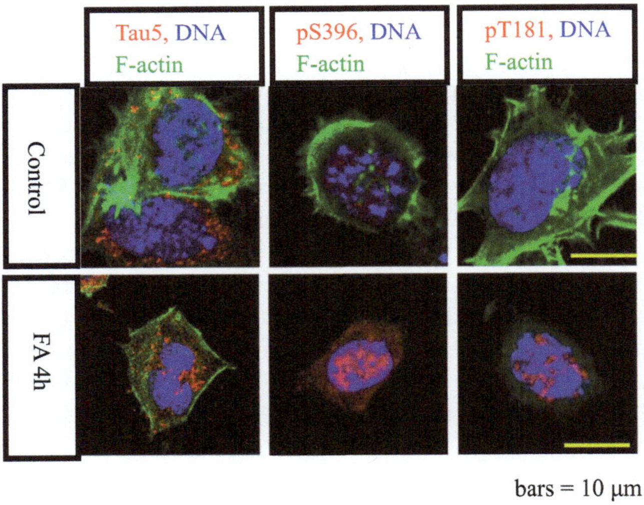

Fig. 9.2 Signals of phosphorylated Tau (pT181 and pS396) and total Tau (Tau5) after cells were treated with formaldehyde. Phosphorylated Tau was detected by anti-pT181 and anti-pS396 antibodies, and total Tau was assayed with Tau5 antibody after cells were treated with formaldehyde for 4 h. Nuclei and F-actin were stained with Hoechst-33258 (blue) and phalloidin (green), respectively. The signals of anti-pT181 and anti-pS396 strikingly increased in nuclei of N2a cells, and most of them were rarely co-localized with the DNA staining. A small part of Tau5 signal was co-localized with DNA except for most of them (Modified from (Lu et al. 2011). The usage of this figure was authorized by the authors and chief editor of Prog. Biochem. Biophys)

disappeared 8 h after formaldehyde treatment, illustrating that these polymers were reversible. Tau polymerization occurred in synchrony with its hyperphosphorylation, and these Tau polymers were shown to be ThS positive in N2a cells under formaldehyde treatment. These data suggested that the polymers should be hyperphosphorylated Tau as they are co-localized with phosphorylation staining and reversible. Otherwise, the cross-linked polymers by formaldehyde could not be so easy to disassemble, and a higher concentration of formaldehyde is also required (Luo and He 1999).

Using high-performance liquid chromatography measurement, Lu and coworkers observed that formaldehyde concentration decreased from 0.5 mM to around 0.1 mM at 8 h and around 0.02 mM at 24 h in the medium with cell cultures, while the concentration only declined to 0.45 mM at 24 h in the cell-free medium. This experiment exhibited that cells were effectively metabolizing formaldehyde when there was formaldehyde in excess. During the formaldehyde treatment, Tau hyperphosphorylation progressed in both concentration-dependent and time-dependent manners. As hyperphosphorylation of Tau leads to microtubule system disruption according to many previous reports (Alonso et al. 2016), this presumably explained the atrophy of cellular processes and neurites observed under formaldehyde treatment. In order to further determine the roles formaldehyde played in Tau protein functionality in vivo, formaldehyde was administrated to mouse as shown in the following paragraph.

2.2 *Tau Hyperphosphorylation in Mouse Brain Administrated with Methanol/Formaldehyde*

Short-term formaldehyde administration was shown to damage the cognition abilities in both mouse and rat (Lu et al. 2008; Tang et al. 2013). In order to investigate the long-term toxicity of industrial alcohol which contains additional methanol, Yang and colleagues fed mice (male, ICR) with methanol through drinking water with concentrations of 0%, 2%, and 3.8% over a 6-week period. Their results showed that the spatial recognition and olfactory memory were impaired in the methanol-fed mice through Y-maze and olfactory memory paradigms (Yang et al. 2014a). Meanwhile, pathogenically examination showed that Tau hyperphosphorylation and neuronal death were observed in the hippocampus of those mice. In addition, the severity of those impairments was correlated with doses of methanol given.

Similarly, Tau hyperphosphorylation was also detected in the hippocampus and cortex of mouse brains when the mouse was administrated with formaldehyde through its tail vein. During this treatment, Tau hyperphosphorylation lasted much longer up to 7 days than that in N2a cells for 24 h (Lu et al. 2013a). Likewise, the phosphorylation of Tau was not only observed in the cytoplasm but also in the nucleus (Lu et al. 2011). Furthermore, the phosphorylated Tau was not overlapped with DNA staining, similar to the results in N2a cells. Phosphorylation at sites Thr181 and Ser396 of Tau protein was also detected with Western blotting. Tau epitopes, such as Thr181 and Ser396, are sensitive to phosphorylation (Gong et al.

2005; Li and Paudel 2006). Hyperphosphorylation of Thr181 in the cerebrospinal fluid (CSF) is employed as a biomarker to predict AD (Blennow 2005; Olsson et al. 2005). Formaldehyde was also detected high in the CSF of AD patients provided by Beijing Hospital of Elderly (our unpublished data). Hyperphosphorylation of Ser396 results in decreased Tau solubility and is believed to be a crucial step in the development of a neurofibrillary pathology in AD patients' brain (Abraha et al. 2000; Michel et al. 1998).

Note that under the same experimental conditions, neither neurofibrillary tangles nor Aβ aggregates were detected in the brain of formaldehyde-treated mouse for 1–3 months. Thus, it is suggested that the authors prolong the administration timing of mouse with methanol, for instance, 6 months or longer, like the administration of mouse with D-ribose that showed Aβ deposition and Tau hyperphosphorylation in neurofibrillary-like tangles in the mouse brain with 6-month treatment (Wu et al. 2015). More phosphorylation sites should be detected although Thr181 and Ser396 are found critical in the brain of Alzheimer's patients, so that more evidence could be given to support formaldehyde-related dementia in the aging population.

2.3 Tau Hyperphosphorylation in Monkey Brain Administrated with Formaldehyde

Though Tau hyperphosphorylation was detected in the mouse brain by administration of methanol/formaldehyde, mouse is much more efficient in degradation of formaldehyde than human beings. Excess methanol is seriously toxic to primates, leading to severe lesions to the eye and neural cells where methanol and metabolite formic acid accumulated (MacAllister et al. 2011). Rodents, however, are able to synthesize enough folate to degrade formic acid so that methanol-induced brain damages are compromised in rodents (Bruckner and Warren 2001). On the other hand, Alzheimer's disease develops in a chronic pathological way; thus, a long-term trial should be carried out to simulate the disease. To determine the role of formaldehyde, a methanol metabolite, in (AD) pathology, Yang and colleagues provided an impetus to investigate chronic effects of methanol exposure by feeding methanol, the precursor of formaldehyde. They employed rhesus macaque to establish a non-human primate model in age-related cognitive impairment through a chronic feeding of young male monkeys with 3% methanol ad libitum (Yang et al. 2014b). As a small molecule, formaldehyde can freely pass the blood-brain barrier (Shcherbakova et al. 1986).

Cognitive ability (working memory) of monkeys was assessed with a variable spatial delayed response task (VSDRT) as used previously (Arnsten et al. 1988; Wang et al. 2013a). As the task results showed, a long term of methanol feeding leads to persistent memory impairment in the monkeys. The working memory impairment lasted at least 6 months beyond the feeding regimen, suggesting a permanent-like cognitive impairment for the employed monkeys. The cognitive impairment coincided with the increase in Tau phosphorylation at residues of Thr181 and Ser396 in cerebrospinal fluid during the feeding. Distinct increases in

Tau-phosphorylated aggregates (some of them look like neurofibrillary tangles) were also observed in four brain regions postmortem: frontal lobe, parietal lobe, temporal lobe, and hippocampus (Yang et al. 2014b). Tau hyperphosphorylation in the brain maintained and persisted even 6 months after a feed of methanol ceased. That is to say, methanol feeding caused long-lasting and persistent pathological changes that were related to AD-like behavior in monkeys. In addition, levels of phosphorylated Tau in cerebrospinal fluid were dependent on doses of methanol being fed.

Further cytotoxicity test, using mouse the primary neurons with methanol, formaldehyde, and formic acid, showed that formaldehyde was the most toxic with the least neurites and highest Tau phosphorylation levels (Yang et al. 2014a).

2.4 *Activation of GSK-3β in the Presence of Formaldehyde*

As mentioned above, Tau was phosphorylated at Thr181 and Ser396 under formaldehyde treatment both in the nucleus and cytoplasm. So far, more than 20 kinases are involved in Tau phosphorylation (Wang et al. 2013b). Glycogen synthase kinase 3 beta (GSK-3β) that recognizes and reacts with Tau protein at more than 11 phosphorylation sites including Thr181 was assessed in formaldehyde-participated Tau hyperphosphorylation. The phosphorylation of Tyr216, which signified the activation of GSK-3β, was markedly elevated in N2a cells in the presence of formaldehyde (He et al. 2016). In the meantime, GSK-3β was found to be increased in the nucleus with the addition of formaldehyde which might contribute to the phosphorylation of nuclear Tau (Lu et al. 2013a). On the other hand, formaldehyde-induced Tau phosphorylation was greatly reduced when activation of GSK-3β was inhibited by its inhibitor LiCl. In order to confirm the effect, siRNA targeted to GSK-3β was transfected into N2a cells, which resulted in decreased Tau phosphorylation levels in the same formaldehyde treatment. In fact, GSK-3β is regarded as one of the most important kinases to produce hyperphosphorylation of Tau in AD (Jembrek et al. 2013). Intracellular accumulation of Aβ also likely induces increase in hyperphosphorylated Tau by a mechanism dependent on GSK-3β (Alvarez et al. 1999). Other kinases such as CDK5, EARK1/EARK2, and 14-3-3ε were also tested, and none showed strong effect as GSK-3β. Thus, these experiments indicated formaldehyde can activate GSK-3β and further lead to Tau multiple phosphorylation and aggregation in vitro and in vivo (Lu et al. 2013a).

2.5 *Activation of CaMKII in the Presence of Formaldehyde*

Calmodulin-dependent protein kinase II (CaMKII) is demonstrated to participate in Tau phosphorylation (Wang et al. 2013b). It is a remarkably complex protein kinase, involving in synaptic plasticity and memory formation. Dysregulation of CaMKII may thus be a trigger of pathology in AD, a dementia characterized by aberrant calcium signaling, synapse and neuronal loss, and impaired memory (Ghosh and

Giese 2015). CaMKII prefers phosphorylation of Tau at Ser396 and six other residues (Wang et al. 2013b). Thus, He and colleagues determined expression levels of CaMKII in N2a cells which were incubated with formaldehyde. They observed upregulation of CaMKII protein levels induced by formaldehyde (He et al. 2016). That is to say, the activation of CaMKII may contribute to Tau phosphorylation at Ser396, too. CaMKII dysregulation critically contributes to neurodegeneration and memory impairment in AD (Sanhueza and Lisman 2013).

As described by Sanhueza and Lisman, synaptic strength is stored stably through the combined actions of the CaMKII/NMDAR (N-methyl-D-aspartate receptor) complex and N-cadherin dimers (Sanhueza and Lisman 2013). Storage of synaptic information may instead be mediated by regulated interaction of CaMKII with NMDAR complex. Recent findings also implicated roles of CaMKII in long-term depression (LTD), as well as functional roles at inhibitory synapses (Coultrap and Bayer 2012). Tong and colleagues found that elevation of formaldehyde markedly suppressed LTP in rat hippocampus (Tong et al. 2011). As their Western blotting results showed, formaldehyde also induced downregulation of NMDA receptors, such as NR2B (NMDAR subtype 2B) and NR1 (NMDAR subtype 1), in a concentration-dependent manner (Tong et al. 2013). Thus, it appears that upregulation of CaMKII in N2a cells may be due to the suppression of NMDA receptors in the presence of formaldehyde.

2.6 Changes of Dyrk1A and CDK5 in the Presence of Formaldehyde

Dual specificity tyrosine-phosphorylation-regulated kinase 1A (Dyrk1A) has 11 phosphorylation sites on Tau including Thr181 and Ser396 (Liu et al. 2008; Wang et al. 2013b). According to Lu and coworkers, expression of Dyrk1A, however, decreased in N2a cells during formaldehyde stress (Lu et al. 2013a). That is to say, Dyrk1A did not participate the phosphorylation of Tau in the presence of formaldehyde. Cyclin-dependent kinase 5 (CDK5) functioning with its coactivator p35 and truncated coactivator p25 (p35/p25) phosphorylates Tau protein at multiple sites including Thr181 and Ser396 (Kesavapany et al. 2004; Patrick et al. 1999; Tsai et al. 1994; Wang et al. 2013b). Compared with the control group, formaldehyde caused no significant increase in levels of p35/p25 or transcription of CDK5 in N2a cells under the same experimental conditions (Lu et al. 2013a).

2.7 Downregulation of PP2A in the Presence of Formaldehyde

Phosphoseryl/phosphothreonyl protein phosphatase-2A (PP2A) is a large family of enzymes that account for the majority Ser/Thr phosphatase activity in the brain (Wang et al. 2013b). Dysfunction of PP2A is closely related to Tau hyperphosphorylation, amyloidogenesis, and synaptic deficits. It is a particularly interesting field to study changes in PP2A catalytic activities, regulators of PP2A, expressions of its subunits, methylation, and/or phosphorylation modifications in AD-affected brain regions. Deregulation of PP2A affects activity of many Ser/Thr protein kinases implicated in AD (Sontag and Sontag 2014). However, it remains unclear what are the primary events underlying "PP2A" dysfunction in AD. He and coworkers incubated N2a cells with formaldehyde and found that the levels of PP2A protein markedly decreased. However, the decrease of PP2A could be rescued by addition of resveratrol to N2a cells (He et al. 2016).

According to Chan and Sucher, stimulation of NMDARs led to dissociation of PP2A from the complex and reduction of PP2A activity. They believed that NMDARs were allosteric modulators of PP2A, which in turn controlled their own phosphorylation state (Chan and Sucher 2001). Comparing previous work reported by Tong and colleagues, it suggested that the suppression of PP2A of N2a cells in the presence of formaldehyde was due to inactivation and downregulation of NMDARs (He et al. 2016; Tong et al. 2015).

2.8 *Tau Hyperphosphorylation and DNA Damage*

Although Tau is mainly known as an axonal microtubule-associated protein, many studies indicate that it is also located in nucleus (Loomis et al. 1990; Lu et al. 2014; Wang et al. 1993). Tau protein can be stained with immunofluorescence only in human cells such as SH-SY5Y. However, when applied with subcellular fractionation or nuclei isolation method, Tau was able to detect in rodent nucleus, too (Lu et al. 2014). Nuclear Tau is usually hypophosphorylated, which means the phosphorylation may also regulate the transportation of Tau across the nuclear envelope (Sultan et al. 2011). As described previously, Tau binding to DNA enhanced melting temperature of DNA and prevented DNA from attacks by free radicals (Hua and He 2002, 2003; Hua et al. 2003). Tau protein and DNA complex displayed a beads-on-string-like structure (Hua and He 2003). The beads were composed of Tau and folded DNA called DNA-Tauosome (Lu et al. 2013b). Hyperphosphorylation promoted Tau to separate from not only microtubules but also DNA double strands. Thus, the DNA-Tauosome will be disrupted when the bound Tau protein was dissociated from the complex (Lu et al. 2011). As mentioned above, formaldehyde triggers Tau hyperphosphorylation in neural cells (Lu et al. 2013a; Yang et al. 2014a). Therefore, DNA could be suffered from the loss of protection by Tau protein when Tau is hyperphosphorylated in the presence of formaldehyde.

2.9 *Tau Phosphorylation in Accumulation at the Pathological Concentration of Formaldehyde*

In fact, age-related cognitive impairment is a chronic and relentless progression. A chronic exposure to formaldehyde, which can cross the blood-brain barrier freely (Shcherbakova et al. 1986), leads to impairment of human memory (Kilburn 1994;

Tong et al. 2013). On clinical, the average pathological concentration of FA in the urine of AD patients is 13.70 ± 5.17 μM, which is ~1.4 times as high as that of the physiological FA concentration (9.61 ± 2.90 μM) from age-matched normal participants (Li et al. 2016). However, concentrations of formaldehyde used in previous reports were much higher than concentrations of endogenous FA in aging people under either physiological or pathological states (Sun et al. 2016). Therefore, long-term cellular effects of exposure to low concentrations of formaldehyde, which is consistent with pathological concentration, should be investigated. The concentration of formaldehyde used to treat cells was based on the average levels of endogenous formaldehyde in AD patients, and the level of formaldehyde in age-matched normal aging participants was used as control. The experimental process was attempted to clarify whether formaldehyde at as low as the pathological concentration affects the morphology and connectivity of neurons.

Wang and colleagues established a cell culture system with a long-term treatment with formaldehyde to simulate the chronic exposure to pathological concentration of formaldehyde. This experimental protocol was to clarify whether low concentration and long-term exposure can cause formaldehyde accumulation for cells (Wang et al. 2017). The answer was positive. The response of N2a cells to the low concentration formaldehyde exposure was examined (Wang et al. 2017). They found that the pathological concentration of formaldehyde decreased cell viability, altered cell morphology, and increased Tau hyperphosphorylation. They also monitored the changes of cellular morphology using digital holographic microscopy after a 6-day incubation with formaldehyde. As shown in Fig. 9.3, changes in the morphological indexes can be observed, which could directly reflect the effect of formaldehyde accumulation on cell growth, cell adhesion, and the spreading of cell processes (Wang et al. 2017). They adjusted thickness value to exclude interference of dimension of cell volume on cell thickness, making it more reliable for describing the adhesive capacity of cells. After a long-term treatment with pathological level of formaldehyde, cells became rounder, less attached to the bottom of the culture dish, and fewer cell processes compared with the control cells. To demonstrate the effect of formaldehyde accumulation (Fig. 9.4), they also performed the primary culture of hippocampal neurons from mouse and observed the decreases in neurites under the same experimental conditions (Wang et al. 2017). These results suggested that low concentration of formaldehyde accumulation chronically suppresses the connectivity among neurons in culture.

The cytoskeletal system, for instance, the microtubule system, is essential to maintain cellular morphology. As a microtubule-associated protein, Tau functions in the promotion and stabilization of the microtubule system and actin filaments (Weingarten et al. 1975; Drubin and Kirschner 1986; Avila 2006), and it is regulated by phosphorylation and dephosphorylation (Caceres and Kosik 1990). When Tau is hyperphosphorylated, Tau disassociates from microtubules or actin filaments, followed by the disruption of the cytoskeleton. In other words, Tau hyperphosphorylation causes the disruption of cellular processes, especially the terminal processes. Tau was markedly phosphorylated during the long-term treatment with low concentration of formaldehyde, followed by cell shrinkage and even cell death.

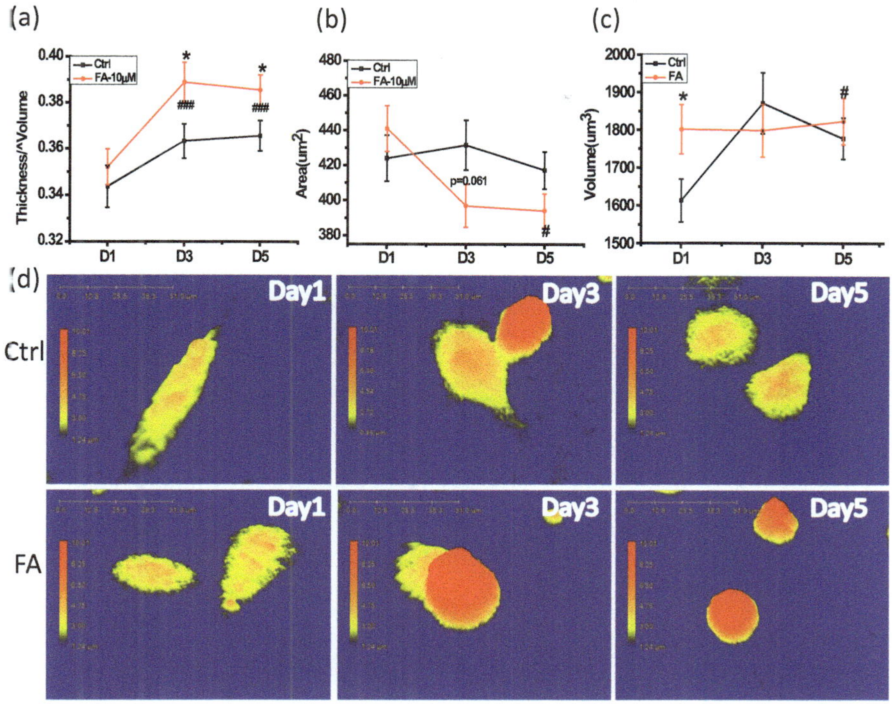

Fig. 9.3 Morphological changes of N2a cells in the presence of formaldehyde. Cell culture and inoculation were performed as described for this figure. Aliquots of N2a cells were taken daily for measurements of their morphologies on a holographic imaging system (Phase Holographic Imaging AB, Sweden) in order to analyze the thickness/^volume (panel **a**), average area (panel **b**), and average volume (panel c) at 4 h, when N2a cells were adherent. Values are expressed as the means ± SE; *P<0.05, **P<0.01,***P<0.001 vs the control group; ##P<0.01, ###P<0.001 vs the control group on the first day. The P values were obtained based on comparative analysis of the indicated group with the controls. Inverted phase-contrast microscopy showed the morphology of N2a cells in the presence (panel **d**) or absence (panel **e**) of formaldehyde on day 5 (Wang et al. 2017). The usage of this figure was authorized by the authors and chief editor of Prog. Biochem. Biophys

Tau hyperphosphorylation should be one of the explanations for the distinct change of cellular morphologies in the presence of low concentration of formaldehyde. Alzheimer's disease is characterized by an almost complete redistribution of Tau protein pool from a microtubule-bound to an aggregated state in the form of pathological amorphous oligomers and PHFs (Geodert et al. 1988; Harrington et al. 1991). Tau hyperphosphorylation usually occurred at early stage in the process of neurofibrillary degeneration of AD and positively correlated with progression and diagnosis of dementia in AD patients (Kopke et al. 1993). In other words, accumulation of the pathological concentration of endogenous formaldehyde referred to AD patients induced chronic damages to N2a cells and primary cultured murine hippocampal cells, especially the neuronal processes and neurites resulted from Tau hyperphosphorylation.

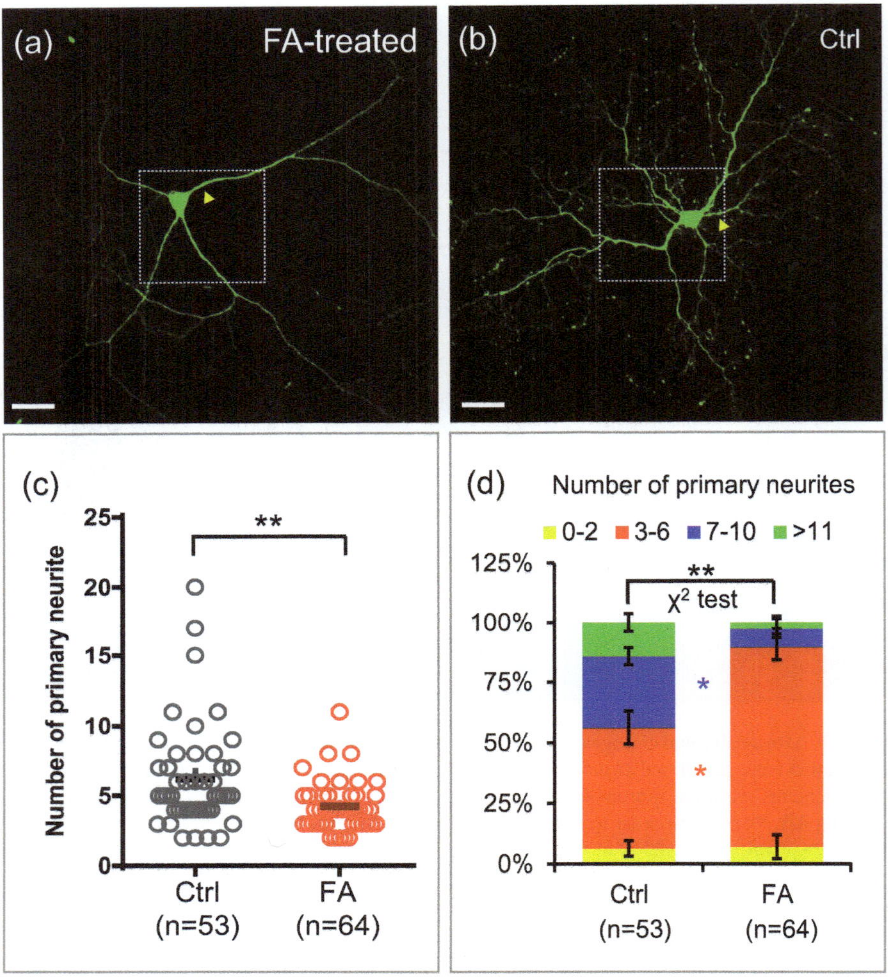

Fig. 9.4 Changes in neurites in the presence of formaldehyde accumulation. The representative morphology of the neurites of primary hippocampal neurons cultured in the presence of 10 μM formaldehyde (panel **a**) and control neurons cultured without the presence of formaldehyde (panel **b**) was displayed. The primary neurons were transfected with EGFP plasmid at 13DIV, and the transfected neurons were then treated with FA at the concentration of 10 μM for 3 days. The medium was changed every day. Images of EGFP-positive neurons were observed with fluorescence confocal microscopy. *Scale bar* = 10 μm. The number of primary neurites was quantified by the ImageJ software. The number of primary neurites of the FA-treated neurons ($n = 53$) was significantly smaller than the control neurons ($n = 64$). (**d**) The distribution of primary neurites numbers in the FA-treated and control neuron populations differed significantly. The distribution of neurite number was analyzed by chi-square test. Data in (panels **c** and **d**) were collected from three independent experiments. Data are shown as mean ± s.e.m. Means were compared with one-way ANOVA $*P < 0.05$; $**P < 0.01$ (Wang et al. 2017). The usage of this figure was authorized by the authors and chief editor of Prog. Biochem. Biophys

3 Aβ Deposition in the Presence of Formaldehyde

Amyloid-β peptides are critically involved in AD as the main component of senile plaques found in the brains of Alzheimer's patient. The amyloid precursor protein (APP) is cleaved by γ-secretase and β-secretase to yield Aβ (Shu et al. 2015). A large body of literatures reported that vascular disorders are involved in the pathogenesis of vascular dementia and AD (Wallin et al. 2016a, b). In 2001, Yu hypothesized that SSAO-mediated generation of formaldehyde can induce protein $A\beta_{1-40}$ cross-linkage, deposition, and subsequently plaque formation in the compartment adjacent to the cerebral vessels (Yu 2001). As described, SSAO-mediated deamination of methylamine or amino acetone produces toxic formaldehyde and methylglyoxal, respectively (Yu 2001). These aldehydes can cause intramolecular and intermolecular protein cross-linkages. As Chen and coworkers have reported, aldehydes such as formaldehyde, methylglyoxal, and malondialdehyde at low concentrations accelerated the formation of β-amyloid, β-sheets, oligomers, and protofibrils as well as increased size of the aggregates (Chen et al. 2007). So far, we believed that imbalance of formaldehyde metabolism, originated from many metabolic pathways, is one of the risk factors to age-related cognitive impairment (He et al. 2010). Sun and coworkers observed that treatment on human neuroblastoma cells with *geniposide*, a traditional Chinese drug, resisted to Aβ aggregation induced by formaldehyde (Sun et al. 2013).

To demonstrate whether formaldehyde is able to induce amyloid plaques in vivo, Yang and collaborators have studied the brain of monkey which was fed with methanol as described above (Yang et al. 2014a, b). In their observations, senile plaques were found in four brain regions postmortem: frontal lobe, parietal lobe, temporal lobe, and hippocampus. It seems as if formaldehyde induced Aβ deposition by activation of β- and γ-secretase. The presence of amyloid plaques in monkeys highlighted a marked difference in animal systems used in AD investigations, suggesting that the innate defense in mouse against methanol toxicity makes it unable to completely mimicking AD pathology. In addition, senile plaques were also observed by intracerebroventricular injection of low concentrations of formaldehyde in monkeys (unpublished data). Nonetheless, these findings support a growing body of evidence to a complicated link between dysmetabolism of formaldehyde and AD pathology.

4 Aggregation of APOE4 in the Presence of Formaldehyde

In humans, there are three variant alleles at the APOE locus: ε2, ε3, and ε4. ApoE ε4 is the most risk factor in the development of late-onset AD (LOAD) (Castellano et al. 2011). Apolipoprotein E is an essential cholesterol carrier in the brain and helps in neuronal maintenance and repair. People who develop AD are more likely to have an APOE ε4 allele than people who do not develop the disease. APOE ε4 is present in about 25–30% of the population and in about 40% of people with

LOAD. Although research supports the correlation of APOE ε4 variant and the occurrence of LOAD, full mechanism of the pathophysiology is not known (Huang et al. 2004). Formaldehyde is also a risk factor increasing with age that is similar to APOE ε4. Epidemiological investigation of aging population in Beijing communities showed that 40% of the LOAD patients got high levels of endogenous formaldehyde. This rate is coincidently similar to that of ApoE ε4 cases. APOE protein is believed to involve in aggregation and/or clearance of Aβ from extracellular space, where APOE ε4 leads to prominent Aβ accumulation (Verghese et al. 2011; Youmans et al. 2012). Rizak and his colleagues combined these two factors and investigated their functions in Aβ aggregation. Their results showed that formaldehyde could induce aggregation of $A\beta_{1-40}$ and ApoE (ε4, ε3, and ε2) in vitro (Rizak et al. 2014). Further mixed the three factors ApoE isoforms with $A\beta_{1-40}$ in high formaldehyde concentrations (10 mM) reflected the trend of ApoE isoform genetic vulnerability to AD (APOE ε4 > ε3 > ε2). The higher formaldehyde with ApoE ε4 showed higher molecular bands in Western blotting (>250 kDa). These data provided an insight of the APOE genetic variability in AD pathology.

5 Conclusion

Formaldehyde acts as an effective trigger of Tau protein hyperphosphorylation in neural cell lines and primary cultured neurons. As shown in the schematic Fig. 9.5., with a short-term formaldehyde treatment, Tau phosphorylation and polymerization occur, followed by dephosphorylation and depolymerization as formaldehyde levels decreased. Maintaining a high level of formaldehyde for a longer period, the Tau hyperphosphorylation lasts even longer and leads to irreversible aggregation and cell death. Administration of methanol to rodents by gavage, feeding in water, intraperitoneal injection, or cerebral ventricle injection of formaldehyde can also induce Tau hyperphosphorylation in neural cells from the hippocampus and brain cortex accompanied with cognitive impairment, but not Aβ deposition or plaques. The activation of GSK-3β and CaMKII as well as suppression of PP2A might be a mechanism under formaldehyde-induced Tau pathologies. Furthermore, primate model rhesus monkeys bearing both Tau hyperphosphorylation and Aβ deposition in the hippocampus and cerebral cortex suffered from decline of cognitive ability through a long-term feeding with water containing formaldehyde. Aβ deposition in monkey brain may be due to the activation of β- and γ-secretase, but the mechanism of activation needs further investigating. Amyloid β deposition and Tau hyperphosphorylation occurs early (preclinical) in age-related cognitive impairment. Cells may respond to overloaded formaldehyde in diverse manners. However, as a trigger of Aβ deposition and Tau phosphorylation, it suggests that the onset of dysmetabolism of endogenous formaldehyde should process before Aβ deposition and Tau phosphorylation in the early onset of preclinical stage of AD.

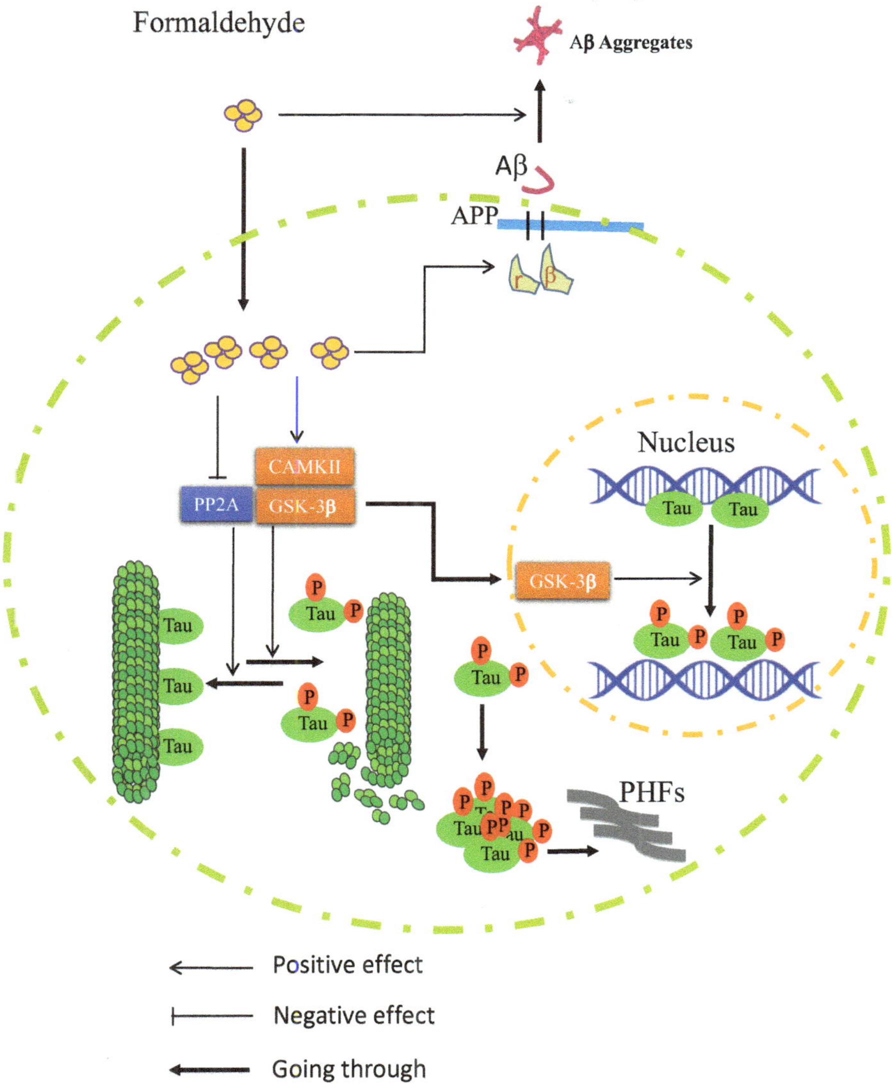

Fig. 9.5. A putative mechanism for Tau hyperphosphorylation and polymerization and Aβ deposition in the presence of formaldehyde overload. Formaldehyde overload causes activation GSK-3β and CAMKII and nuclear accumulation of GSK-3β, then Tau protein, especially nuclear Tau, becomes highly phosphorylated; meanwhile, PP2A was inhibited and leading to the accumulation of hyperphosphorylated Tau and the subsequent formation of ThS-positive polymers, even PHFs. Hyperphosphorylated Tau in the nucleus may disassociate from DNA and lose its protective function of the microtubule system. Levels of Tau phosphorylation in the cytoplasm increase and reduce the ability of Tau protein to bind to microtubules, resulting in destabilization of microtubules. Formaldehyde leads to the activation of β- and γ-secretases, increasing the production of Aβ and secretion of Aβ. Formaldehyde extracellular induces the accumulation and deposition

Acknowledgment This project was supported by grants from the National Key Research and Development Program of China (2016YFC1306300), the National Basic Research Program of China (973 Program) (2012CB911004), the Natural Scientific Foundation of China (NSFC 31270868), Foundation of Chinese Academy of Sciences CAS-20140909, and the Queensland-Chinese Academy of Sciences Biotechnology Fund (GJHZ201302).

Competing Financial Interests The authors declare no competing financial interests.

References

Abraha A, Ghoshal N, Gamblin TC, Cryns V, Berry RW, Kuret J, Binder LI (2000) C-terminal inhibition of tau assembly in vitro and in Alzheimer's disease. J Cell Sci 113(Pt 21):3737–3745

Allison SL, Fagan AM, Morris JC, Head D (2016) Spatial navigation in preclinical Alzheimer's disease. J Alzheimer's Dis JAD 52:77–90

Alonso AD, Beharry C, Corbo CP, Cohen LS (2016) Molecular mechanism of prion-like tau-induced neurodegeneration. Alzheimer's Dementia J Alzheimer's Assoc 12:1090–1097

Alvarez G, Munoz-Montano JR, Satrustegui J, Avila J, Bogonez E, Diaz-Nido J (1999) Lithium protects cultured neurons against beta-amyloid-induced neurodegeneration. FEBS Lett 453:260–264

Alzheimer A (1907) Über eine eigenartige Erkrankung der Hirnrinde [about a peculiar disease of the cerebral cortex]. Allgemeine Zeitschrift fur Psychiatrie und Psychisch-Gerichtlich Medizin 64:146–148

Alzheimer's A (2015) 2015 Alzheimer's disease facts and figures. Alzheimer's Demen 11:332–384

Arnsten AF, Cai JX, Goldman-Rakic PS (1988) The alpha-2 adrenergic agonist guanfacine improves memory in aged monkeys without sedative or hypotensive side effects: evidence for alpha-2 receptor subtypes. J Neurosci 8:4287–4298

Avila J (2006) Tau phosphorylation and aggregation in Alzheimer's disease pathology. FEBS Lett 580:2922–2927

Ballard C, Khan Z, Clack H, Corbett A (2011) Nonpharmacological treatment of Alzheimer disease. Can J Psychiatr 56:589–595

Barron M, Gartlon J, Dawson LA, Atkinson PJ, Pardon MC (2016) A state of delirium: deciphering the effect of inflammation on tau pathology in Alzheimer's disease. Exp Gerontol 94:103–107

Bird TD (1993) Early-onset familial Alzheimer disease. In: GeneReviews(R) editors: Pagon RA, Adam MP, Ardinger HH, Wallace SE, Amemiya A, Bean LJH, Bird TD, Fong CT, Mefford HC, Smith RJH et al. (eds) Seattle (WA)

Blennow K (2005) CSF biomarkers for Alzheimer's disease: use in early diagnosis and evaluation of drug treatment. Expert Rev Mol Diagn 5:661–672

Bruckner J, Warren D (2001) Toxic effects of solvents and vapors in Casarett & Doull's toxicology. The basic science of poisons, 6th edn. Klaassen CD, McGraw-Hill, Kansas, USA, pp 894–895

Bufill E, Blesa R, Augusti J (2013) Alzheimer's disease: an evolutionary approach. J Anthropol Sci 91:135–157

Caceres A, Kosik KS (1990) Inhibition of neurite polarity by tau antisense oligonucleotides in primary cerebellar neurons. Nature 343:461–463

Cai T, Che H, Yao T, Chen Y, Huang C, Zhang W, Du K, Zhang J, Cao Y, Chen J et al (2011) Manganese induces tau hyperphosphorylation through the activation of ERK MAPK pathway in PC12 cells. Toxicol Sci 119:169–177

Castellano JM, Kim J, Stewart FR, Jiang H, DeMattos RB, Patterson BW, Fagan AM, Morris JC, Mawuenyega KG, Cruchaga C et al (2011) Human apoE isoforms differentially regulate brain amyloid-beta peptide clearance. Sci Transl Med *3*:89ra57

Chan SF, Sucher NJ (2001) An NMDA receptor signaling complex with protein phosphatase 2A. J Neurosci 21:7985–7992
Chen XX, Su T (2015) Microcirculation dysfunction in age-related cognitive impairment. Prog Biochem Biophys 42(12):1077–1083
Chen YH, Luo JY, Li W, He RQ (1999) Effect of acetaldehyde on phosphorylation of human neuronal tau. J Biochem Mol Biol & Biophys 3:197–202
Chen K, Kazachkov M, Yu PH (2007) Effect of aldehydes derived from oxidative deamination and oxidative stress on beta-amyloid aggregation; pathological implications to Alzheimer's disease. J Neural Trans (Vienna, Austria: 1996) 114:835–839
Chen N, Dai LF, Jiang WY, Wu Y (2012a) Pathogenic role of UPR (unfolded protein response) among hereditary Leukoencephalopathy and neurodegenerative disorders after endoplasmic reticulum stress. Prog Biochem Biophys 39:764–770
Chen NN, Luo DJ, Yao XQ, Yu C, Wang Y, Wang Q, Wang JZ, Liu GP (2012b) Pesticides induce spatial memory deficits with synaptic impairments and an imbalanced tau phosphorylation in rats. J Alzheimer's Dis 30:585–594
Coultrap SJ, Bayer KU (2012) CaMKII regulation in information processing and storage. Trends Neurosci 35:607–618
Cui Y, Su T, Zhang SD, Huang P, He YG, Liu Y, Zhang C, Ritch R, He RQ (2016) Elevated urine formaldehyde in elderly patients with primary open angle glaucoma. Int J Ophthalmol 9:411–416
Drubin DG, Kirschner MW (1986) Tau protein function in living cells. J Cell Biol 103:2739–2746
Engelborghs S, De Deyn PP (2001) Biological and genetic markers of sporadic Alzheimer's disease. Acta Med Okayama 55:55–63
Evans AM, Fameli N, Ogunbayo OA, Duan J, Navarro-Dorado J (2016) From contraction to gene expression: nanojunctions of the sarco/endoplasmic reticulum deliver site- and function-specific calcium signals. Sci China Life Sci 59(8):749–763
Fu HJ, Rodriguez GA, Herman M, Emrani S, Nahmani E, Barrett G, Figueroa HY, Goldberg E, Hussaini SA, Duff KE (2017) Tau pathology induces excitatory neuron loss, grid cell dysfunction, and spatial memory deficits reminiscent of early Alzheimer's disease. Neuron 93(3):533–541
Fyhn M, Molden S, Witter MP, Moser EI, Moser MB (2004) Spatial representation in the entorhinal cortex. Science 305:1258–1264
Ghosh A, Giese KP (2015) Calcium/calmodulin-dependent kinase II and Alzheimer's disease. Mol Brain 8(78). doi:10.1186/s13041-015-0166-2
Goedert M, Wischik CM, Crowther RA et al (1988) Cloning and sequencing of the cDNA encoding a core protein of the paired helical filament of Alzheimer disease: identification as the microtubule-associated protein tau. Proc Natl Acad Sci U S A 85(11):4051–4055
Gong CX, Singh TJ, Grundke-Iqbal I, Iqbal K (1993) Phosphoprotein phosphatase activities in Alzheimer disease brain. J Neurochem 61:921–927
Gong CX, Liu F, Grundke-Iqbal I, Iqbal K (2005) Post-translational modifications of tau protein in Alzheimer's disease. J Neural Trans (Vienna, Austria: 1996) 112:813–838
Grundke-Iqbal I, Iqbal K, Quinlan M, Tung YC, Zaidi MS, Wisniewski HM (1986) Microtubule-associated protein tau. A component of Alzheimer paired helical filaments. J Biol Chem 261:6084–6089
Hafting T, Fyhn M, Molden S, Moser MB, Moser EI (2005) Microstructure of a spatial map in the entorhinal cortex. Nature 436:801–806
Harrington CR, Mukaetova-Ladinska EB, Hills R et al (1991) Measurement of distinct immunochemical presentations of tau protein in Alzheimer disease. Proc Natl Acad Sci U S A 88(13):5842–5846
He R (2016) Abnormal lysosome, formaldehyde Dysmetabolism and age-related cognitive impairment. Prog Biochem Biophys 43(12):1197
He R, Luo J, Li W (1998) Effect of ethanol on the aggregation of human neuronal tau protein. Protein Pept Lett 5(5):279–285

He R, Lu J, Miao JY (2010) Formaldehyde stress. Sci China Life Sci 53(12):1399–1404
He X, Li Z, Rizak JD, Wu S, Wang Z, He R, Su M, Qin D, Wang J, Hu X (2016) Resveratrol attenuates formaldehyde induced hyperphosphorylation of Tau protein and cytotoxicity in N2a cells. Front Neurosci 10(598). doi:10.3389/fnins.2016.00598
Hu X, Wang T, Jin F (2016) Alzheimer's disease and gut microbiota. Sci China Life Sci 59(10):1006–1023
Hua Q, He RQ (2002) Effect of phosphorylation and aggregation on tau binding to DNA. Protein Pept Lett 9(4):349–357
Hua Q, He RQ (2003) Tau could protect DNA double helix structure. Biochim Biophys Acta 1645(2):205–211
Hua Q, He RQ, Haque N, Qu MH, del Carmen Alonso A, Grundke-Iqbal I, Iqbal K (2003) Microtubule associated protein tau binds to double-stranded but not single-stranded DNA. Cell Mol Life Sci CMLS 60(2):413–421
Huang W, Qiu C, von Strauss E, Winblad B, Fratiglioni L (2004) APOE genotype, family history of dementia, and Alzheimer disease risk: a 6-year follow-up study. Arch Neurol 61(12):1930–1934
Jembrek MJ, Babic M, Pivac N, Hof PR, Simic G (2013) Hyperphosphorylation of tau by gsk-3 beta in alzheimer's disease: the interaction of a beta and sphingolipid mediators as a therapeutic target. Transl Neurosci 4(4):466–476
Jian X, Zhu MX (2016) Regulation of lysosomal ion homeostasis by channels and transporters. Sci China Life Sci 59(8):777–791
Kar A, Kuo D, He R, Zhou J, Wu JY (2005) Tau alternative splicing and frontotemporal dementia. Alzheimer Dis Assoc Disord 19(Suppl 1):S29–S36
Kesavapany S, Li BS, Amin N, Zheng YL, Grant P, Pant HC (2004) Neuronal cyclin-dependent kinase 5: role in nervous system function and its specific inhibition by the Cdk5 inhibitory peptide. Biochim Biophys Acta 1697(1–2):143–153
Kilburn KH (1994) Neurobehavioral impairment and seizures from formaldehyde. Arch Environ Health 49(1):37–44
Kim B, Backus C, Oh S, Hayes JM, Feldman EL (2009) Increased tau phosphorylation and cleavage in mouse models of type 1 and type 2 diabetes. Endocrinology 150(12):5294–5301
Kim DH, Huh JW, Jang M, Suh JH, Kim TW, Park JS, Yoon SY (2012) Sitagliptin increases tau phosphorylation in the hippocampus of rats with type 2 diabetes and in primary neuron cultures. Neurobiol Dis 46(1):52–58
Kopke E, Tung YC, Shaikh S et al (1993) Microtubule-associated protein tau. Abnormal phosphorylation of a non-paired helical filament pool in Alzheimer disease. J Biol Chem 268(32):24374–24384
Ksiezak-Reding H, Liu WK, Yen SH (1992) Phosphate analysis and dephosphorylation of modified tau associated with paired helical filaments. Brain Res 597(2):209–219
Li T, Paudel HK (2006) Glycogen synthase kinase 3beta phosphorylates Alzheimer's disease-specific Ser396 of microtubule-associated protein tau by a sequential mechanism. Biochemistry 45(10):3125–3133
Li XH, Lv BL, Xie JZ, Liu J, Zhou XW, Wang JZ (2012) AGEs induce Alzheimer-like tau pathology and memory deficit via RAGE-mediated GSK-3 activation. Neurobiol Aging 33(7):1400–1410
Lindwall G, Cole RD (1984) Phosphorylation affects the ability of tau protein to promote microtubule assembly. J Biol Chem 259(8):5301–5305
Lithfous S, Dufour A, Despres O (2013) Spatial navigation in normal aging and the prodromal stage of Alzheimer's disease: insights from imaging and behavioral studies. Ageing Res Rev 12(1):201–213
Liu F, Liang Z, Wegiel J, Hwang YW, Iqbal K, Grundke-Iqbal I, Ramakrishna N, Gong CX (2008) Overexpression of Dyrk1A contributes to neurofibrillary degeneration in down syndrome. FASEB J 22(9):3224–3233
Liu K, He Y, Yu L, Chen Y, He R (2017) Markedly elevated formaldehyde in the cecum of APP/PS1 mouse. Microbiol China 44(8):1761–1766

Loomis PA, Howard TH, Castleberry RP, Binder LI (1990) Identification of nuclear tau isoforms in human neuroblastoma cells. Proc Natl Acad Sci U S A 87(21):8422–8426
Lu Z, Li CM, Qiao Y, Yan Y, Yang X (2008) Effect of inhaled formaldehyde on learning and memory of mice. Indoor Air 18(2):77–83
Lu J, Miao JY, Pan R, He RQ (2011) Formaldehyde-mediated hyperphosphorylation disturbs the interaction between tau protein and DNA. Prog Biochem Biophys 38(12):1113–1120
Lu J, Miao J, Su T, Liu Y, He R (2013a) Formaldehyde induces hyperphosphorylation and polymerization of tau protein both in vitro and in vivo. Biochim Biophys Acta 1830(8):4102–4116
Lu Y, He HJ, Zhou J, Miao JY, Lu J, He YG, Pan R, Wei Y, Liu Y, He RQ (2013b) Hyperphosphorylation results in tau dysfunction in DNA folding and protection. J Alzheimers Dis 37(3):551–563
Lu J, Li T, He R, Bartlett PF, Gotz J (2014) Visualizing the microtubule-associated protein tau in the nucleus. Sci China Life Sci 57(4):422–431
Luo J, He R (1999) Effect of acetaldehyde on aggregation of neuronal tau. Protein Pept Lett 6(2):105–110
MacAllister SL, Choi J, Dedina L, O'Brien PJ (2011) Metabolic mechanisms of methanol/formaldehyde in isolated rat hepatocytes: carbonyl-metabolizing enzymes versus oxidative stress. Chem Biol Interact 191(1–3):308–314
Miao J, Lu J, Zhang Z, Tong Z, He R (2013) The effect of formaldehyde on cell cycle is in a concentration-dependent manner. Prog Biochem Biophys 40(7):641–651
Michel G, Mercken M, Murayama M, Noguchi K, Ishiguro K, Imahori K, Takashima A (1998) Characterization of tau phosphorylation in glycogen synthase kinase-3beta and cyclin dependent kinase-5 activator (p23) transfected cells. Biochim Biophys Acta 1380(2):177–182
Monte WC (2012) While science sleeps (SC, USA: Charleston)
Mudher A, Lovestone S (2002) Alzheimer's disease-do tauists and baptists finally shake hands? Trends Neurosci 25(1):22–26
Nie CL, Wei Y, Chen XY, Liu YY, Dui W, Liu Y, Davies MC, Tendler SJB, He RG (2007) Formaldehyde at low concentration induces protein tau into globular amyloid-like aggregates in vitro and in vivo. PLoS One 2(7). doi:10.1371/journal.pone.0000629
Nobutoki T, Ihara T (2015) Early disruption of neurovascular units and microcirculatory dysfunction in the spinal cord in spinal muscular atrophy type I. Med Hypotheses 85(6):842–845
Oliveira JM, Henriques AG, Martins F, Rebelo S, Silva OABDE (2015) Amyloid-beta modulates both a beta PP and tau phosphorylation. J Alzheimers Dis 45(2):495–507
Olsson A, Vanderstichele H, Andreasen N, De Meyer G, Wallin A, Holmberg B, Rosengren L, Vanmechelen E, Blennow K (2005) Simultaneous measurement of beta-amyloid(1-42), total tau, and phosphorylated tau (Thr181) in cerebrospinal fluid by the xMAP technology. Clin Chem 51(2):336–345
Patrick GN, Zukerberg L, Nikolic M, de la Monte S, Dikkes P, Tsai LH (1999) Conversion of p35 to p25 deregulates Cdk5 activity and promotes neurodegeneration. Nature 402(6762):615–622
Planel E, Tatebayashi Y, Miyasaka T, Liu L, Wang L, Herman M, Yu WH, Luchsinger JA, Wadzinski B, Duff KE et al (2007) Insulin dysfunction induces in vivo tau hyperphosphorylation through distinct mechanisms. J Neurosci 27(50):13635–13648
Qu Z, Jiao Z, Sun X, Zhao Y, Ren J, Xu G (2011) Effects of streptozotocin-induced diabetes on tau phosphorylation in the rat brain. Brain Res 1383(300–6):300–306
Rahman A, Ting K, Cullen KM, Braidy N, Brew BJ, Guillemin GJ (2009) The Excitotoxin Quinolinic acid induces tau phosphorylation in human neurons. PLoS One 4(7). doi:10.1371/journal.pone.0006344
Rhein V, Song X, Wiesner A, Ittner LM, Baysang G, Meier F, Ozmen L, Bluethmann H, Drose S, Brandt U et al (2009) Amyloid-beta and tau synergistically impair the oxidative phosphorylation system in triple transgenic Alzheimer's disease mice. Proc Natl Acad Sci U S A 106(47):20057–20062
Rizak JD, Ma Y, Hu X (2014) Is formaldehyde the missing link in AD pathology? The differential aggregation of amyloid-beta with APOE isoforms in vitro. Curr Alzheimer Res 11(5):461–468

Run X, Liang Z, Zhang L, Iqbal K, Grundke-Iqbal I, Gong CX (2009) Anesthesia induces phosphorylation of tau. J Alzheimer's Dis 16(3):619–626

Sanhueza M, Lisman J (2013) The CaMKII/NMDAR complex as a molecular memory. Mol Brain 6(10). doi:10.1186/1756-6606-6-10

Scheuner D, Eckman C, Jensen M, Song X, Citron M, Suzuki N, Bird TD, Hardy J, Hutton M, Kukull W et al (1996) Secreted amyloid beta-protein similar to that in the senile plaques of Alzheimer's disease is increased in vivo by the presenilin 1 and 2 and APP mutations linked to familial Alzheimer's disease. Nat Med 2(8):864–870

Serrano J, Fernandez AP, Martinez-Murillo R, Martinez A (2010) High sensitivity to carcinogens in the brain of a mouse model of Alzheimer's disease. Oncogene 29(15):2165–2171

Shcherbakova LN, Tel'pukhov VI, Trenin SO, Bashilov IA, Lapkina TI (1986) Permeability of the blood-brain barrier to intra-arterial formaldehyde. Biull Eksp Biol Med 102(11):573–575

Shu R, Wong W, Ma QH, Yang ZZ, Zhu H, Liu FJ, Wang P, Ma J, Yan S, Polo JM, et al (2015) APP intracellular domain acts as a transcriptional regulator of miR-663 suppressing neuronal differentiation. Cell Death Dis 19(6). doi:10.1038/cddis.2015.10

Sontag JM, Sontag E (2014) Protein phosphatase 2A dysfunction in Alzheimer's disease. Front Mol Neurosci 7(16). doi:10.3389/fnmol.2014.00016

Stoothoff WH, Johnson GV (2005) Tau phosphorylation: physiological and pathological consequences. Biochim Biophys Acta 1739(2–3):280–297

Su T, Monte WC, Hu X, He Y, He R (2016) Formaldehyde as a trigger for protein aggregation and potential target for mitigation of age-related, progressive cognitive impairment. Curr Top Med Chem 16(5):472–484

Sultan A, Nesslany F, Violet M, Begard S, Loyens A, Talahari S, Mansuroglu Z, Marzin D, Sergeant N, Humez S et al (2011) Nuclear tau, a key player in neuronal DNA protection. J Biol Chem 286(6):4566–4575

Sun P, Chen JY, Li J, Sun MR, Mo WC, Liu KL, Meng YY, Liu Y, Wang F, He RQ et al (2013) The protective effect of geniposide on human neuroblastoma cells in the presence of formaldehyde. BMC Complement Altern Med 13(152). doi:10.1186/1472-6882-13-152

Tang XQ, Zhuang YY, Zhang P, Fang HR, Zhou CF, Gu HF, Zhang H, Wang CY (2013) Formaldehyde impairs learning and memory involving the disturbance of hydrogen sulfide generation in the hippocampus of rats. J Mol Neurosci 49(1):140–149

Tong Z, Zhang J, Luo W, Wang W, Li F, Li H, Luo H, Lu J, Zhou J, Wan Y et al (2011) Urine formaldehyde level is inversely correlated to mini mental state examination scores in senile dementia. Neurobiol Aging 32(1):31–41

Tong Z, Han C, Luo W, Li H, Luo H, Qiang M, Su T, Wu B, Liu Y, Yang X et al (2013) Aging-associated excess formaldehyde leads to spatial memory deficits. Sci Rep 3(1807). doi:10.1038/srep01807

Tong Z, Han C, Qiang M, Wang W, Lv J, Zhang S, Luo W, Li H, Luo H, Zhou J et al (2015) Age-related formaldehyde interferes with DNA methyltransferase function, causing memory loss in Alzheimer's disease. Neurobiol Aging 36(1):100–110

Tsai LH, Delalle I, Caviness VS Jr, Chae T, Harlow E (1994) p35 is a neural-specific regulatory subunit of cyclin-dependent kinase 5. Nature 371(6496):419–423

Tsuji S (2010) Genetics of neurodegenerative diseases: insights from high-throughput resequencing. Hum Mol Genet 19(R1):R65–R70

Verghese PB, Castellano JM, Holtzman DM (2011) Apolipoprotein E in Alzheimer's disease and other neurological disorders. Lancet Neurol 10(3):241–252

Wallin A, Nordlund A, Jonsson M, Blennow K, Zetterberg H, Ohrfelt A, Stalhammar J, Eckerstrom M, Carlsson M, Olsson E et al (2016a) Alzheimer's disease--subcortical vascular disease spectrum in a hospital-based setting: overview of results from the Gothenburg MCI and dementia studies. J Cereb Blood Flow Metab 36(1):95–113

Wallin A, Nordlund A, Jonsson M, Lind K, Edman A, Gothlin M, Stalhammar J, Eckerstrom M, Kern S, Borjesson-Hanson A, Carlsson M, Olsson E, Zetterberg H, Blennow K, Svensson J, Öhrfelt A, Bjerke M, Rolstad S, Eckerström C (2016b) The Gothenburg MCI study: design and distribution of Alzheimer's disease and subcortical vascular disease diagnoses from baseline to 6-year follow-up. J Cereb Blood Flow Metab 36(1):114–131

Wang Y, Loomis PA, Zinkowski RP, Binder LI (1993) A novel tau transcript in cultured human neuroblastoma cells expressing nuclear tau. J Cell Biol 121(2):257–267

Wang JH, Rizak JD, Chen YM, Li L, Hu XT, Ma YY (2013a) Interactive effects of morphine and dopaminergic compounds on spatial working memory in rhesus monkeys. Neurosci Bull 29(1):37–46

Wang JZ, Xia YY, Grundke-Iqbal I, Iqbal K (2013b) Abnormal hyperphosphorylation of tau: sites, regulation, and molecular mechanism of neurofibrillary degeneration. J Alzheimer's Dis 33(Suppl 1):S123–S139

Wang J, Zhou J, Mo W, He Y, Wei Y, He R, Yi F (2017) Pathological level of formaldehyde decreases cell viability and adhesive morphology in murine neuroblastoma cells. Prog Biochem Biophys 44(7):601–614

Weingarten MD, Lockwood AH, Hwo SY, Kirschner MW (1975) A protein factor essential for microtubule assembly. Proc Natl Acad Sci U S A 72(5):1858–1862

Whittington RA, Virag L, Marcouiller F, Papon MA, El Khoury NB, Julien C, Morin F, Emala CW, Planel E (2011) Propofol directly increases tau phosphorylation. PLoS One 6(1):e16648. doi:10.1371/journal.pone.0016648

Wu L, Rosa-Neto P, Hsiung GY, Sadovnick AD, Masellis M, Black SE, Jia J, Gauthier S (2012) Early-onset familial Alzheimer's disease (EOFAD). Can J Neurol Sci Le journal canadien des sciences neurologiques 39(4):436–445

Wu B, Wei Y, Wang Y, Su T, Zhou L, Liu Y, He R (2015) Gavage of D-Ribose induces Abeta-like deposits, Tau hyperphosphorylation as well as memory loss and anxiety-like behavior in mice. Oncotarget 6(33):34128–34142

Yanagisawa M, Planel E, Ishiguro K, Fujita SC (1999) Starvation induces tau hyperphosphorylation in mouse brain: implications for Alzheimer's disease. FEBS Lett 461(3):329–333

Yang M, Lu J, Miao J, Rizak J, Yang J, Zhai R, Zhou J, Qu J, Wang J, Yang S et al (2014a) Alzheimer's disease and methanol toxicity (part 1): chronic methanol feeding led to memory impairments and tau hyperphosphorylation in mice. J Alzheimer's Dis JAD 41(4):1117–1129

Yang M, Miao J, Rizak J, Zhai R, Wang Z, Huma T, Li T, Zheng N, Wu S, Zheng Y et al (2014b) Alzheimer's disease and methanol toxicity (part 2): lessons from four rhesus macaques (*Macaca mulatta*) chronically fed methanol. J Alzheimer's Dis 41(4):1131–1147

Youmans KL, Tai LM, Nwabuisi-Heath E, Jungbauer L, Kanekiyo T, Gan M, Kim J, Eimer WA, Estus S, Rebeck GW et al (2012) APOE4-specific changes in Abeta accumulation in a new transgenic mouse model of Alzheimer disease. J Biol Chem 287(50):41774–41786

Yu PH (2001) Involvement of cerebrovascular semicarbazide-sensitive amine oxidase in the pathogenesis of Alzheimer's disease and vascular dementia. Med Hypotheses 57(2):175–179

Yu J, Su T, Zhou T, He Y, Lu J, Li J, He R (2014) Uric formaldehyde levels are negatively correlated with cognitive abilities in healthy older adults. Neurosci Bull 30(2):172–184

Zhao B, Wan L (2012) Metal metabolic homeostasis disruption and early initiation of mechanism for Alzheimer's disease. Prog Biochem Biophys 39(8):756–763

Chapter 10
Formaldehyde Exposure and Neuropsychiatric Disorders

Xiumei Wang and Rongqiao He

Abstract Formaldehyde is getting more widely used in modern industrial and information society. Exposure to ambient air pollution is a serious and common public health concern. The high risk of formaldehyde exposure often occurs in the occupational settings, including scientific laboratories in hospitals and universities, particle board/plywood plants, fire sites, etc. Despite the data showing workplace formaldehyde exposures well below those typically considered risks to health, workers complained psychiatric disorders more frequently, and the syndromes could be rescued after leaving the workplace for a period of time. In addition to the occupational formaldehyde exposure sites, urea-formaldehyde resins in building and furnishing materials contributes to the major component of indoor air pollution where people act and live in newly decorated houses and rooms. More people are at high risk of long-term and low-level formaldehyde exposure because of the low ventilation rate indoor. Epidemiological studies show that people complain a series of neuropsychiatric symptoms, such as depression, anxiety, sleep disorders, malaise, balance dysfunctions, headache, indigestion, lethargy, decrease in motor activity and loss of appetite. All those further confirmed that the neuropsychiatric symptoms are highly related to the long-term formaldehyde exposure in the air. In the case of long-term formaldehyde exposure, the victims (17 males, 20 females; average age of 38 years old) mainly showed anxiety symptoms. Around 60.7% of them had elevated levels of urine formaldehyde compared with the normal control. In other words, it is necessary to determine and monitor endogenous formaldehyde for the victims suffering a long-term exposure. Although exogenous formaldehyde causes depression, anxiety and circadian rhythm disorders, whether endogenous formaldehyde induces those symptoms is still unclear. Here, we discuss the effects of formaldehyde exposure on psychosomatic behaviours such as rhythm

---

X. Wang
State Key Laboratory of Brain and Cognitive Science, Institute of Biophysics, Chinese Academy of Sciences, Beijing 100101, China

R. He (✉)
State Key Laboratory of Brain and Cognitive Science, Institute of Biophysics, Chinese Academy of Sciences, Beijing 100101, China

University of Chinese Academy of Sciences, Beijing 100049, China
e-mail: rongqiaohe@163.com

R. He, *Formaldehyde and Cognition*, DOI 10.1007/978-94-024-1177-5_10

disorders, depression, anxiety and other behavioural disorders except for cognitive impairment.

Keywords Formaldehyde • Exposure • Neuropsychiatric disorder • Anxiety • Depression • Circadian rhythm

1 Introduction

Formaldehyde is ubiquitous, which enters the human body through different ways such as respiratory tract, skin absorption and food intake. Due to the high water solubility and reactivity, the airborne formaldehyde is absorbed mainly in the upper airways (>90%) (Nielsen et al. 2013). Low level of chronic formaldehyde exposure in the airborne environment mainly induces disorders of respiratory symptoms such as rhinitis and anaphylaxis, as well as the effects on neurological, musculoskeletal, dermatological, gastrointestinal, cardiac and endocrinological systems (Gibson and Vogel 2009). High-level exposure has the potential risk to cause tumours in the upper respiratory tract (Kilburn et al. 1985a) and leukaemia (Cogliano et al. 2005). The International Agency for Research on Cancer (IARC) has evaluated: "There is sufficient evidence in humans for the carcinogenicity of formaldehyde. Formaldehyde causes cancer of the nasopharynx and leukaemia" (IARC2012).

Generally speaking, exposure to formaldehyde is higher indoors than outdoors. This is mainly due to the multiple formaldehyde sources and low air exchange rates in the indoor environment. Persistent central nervous system (CNS) symptoms and impairments have been reported in occupationally or nonoccupationally formaldehyde-exposed humans (Sparks et al. 1990; Lu et al. 2008). Epidemiological studies have shown that work-related exposure to formaldehyde results in headaches, anxiety, depression, panic, fatigue, sleep disorders, cognition and memory decline (Kilburn et al. 1987). Formaldehyde exposure to murine has effect on anxiety, depression, cognitive ability, exhibiting the same behaviour symptoms like humans (Bhatt and Panchal 1992; Aslan et al. 2006; Li et al. 2016a).

Because occupational formaldehyde places have protective measures, the people could reduce the risk of high-concentration formaldehyde exposure in working place except for accident. But materials containing formaldehyde resin are widely used in house decoration, and more and more people are subjected in low-level and long-term formaldehyde exposure environment. Coincidentally, clinical patients with melatonin (MT) deficiency also complain cognitive problems associated with the mental disorders discribed above. Age-related cognitive impairment is often accompanied with neuropsychiatric disorders such as sleep disorders, depression, anxiety and aggression. The indoor symptom, named sick building syndrome (SBS), affects the normal life and work on people. Mental disorders should be paid attention to and deserved research (Hu et al. 2016), in order to avoid cell dysfunction, substantive organ damage and even cancer caused by formaldehyde. We have reviewed cognitive dysfunction and formaldehyde in Chap. 8. Here, we are going to discuss the effects of formaldehyde exposure on psychosomatic behaviours such as biological rhythm, depression, anxiety and the other behavioural disorders except for cognitive impairment.

2 Formaldehyde and Circadian Rhythm

2.1 Formaldehyde Exposure Disturbing Sleep

It is known that there is disturbed circadian rhythm in Alzheimer's patients, which is related to the declined cognitive function. In 1985, Kilburn and colleagues recruited the women working in histology who had daily exposure to formaldehyde, xylene and toluene and the unexposed female clerical workers working in the same hospitals. They compared the disturbances of sleep, memory, mood and equilibrium that occurred simultaneously with headache and indigestion. They found that formaldehyde exposure correlated better with neurobehavioural symptoms and with respiratory and mucous membrane symptoms than did exposure to xylene/toluene or to other agents (Kilburn et al. 1985a). Weber and colleagues reported a family with chronic exposure to formaldehyde in a renovated apartment. The family members suffered from sleeping disturbances, malaise, headache and nausea, besides eye and upper airway irritation and lack of appetite (Weber et al. 1988). In fact, formaldehyde has long been used as a pain agent for disturbance on sleeping (Coderre et al. 1984; Park et al. 2011). Zaman and collaborators performed formaldehyde-induced paw oedema with pain to disturb sleeping in the study of Nidrakar Bati, an herbal remedy used to cure somnifacient (sleeping aid) in South Asia as Ayurvedic medicinal system (Zaman et al. 2015). To mimic the occupational formaldehyde exposure environment, Mei and collaborators employed 16 healthy adult male mice which were exposed to gaseous formaldehyde (3 mg/m^3) for 7 consecutive days. Formaldehyde exposure elicits an intensive oxidative stress by reducing systemic glutathione levels, in particular, decreasing the concentrations of brain melatonin (MT), accompanied with the impairment of spatial memory associated with hippocampal neuronal death (Mei et al. 2016). Zhou and colleagues have reported the disturbance of circadian rhythm and melatonin levels as an early event in the process of AD, which is accompanied with the progression of AD neuropathology. They employed melatonin as a drug to treat AD patients combined with light regulation, and delayed the impairment of their cognitive function (Zhou and Liu 2012). So far, however, evidence should be obtained to support the hypothesis that formaldehyde is involved in circadian rhythm. Whether MT ameliorates the disorders in circadian rhythm through regulation of the metabolism of endogenous formaldehyde needs to be investigated.

2.2 Formaldehyde Exposure Affecting Water Intake Habit

In order to investigate the relationship between formaldehyde and circadian rhythm, Li and her colleagues first compared the water-intake frequency and volume of 1-month-old mice with 10-month-old mice. The water intake of old mice decreased in both the frequency and volume (Li et al. 2012). Then, they intraperitoneally injected the mice with formaldehyde (0.5 mg/kg, once daily) for a week, and then they monitored the mice in their water-intake frequency and volume (Li et al. 2016c). Both the frequency and volume of water intake significantly decreased compared with those of mice injected with saline as control. Furthermore, changes in the frequency of water intake per hour were also observed in the formaldehyde-treated group. In other words, administration with formaldehyde could change the daily rhythm of the mice in their water intake habit. Furthermore, Mei and colleagues' work shows that formaldehyde affects endogenous MT metabolism and intraperitoneal injection of MT markedly attenuated formaldehyde-induced hippocampal neuronal death, restored brain MT levels and reversed memory decline. Formaldehyde directly inactivates MT in vitro and in vivo (Mei et al. 2016). Melatonin supplementation could contribute to the rescue of the biological rhythmic disorders and age-related cognitive impairment (Maurizi 1987; Skene et al. 1990). These data support the hypothesis that endogenous formaldehyde is involved in daily biological rhythm of animals' behaviours (Li et al. 2016c).

3 Changes in Neuropsychological Behaviours

3.1 Formaldehyde and Depression

Formaldehyde exposure causes emotional and behavioural symptoms. In a case series study of composite-material workers who were in the phenol-formaldehyde exposure condition, ~19% of them had antecedent depression disease (Parks and Pilisuk 1991). The hippocampus is in situ at the limbic area involved in emotion, memory and learning. It is known to possess the greatest levels of adrenocorticosteroid receptor binding and mRNA expression (Reul and de Kloet 1985; Aronsson et al. 1988; Reul et al. 1989), which more likely underlie the deleterious effects of glucocorticoids on learning and memory and long-term potentiation (Diamond and Rose 1994; Bodnoff et al. 1995).

Formaldehyde exposure can activate the hypothalamic-pituitary-adrenal gland (HPA) axis and subsequently increase levels of glucocorticoids (Dallman et al. 2003; Makino et al. 2002; Sorg et al. 2001a) that regulate the HPA response via negative feedback of glucocorticoid receptors (GRs) binding in the hippocampus (Raone et al. 2007). Repeated stress exposure and long-term glucocorticoid treatment downregulate levels of brain GRs, in particular of the hippocampus (Huot et al. 2004; Kitraki et al. 2004). Li and colleagues used open-field tests (OFT), elevated plus-maze tests

(EPM) and forced swimming tests (FST) to assess levels of anxiety- and depression-like behaviours following repeated formaldehyde exposure. They observed that repeated formaldehyde exposure reduces levels of GRs in the hippocampus (Li et al. 2016a). As we know, low levels of GRs in the hippocampus induce less negative feedback to regulate levels of corticosterone (CORT) (Kitraki et al. 2004), subsequently increase levels of CORT and lead to higher levels of anxiety and depression.

Acute formaldehyde (5 ppm) exposure resulted in decreased motor activity and increased levels of dopamine (DA) together with its enzymatic metabolites in the hypothalamus of rats (Boja et al. 1985). Inhalative formaldehyde treatment (13.5 ± 1.5 ppm) for 2 weeks enhances aggressive behaviour and increases DA in the frontal cortex synaptosome (Liu et al. 2009). In other words, changes in behaviour induced by formaldehyde exposure could be associated with alteration in DA levels because DA plays a role in the control motor activity and emotional behaviour (Gainetdinov et al. 1999; Rodgers et al. 1994; Zhuang et al. 1999; Zhou and Palmiter 1995). Tyrosine hydroxylase (TH), a rate-limiting enzyme for DA synthesis, as an indicator of DA production, is distributed in many areas of the rat brain. Li and her colleagues found that different concentrations of gaseous formaldehyde exposure result in different effects on anxiety, depression-like behaviour and cognition ability, which may be associated with alterations in hippocampal glucocorticoid receptors and brain tyrosine hydroxylase levels (Li et al. 2016a).

3.2 *Formaldehyde and Anxiety*

Though anxiety, another nervous disorder, is different from depression, many people who develop depression have a history of an anxiety disorder earlier in their lives. It is usual for someone with an anxiety disorder to also suffer from depression. The two psychosomatic disorders are often co-occurrence. Actually, the anxiety-like behaviour was also observed in the 1 week formaldehyde-exposed mice. The data from open-field test and elevated plus-maze test supported the relation between formaldehyde exposure and anxiety. Even if mice were exposed to the formaldehyde vapour (1.1, 2.3 and 2.5 ppm) for 2 h, their locomotor and explorative activity in the open field would be affected (Malek et al. 2004). They could still observe the effects on the mice after 24 h. These data suggested that formaldehyde can induce emotional and mood changes, even disorders.

Formaldehyde has been used as an inducer in pain-associated anxiety (Rahman et al. 1994a), namely, the animal model of experimental pain/anxiety (Rahman et al. 1994b). In the study of chemical intolerance, Sorg and colleagues observed that repeated low-level formaldehyde exposure produces and increases anxiety, besides sleep disturbance and/or fatigue (Sorg et al. 2001b). Li and her colleagues intraperitoneally injected mice with formaldehyde (once daily) for 7 days, leading to a decrease of brain 5-hydroxytryptamine (5-TH) compared with the control group. Simultaneously, the formaldehyde-treated mice became markedly sensitive and easily irritated, besides slower learning in "Shuttle box" behaviour assay than those fed with water as controls (Li et al. 2016a, b, c). Behavioural sensitization occurred after the mice were injected with formaldehyde for 7 days, suggesting anxiety induced by formaldehyde, similar to the results reported by Sorg and colleagues previously (Sorg et al. 2001b). Behavioural sensitization, easy irritation and anxiety also were observed in monkey trials after the animal had drunk 3% methanol for 3 months.

Herewith is an unpublished data about a case of about 37 staffs (17 males, 20 females; 28–55 years old, average age of 38 years old) who worked in a company complaining of the air pollution with formaldehyde or other organic compounds in their workplace. They were worried about their health in the presence of formaldehyde odour. Some of them showed anxiety-like behaviours such as headache, vertigo, nausea, indigestion, dry mouth, fatigue and itchy skin (World Health Organization 2009; Testa et al. 2013). The authors collected their urine (28 of them) and measured the concentrations of urine formaldehyde by using HPLC coupled with dinitrophenylhydrazone (DNPH). Seventeen (60.7%) of them had high levels of urine formaldehyde compared with the age-matched normal control (5.9 ± 0.62 μM). Among the participants with high formaldehyde levels, 73.33% (11/15) were males, and 46.15% (6/13) were females. The most irrational and excessively worried person who had the highest concentration of endogenous formaldehyde (18.28 μM) was working in the same office with the other two persons whose formaldehyde levels were 15.13 μM and 12.26 μM. They complained of the formaldehyde odour in their office and felt agitated and restless. It was possible that the wood desk they sat at contained excess formaldehyde released into the office air. However, it was difficult to make a conclusion that their anxious symptoms are directly resulted from the formaldehyde toxicosis because their behaviours were also affected by improper emotions. The urine samples were only 28, too small to get a solid conclusion. Referred to the reports by Kilburn and colleagues (1985a, b, 1998), these data suggested that a long-term exposure to formaldehyde affects human's mood including anxiety and sensitization. In addition, determination of endogenous formaldehyde should be recommended for the victims who have long been exposed to a polluted environment.

4 Changes in Neurobiological Behaviours

4.1 Olfaction Impairment

Formaldehyde is colourless, with a particular pungent smell of gas. Even in very low concentrations, it can be smelt out and induce lesions in the nasal area. It has been previously established that the low-level formaldehyde exposure (0.25 ppm for 130 days) can influence odour sensitivity (Apfelbach and Weiler 1991). Zhang and his colleagues found that in the higher concentration for short time (13.5 ± 1.5 ppm, twice 30 minutes/day for 14 days), the rat olfactory function was impaired by buried food pellet test (Zhang et al. 2014). As described by Lang and colleagues, olfactory symptoms occurred at concentrations as low as 0.3 ppm. Long-term formaldehyde inhalation can change cellular morphologies in the olfactory bulb (Lang et al. 2008). Therefore, we suggest the effects of formaldehyde on the olfactory system should be urgently paid attention to.

MicroRNA (miRNA) is a small non-coding RNA molecule involved in a wide range of biological processes such as cell cycle control, apoptosis and several developmental and physiological processes including the differentiation of olfactory precursors, olfactory neuron morphogenesis and neurogenesis. Recently, Li and coworkers have tried to study the molecular mechanisms of formaldehyde-induced olfactory dysfunction (Li et al. 2015). They found that formaldehyde inhalation markedly alters the miRNA expression profile of olfactory bulb. With high-throughput microarray technology, 18 upregulated miRNAs and 7 downregulated miRNAs were identified (Li et al. 2015). Among them, miR-199a, miR-146b and miR-200a were changed most. A long-term formaldehyde exposure resulted in an increase in the number of tyrosine hydroxylase-immunopositive periglomerular cells of the main olfactory bulb (Hayashi et al. 2004). As we know, tyrosine hydroxylase uses tetrahydrobiopterin and molecular oxygen to convert tyrosine to dopamine. The olfactory bulb of mammals has a large population of dopaminergic neurons within the glomerular layer. Dopamine has been proved to modulate olfactory information processing. Synaptosomal-associated protein 25 (SNAP25) is a presynaptic plasma membrane protein involved in the regulation of neurotransmitter release. The olfactory bulbs exhibit high levels of SNAP25. After formaldehyde exposure, the protein expression decreased (Zhang et al. 2014). These results show that formaldehyde may impact the olfaction by affecting neurotransmitter system.

4.2 *Algesthesia*

Formalin, an aqueous solution of formaldehyde, is commonly employed to study algesthesia. The rodent formalin model is utilized for evaluating the effects of pain and analgesic compounds in laboratory animals (Tajolsen et al. 1992). When formalin is injected into the paw of a rodent, it will induce biphasic nociceptive behavioural responses including early and late phases. The early-phase response seems to be caused by direct activation of the small primary afferents, while the late phase is dependent on the combination of an inflammatory reaction in the peripheral tissue and functional changes in the dorsal horn of the spinal cord. It is a valuable test available to study nociception. A lot of research has been done to uncover the mechanisms. Transient receptor potential cation channel, subfamily A, member 1 (TRPA1) is identified as the principal site of formalin's pain-producing action (McNamara et al. 2007). The activation of this excitatory channel underlies the physiological and behavioural responses associated with this model of pain hypersensitivity.

The other pain-associated molecules play roles in the model of nociceptive and inflammatory pain and inflammatory oedema (Dutra et al. 1994; Godin et al. 2015). Mogil has reviewed animal pain models and considered pain is so complicated that the researchers should carefully pay attention to the models and must take into account when using an animal model (Mogil 2009). One interesting discovery is that the cancer tissues can directly secrete endogenous formaldehyde and so forth. Formaldehyde at low concentration induces metastatic bone cancer pain through TRPV1 activation especially under tumour acidic environment (Tong et al. 2010).

4.3 *Effect on Motor Activities*

Amyotrophic lateral sclerosis (ALS) is a progressive neurological disease involving the death of upper and lower motor neurons. It causes muscle weakness and impacts physical function. At the late stages, the condition affects nerves that control breathing and other vital bodily functions, resulting in death. Currently, experts do not know what cause ALS precisely.

The neurotoxic effects of formaldehyde from animal models and in vitro experiments indicate that it may be relevant for ALS. Several epidemiological researches were performed to determine the relationship between ALS and formaldehyde exposure. Weisskopf assessed the association between exposure to chemicals and risk of ALS in a prospective cohort study including over 1 million participants in the American Cancer Society's Cancer Prevention Study II. A strongly significant dose-response relation with increasing years of formaldehyde exposure was concluded (Weisskopf et al. 2009). It was reported that certain occupations and workplace exposures may be associated with increased risk of ALS (Fang et al. 2009). A cohort research of formaldehyde-exposed garment workers does not suggest that ALS is associated with formaldehyde exposure (Pinkerkon et al. 2013). In 2015, Roberts and colleagues examined the association of ALS mortality with job-related formaldehyde exposure in the National Longitudinal Mortality Study (NLMS). The results indicated that workers like funeral directors, with a high probability of exposure to formaldehyde, have an almost threefold greater rate of death related to ALS (Roberts et al. 2016).

4.4 *Primary Open-Angle Glaucoma*

Primary open-angle glaucoma (POAG) is a leading cause of irreversible blindness worldwide. Growing evidence indicates a relation between POAG and Alzheimer's disease (Bayer et al. 2002). Recently, an increased occurrence rate of POAG could be found in AD patients (Tislis et al. 2014). In order to reveal the risk factor of POAG, Cui and his colleagues investigated the correlation of the level of endogenous formaldehyde with POAG by high-performance liquid chromatography (HPLC) to detect the endogenous aldehyde in a double-blind manner (Cui et al. 2016). They found that endogenous formaldehyde level is positively correlated with the neuronal defects of POAG. This suggests again that endogenous formaldehyde is involved in neurodegenerative diseases.

5 The Other Psychosomatic Behaviours

Besides those effects on CNS functions, as mentioned above, formaldehyde can also cause some other nervous symptoms such as seizures, autism and amnesia. Three patients were evaluated for effects of formaldehyde on CNS function by Kilburn. All were disabled, and two had developed seizures (Kilburn 1994). Tracking a case of epilepsy caused by formaldehyde exposure, Perna and colleagues suggested that chronic formaldehyde exposure resulted in heightened sensitivity to formaldehyde, three tonic-clonic seizures and dramatic amnesia as well as other cognitive dysfunction (Perna et al. 2001). Utero exposure of ambient toxics including formaldehyde is considered as a risk of childhood autism (von Ehrenstein et al. 2014). Complaints of headache, nosebleeds and stomachache were observed in the villagers who live in the trailers built with particleboard which have been found contaminated with formaldehyde (Madrid et al. 2008). Of course formaldehyde affects the other sensory organs; for instance, indoor formaldehyde exposure stimulates nasal mucosa. Temporary abnormalities in the olfaction test and increased nasal mucosal hypersensitivity to histamine were observed in a few students with preexisting allergic rhinitis after environmental exposure of high concentrations of formaldehyde though these effects appeared to be transient (Hisamitsu et al. 2011).

6 Potential Mechanisms in Multiple Ways

Over the years, neurotoxicity and cognitive dysfunction have separately been associated with endogenous formaldehyde and reduction of acetylcholine signals. 0.024 to 0.74 ppm formaldehyde-exposed workers with increased alcohol dehydrogenase III $ADH3_{2\text{-}2}$ genotype had higher AChE (acetylcholinesterase) than controls, suggesting that the neurotoxic effects of formaldehyde depend on the AChE activity (Zendehdel et al. 2016). Potential chronic and debilitating effects of solvents and formaldehyde have been attributed to the formation of biologically active epoxides near axons. Such epoxides bind to neurofilamental proteins. These proteins swell and neurofilaments proliferate that are believed to result in demyelination (Kilburn et al. 1987; Savolarnin 1977). Using an animal study of acute low-level formaldehyde exposure with rats, Boja and colleagues have found that formaldehyde exposure resulted in decreased motor activity and neurochemical changes in dopamine and serotonin (5-TH) neurons (Boja et al. 1985). Administration of formaldehyde also decreases norepinephrine (NE) levels in the brain of rodents. Tong and his colleagues provided evidence that accumulated formaldehyde is a critical endogenous factor for ageing-associated NE depletion and cognitive decline (Mei et al. 2015). These neurotransmitters (dopamine, 5-TH and norepinephrine) are involved not only in cognitive function but also in mood and emotion.

Reactive aldehydes have been implicated in the aetiology of several neurological and psychiatric disorders, and increased formaldehyde and upregulation of semicarbazide-sensitive amine oxidase, which forms formaldehyde from methylamine, have been implicated in disorders such as cerebrovascular disorders, alcohol abuse, diabetes and Alzheimer's disease. Glutamate pathway may be the key point, because formaldehyde can reduce the alteration of the second messengers, AKT and p38 (Song et al. 2010). As mentioned above, formaldehyde affects anxiety, depression-like behaviour and cognition ability, through association with alterations in hippocampal glucocorticoid receptors and brain tyrosine hydroxylase levels (Li et al. 2016b).

Chemical modifications of proteins have been well documented to play important roles in the pathophysiology of many human diseases such as cancer, age-related pathology and neurodegenerative disorders. Endogenous and exogenous formaldehydes have a reaction with cysteine residues in proteins, and cysteine residues may serve as a biomarker of formaldehyde exposure (Liu et al. 2016). In addition, most of mechanistic studies related to formaldehyde toxicity have been performed in cytotoxic concentrations enough to trigger cell death. To mimic daily-life formaldehyde exposure level, 200 μM formaldehyde exposed normal human keratinocytes to induce pro-inflammatory responses (Lee et al. 2016). Thus, formaldehyde-induced inflammatory may be another cause to impair neurons and brain function. About these contents, please see Chap. 8.

7 To Clarify Formaldehyde Toxicosis Is a Careful Work

As with many potentially cerebrotoxic chemicals, there are few studies of the effects of chronic formaldehyde exposure on cognitive functioning as described by Pern and collaborators. Williams and Lees-Haley (1998) noted that some research on formaldehyde exposure has often been criticized due to "selection bias in recruitment of research participants and unreliability of participant recall for obtaining data on important background variables and exposure levels". Criticisms such as these are often inherent in the study of any toxic chemical exposure. The difficult situation is that cases involved in litigation may represent more severe outcomes and must be examined critically, with consideration of symptom magnification (Perna et al. 2001). In order to clarify the health impairment resulted from formaldehyde exposure, measures of symptom validity may help clarify this issue and should be included and reported when possible. The second difficulty in field studies and case reports of formaldehyde or other exposure to toxic substances is that individuals are rarely exposed to a single toxin, either because multiple toxins are in the work environment or because toxins are often present in complex solutions or compounds. Another relevant, yet variable, factor is the degree and duration of exposure. The literature provides mixed support regarding potential effects of formaldehyde (Perna et al. 2001). To face these situations, the measurement of endogenous formaldehyde (in blood or urine) may be suggested to clarify some of the cases which are resulted from the formaldehyde exposure as mentioned above. The elevated concentrations of formaldehyde may provide some useful information to support the diagnosis. As with many other toxins, there appears to be large individual differences in reactivity and symptomatology. In fact, psychological factors can worsen the symptoms, which should not be ignored. Neuropsychological symptoms are related to the social psychological factor indicators (spirit) (Sparks et al. 1990). Therefore, although the victim has really exposed to high concentrations of formaldehyde in the air, the effect of psychological factors on the symptoms should be considered. A series of appraisal methods found that the subjects resorted to the formaldehyde concentration have no obvious symptoms and correlation. Because patients exposed to toxic substances recognize themselves in danger, and they may have some physical characterization of discomfort as well, the resulting anxiety depression, insomnia and a series of mental neurological symptoms may not be caused by exposure to pollutants. Application of neurophysiology and neuropsychological assessment of chemical exposure on neurobehavioural damage, age and education level of the synergistic effect should be considered and analysed (Kilburn and Warshaw 1992).

8 Debate and Notable Matters

Hippocampus plays important roles in the consolidation of information from short-term memory to long-term memory and spatial learning in human and other animals (Lieberwirth et al. 2016). Formaldehyde has been found in the cerebrospinal fluid (authors' unpublished data) and thus affects the neuroglia and nerve cell because it can pass easily through the blood-brain barrier (Malek et al. 2003a, b). Exposure to exogenous formaldehyde has been reported to be associated with increased formaldehyde levels in the brain (Tulpule and Dringen 2013; Tulpule et al. 2012). Several recent inhalation studies using labelled formaldehyde (13CD2) did not find DNA adducts outside the nasal tissue in rats and monkeys. Thus, formaldehyde was not considered to reach the internal organs or the blood compartment (Lu et al. 2010; Lu et al. 2011; Swenberg et al. 2011). Formaldehyde accumulation in the brain is considered to come from endogenousness, especially dysmetabolism of formaldehyde. But ultrafine particles matters < 2.5 nanometre ($PM_{2.5}$) can enter the vessel and blood-brain barrier (Ailshire and Clarke 2015; Ailshire and Crimmins 2014). Formaldehyde may be absorbed onto the particles, which may target the receptor tissues and cells.

The results from animals and cells to humans need to be extrapolated carefully (Worek et al. 2002), although epidemiological studies and experimental research are enough to alert us to avoid formaldehyde exposure and reduce exposure levels. There are many factors including the way, concentrations and duration of formaldehyde exposure and different metabolism pathways of formaldehyde in different species significantly influence the outputs of the behavioural components tests. The transport and metabolism in the body gradually become clear. Even though formaldehyde comes from different pathways and goes in different metabolisms, lysosome shows its role as a central organelle to accommodate and transfer formaldehyde in and out of cells (Chen et al. 2017; Xiong and Zhu 2016) (see Chap. 8). Of course, further investigation of neurotoxic mechanisms is required.

9 Precautionary Measures

Formaldehyde is a main compound of the air pollution in workplace and home, and epidemiological and experimental data show that formaldehyde has risks to health. Some precaution should be taken to reduce the risks, such as an increase of ventilation to dilute the levels of formaldehyde and the use of air cleaners to remove formaldehyde. Lime was found to be an efficient reagent to lower the concentration of formaldehyde in highly concentrated effluents (Moussavi et al. 2002). Planting trees reduces formaldehyde pollution in air and water (Su and Liang 2015). Toxics use reduction (TUR) is one part of a comprehensive cancer prevention strategy. TUR emphasizes reducing the use of cancer-causing chemicals by improving manufacturing processes and identifying and adopting safer alternatives. Daily intake of antioxidant food such as grapeseed (resveratrol) reduces formaldehyde level in body to protect health. In fact, disengagement of formaldehyde is the most effective method of protection. More information is described about drug usages to treat the formaldehyde toxicosis in Chap. 12.

10 Conclusion

Excess formaldehyde has multiple harmful effects on different human organs such as kidney, liver, lung and blood vessels, especially CNS. Epidemiological and experimental data support the role of formaldehyde as a risk factor for neuropsychiatric disorders such as depression, anxiety and circadian rhythm dysfunction, including cognitive impairment, though the mechanisms of formaldehyde exposure causing mental disorders still are unclear. Recent studies showed that formaldehyde induces β-amyloid deposition, Tau hyperphosphorylation, dysmetabolism of neurotransmitters and neuronal death. All those may be related to the impairment of brain and neuropsychiatric disorders. It is necessary to strengthen the research and studies on the effects of air pollution on neuropsychological disorders and highly sensitive monitoring methods for formaldehyde and effective protective measures for people's health. Further and more work should be carried out to clarify the relationship between long-term formaldehyde exposure and neuropsychiatric disorders. At the same time, whether endogenous formaldehyde can be used as a biomarker to identify anxiety or depression needs further investigating.

Acknowledgements This project was supported by grants from the National Key Research and Development Program of China (2016YFC1306300), the National Basic Research Program of China (973 Program) (2012CB911004), the Beijing Municipal Science and Technology Project (Z161100000217141; Z161100000216137), the National Natural Science Foundation of China (NSFC 31270868), the Foundation of Chinese Academy of Sciences (CAS-20140909) and the Queensland-Chinese Academy of Sciences Biotechnology Fund (GJHZ201302).

Competing Financial Interests The authors declare no competing financial interests.

References

Ailshire JA, Clarke P (2015) Fine particulate matter air pollution and cognitive function among US older adults. J Gerontol B-Psychol 70:322–328

Ailshire JA, Crimmins EM (2014) Fine particulate matter air pollution and cognitive function among older US adults. Am J Epidemiol 180:359–366

Apfelbach R, Weiler E(1991) Sensitivity to odors in Wistar rats is reduced after low-level formaldehyde-gas exposure.Naturwissenschaften 78:221–223

Aronsson M, Fuxe K, Dong Y, Agnati LF, Okret S (1988) Localization of glucocorticoid receptor mRNA in the male rat brain by in situ hybridization. Proc Natl Acad Sci U S A 85:9331–9335

Aslan H, Songur A, Tunc AT, Ozen OA, Bas O, Yagmurca M, Turgut M, Sarsilmaz M, Kaplan S (2006) Effects of formaldehyde exposure on granule cell number and volume of dentate gyrus: a histopathological and stereological study. Brain Res 1122:191–200

Bayer AU, Ferrari F, Erb C (2002) High occurrence rate of glaucoma among patients with. Alzheimer's Dis 47(3):165–168

Bhatt HV, Panchal GM (1992) Behavioural change in rats due to chronic oral and systemic formaldehyde. Indian J Physiol Pharmacol 36:270–272

Bodnoff SR, Humphreys AG, Lehman JC, Diamond DM, Rose GM, Meaney MJ (1995) Enduring effects of chronic corticosterone treatment on spatial learning, synaptic plasticity, and hippocampal neuropathology in young and mid-aged rats. J Neurosci 15:61–69

Boja JW, Nielsen JA, Foldvary E, Truitt EB Jr (1985) Acute low-level formaldehyde behavioral and neurochemical toxicity in the rat. Prog Neuro Psychopharmacol Biol Psychiat 9:671–674

Chen XX, Su T, He YG, He RQ (2017) Spatial cognition decline caused by excess formaldehyde in the lysosome Prog Biochem Biophys. doi:10.16476/j.pibb.2017.0050

Coderre TJ, Abbott FV, Melzack R (1984) Effects of peripheral antisympathetic treatments in the tail-flick, formalin and autotomy tests. Pain 18(1):13–23

Cogliano VJ, Grosse Y, Baan RA, Straif K, Secretan MB, El Ghissassi F, Working Group for, V (2005) Meeting report: summary of IARC monographs on formaldehyde, 2-butoxyethanol, and 1-tert-butoxy-2-propanol. Environ Health Perspect 113:1205–1208

Cui Y, Su T, Zhang SD, Huang P, He YG, Liu Y, Zhang C, He RQ (2016). The elevated urine formaldehyde level in elderly patients with primary open angle glaucoma. Int J Ophthalmol 18;9(3):411–416

Dallman MF, Akana SF, Laugero KD, Gomez F, Manalo S, Bell ME, Bhatnagar SA (2003) Spoonful of sugar: feedback signals of energy stores and corticosterone regulate responses to chronic stress. Physiol Behav 79:3–12

Diamond DM, Rose GM (1994) Stress impairs LTP and hippocampal-dependent memory. Ann N Y Acad Sci 746:411–414

Dutra MM, Godin AM, Cse GM (1994) Stress impairs LTP and hippocampal-dependent memory. Annals of the new York Academy of Sciences 746, 411-414.Olitan areas: Nashville, Atlanta, Houston, Philadelphia, and Tampa. Environmental science & troup, in the experimental model of pain induced by formaldehyde in mice. Pharmacol Biochem Behav 106:85–90

Fang F, Quinlan P, Ye W, Barber MK, Umbach DM, Sandler DP, Kamel F (2009) Workplace exposures and the risk of amyotrophic lateral sclerosis. Environ Health Perspect 117:1387–1392

Gainetdinov RR, Wetsel WC, Jones SR, Levin ED, Jaber M, Caron MG (1999) Role of serotonin in the paradoxical calming effect of psychostimulants on hyperactivity. Science 283:397–401

Gibson PR, Vogel VM (2009) Sickness-related dysfunction in persons with self-reported multiple chemical sensitivity at four levels of severity. J Clin Nurs 18:72–81

Godin AM, Araújo DP, César IC, Menezes RR, Brito AM, Melo IS, Coura GM, Bastos LF, Almeida MO, Byrro RM, Matsui TC, Batista CR, Pianetti GA, de Fátima Â, Machado RR, Coelho MM (2015) Activities of 2-phthalimidethyl nitrate and 2-phthalimidethanol in the models of nociceptive response and edema induced by formaldehyde in mice and preliminary investigation of the underlying mechanisms.Eur J Pharmacol 756:59–66

Hayashi H, Kunugita N, Arashidani K, Fujimaki H, Ichikawa M (2004) Long-term exposure to low levels of formaldehyde increases the number of tyrosine hydroxylase-immunopositive periglomerular cells in mouse main olfactory bulb. Brain Res 1007:192–197

Hisamitsu M, Okamoto Y, Chazono H, Yonekura S, Sakurai D, Horiguchi S, Hanazawa T, Terada N, Konno A, Matsuno Y et al (2011) The influence of environmental exposure to formaldehyde in nasal mucosa of medical students during cadaver dissection. Allergol Int 60:373–379

Hu X, Wang T, Jin F (2016) Alzheimer's disease and gut microbiota. Sci China Life Sci 59(10):1006–1023

Huot RL, Gonzalez ME, Ladd CO, Thrivikraman KV, Plotsky PM (2004) Foster litters prevent hypothalamic pituitary adrenal axis sensitization mediated by neonatal maternal separation. Psychoneuroendocrinol 29:279–289

Kilburn KH (1994) Neurobehavioral impairment and seizures from formaldehyde. Arch Environ Health 49:37–44

Kilburn KH, Warshaw RH (1992) Neurobehavioral effects of formaldehyde and solvents on histology technicians: repeated testing across time. Environ Res 58:134–146

Kilburn KH, Seidman BC, Warshaw R (1985a) Neurobehavioral and respiratory symptoms of formaldehyde and xylene exposure in histology technicians. Arch Environ Health 40:229–233

Kilburn KH, Warshaw R, Boylen CT, Johnson SJ, Seidman B, Sinclair R, Takaro T Jr (1985b) Pulmonary and neurobehavioral effects of formaldehyde exposure. Arch Environ Health 40:254–260

Kilburn KH, Warshaw R, Thornton JC (1987) Formaldehyde impairs memory, equilibrium, and dexterity in histology technicians: effects which persist for days after exposure. Arch Environ Health 42:117–120

Kilburn KH, Thornton JC, Hanscom B (1998) Population-based prediction equations for neurobehavioral tests. Arch Environ Health 53:257–263

Kitraki E, Kremmyda O, Youlatos D, Alexis MN, Kittas C (2004) Gender dependent alterations in corticosteroid receptor status and spatial performance following 21 days of restraint stress. Neuroscience 125:47–55

Lang I, Bruckner T, Triebig G (2008) Formaldehyde and chemosensory irritation in humans: a controlled human exposure study. Regul Toxicol Pharmacol 50:23–36

Lee E, Kim HJ, Lee M, Jin SH, Hong SH, Ahn S, Kim SO, Shin DW, Lee ST, Noh M (2016) Cystathionine metabolic enzymes play a role in the inflammation resolution of human keratinocytes in response to sub-cytotoxic formaldehyde exposure. Toxicol Appl Pharmacol 310:185–194

Li T, Qiang M, He RQ (2012) Chronic dehydration and regularly drinking water to mitigate age-related cognitive impairment. Acta Neuropharmacologica 2(3):43–51

Li G, Yang J, Ling S (2015) Formaldehyde exposure alters miRNA expression profiles in the olfactory bulb. Inhal Toxicol 27:387–393

Li Y, Song Z, Ding Y, Xin Y, Wu T, Su T, He R, Tai F, Lian Z (2016a) Effects of formaldehyde exposure on anxiety-like and depression-like behavior, cognition, central levels of glucocorticoid receptor and tyrosine hydroxylase in mice. Chemosphere 144:2004–2012

Li T, Su T, He Y-G, He R-Q (2016b) Chronic-dehydrated dysmetabolism of formaldehyde in mouse brain and decline of learning in the shuttle box. Prog Biochem Biophys 43:429–438

Li T, Su T, He Y, Lu J, Mo W, Wei Y, He R (2016c) Brain formaldehyde is related to water intake behavior. Aging and disease 7:561–584

Lieberwirth C, Pan Y, Liu Y, Zhang Z, Wang Z (2016) Hippocampal adult neurogenesis: its regulation and potential role in spatial learning and memory. Brain Res 1644:127–140

Liu Y, Ye ZL, Luo HZ, Sun M, Li M, Fan DS, Chui DH (2009) Inhalative formaldehyde exposure enhances aggressive behavior and disturbs monoamines in frontal cortex synaptosome of male rats. Neurosci Lett 464:113–116

Liu J, Chan KK, Chan W (2016) Identification of protein Thiazolidination as a novel molecular signature for oxidative stress and formaldehyde exposure. Chem Res Toxicol 29:1865–1871

Lu Z, Li CM, Qiao Y, Yan Y, Yang X (2008) Effect of inhaled formaldehyde on learning and memory of mice. Indoor Air 18:77–83

Lu K, Collins LB, Ru H, Bermudez E, Swenberg JA (2010) Distribution of DNA adducts caused by inhaled formaldehyde is consistent with induction of nasal carcinoma but not leukemia. Toxicol Sci 116:441–451

Lu K, Moeller B, Doyle-Eisele M, McDonald J, Swenberg JA (2011) Molecular dosimetry of N2-hydroxymethyl-dG DNA adducts in rats exposed to formaldehyde. Chem Res Toxicol 24:159–161

Madrid PA, Sinclair H, Bankston AQ, Overholt S, Brito A, Domnitz R, Grant R (2008) Building integrated mental health and medical programs for vulnerable populations post-disaster: connecting children and families to a medical home. Prehosp Disaster Med 23:314–321

Makino S, Hashimoto K, Gold PW (2002) Multiple feedback mechanisms activating corticotropin -releasing hormone system in the brain during stress. Phamacol Biochem Behav 73:147–158

Malek FA, Moritz KU, Fanghanel J (2003a) Formaldehyde inhalation & open field behaviour in rats. Indian J Med Res 118:90–96

Malek FA, Moritz KU, Fanghanel J (2003b) A study on the effect of inhalative formaldehyde exposure on water labyrinth test performance in rats. Ann Anat Anatomischer Anzeiger 185:277–285

Malek FA, Moritz KU, Fanghanel J (2004) Effects of a single inhalative exposure to formaldehyde on the open field behavior of mice. Int J Hyg Environ Health 207:151–158

Maurizi CP (1987) Dementia– The failure of hippocampal plasticity and dreams. Is there a preventative role for melatonin? Med Hypotheses 24:59–68

McNamara CR, Mandel-Brehm J, Bautista DM, Siemens J, Deranian KL, Zhao M, Hayward NJ, Chong JA, Julius D, Moran MM et al (2007) TRPA1 mediates formalin-induced pain. Proc Natl Acad Sci U S A 104:13525–13530

Mei Y, Jiang C, Wan Y, Lv J, Jia J, Wang X, Yang X, Tong Z (2015) Aging-associated formaldehyde-induced norepinephrine deficiency contributes to age-related memory decline. Aging Cell 14:659–668

Mei Y, Duan C, Li X, Zhao Y, Cao F, Shang S, Ding S, Yue X, Gao G, Yang H et al (2016) Reduction of endogenous melatonin accelerates cognitive decline in mice in a simulated occupational formaldehyde exposure environment. Int J Environ Res Pub Health 13

Mogil JS (2009) Animal models of pain: progress and challenges. Nat Rev Neurosci 10:283–294

Moussavi M, Mowla D, Edraki H (2002) Chemical pretreatment of formaldehyde-containing effluents. Environ Sci Technol 36:3822–3826

Nielsen GD, Larsen ST, Wolkoff P (2013) Recent trend in risk assessment of formaldehyde exposures from indoor air. Arch Toxicol 87:73–98

Park HJ, Cha DS, Jeon H (2011) Antinociceptive and hypnotic properties of Celastrus orbiculatus. J Ethnopharmacol 137:1240–1244

Parks SH, Pilisuk M (1991) Caregiver burden: gender and the psychological costs of caregiving. Am J Orthopsychiatry 61:501–509

Perna RB, Bordini EJ, Deinzer-Lifrak M (2001) A case of claimed persistent neuropsychological sequelae of chronic formaldehyde exposure: clinical, psychometric, and functional findings. Arch Clin Neuropsychol 16:33–44

Pinkerton LE, Hein MJ, Meyers A, Kamel F (2013) Assessment of ALS mortality in a cohort of formaldehyde-exposed garment workers. Amyotroph Lateral Scler Frontotemporal Degener 14:353–355

Rahman AF, Takahashi M, Kaneto H (1994a) Involvement of pain associated anxiety in the development of morphine tolerance in formalin treated mice. Jpn J Pharmacol 65:313–317

Rahman AFMM, Takahashi M, Kaneto H (1994b) Morphine-dependence with or without tolerance in formalin-treated mice - further evidence for the dissociation. Jpn J Pharmacol 66:277–280

Raone A, Cassanelli A, Scheggi S, Rauggi R, Danielli B, De Montis MG (2007) Hypothalamus-pituitary-adrenal modifications consequent to chronic stress exposure in an experimental model of depression in rats. Neuroscience 146:1734–1742

Reul JM, de Kloet ER (1985) Two receptor systems for corticosterone in rat brain: microdistribution and differential occupation. Endocrinology 117:2505–2511

Reul JM, Pearce PT, Funder JW, Krozowski ZS (1989) Type I and type II corticosteroid receptor gene expression in the rat: effect of adrenalectomy and dexamethasone administration. Mol Endocrinol 3:1674–1680

Roberts AL, Johnson NJ, Cudkowicz ME, Eum KD, Weisskopf MG (2016) Job-related formaldehyde exposure and ALS mortality in the USA. J Neurol Neurosurg Psychiatry 87:786–788

Rodgers RJ, Nikulina EM, Cole JC (1994) Dopamine D1 and D2 receptor ligands modulate the behaviour of mice in the elevated plus maze. Pharmacol Biochem Behav 49:985–995

Savolarnin H (1977) Some aspects of the mechanisms by which industrial solvents produce neurotoxic effects. Chem Biol Interact 18:1–10

Skene DJ, Vivien-Roels B, Sparks DL, Hunsaker JC, Pevet P, Ravid D, Swaab DF (1990) Daily variation in the concentration of melatonin and 5-methoxytryptophol in the human pineal gland: effect of age and Alzheimer's disease. Brain Res 528:170–174

Song MS, Baker GB, Dursun SM, Todd KG (2010) The antidepressant phenelzine protects neurons and astrocytes against formaldehyde-induced toxicity. J Neurochem 114:1405–1413

Sorg BA, Bailie TM, Tschirgi ML, Li N, Wu WR (2001a) Exposure to repeated low-level formaldehyde in rats increased basal corticosterone levels and enhances the corticosterone response to subsequent formaldehyde. Brain Res 898:314–320

Sorg BA, Tschirgi ML, Swindell S, Chen L, Fang J (2001b) Repeated formaldehyde effects in an animal model for multiple chemical sensitivity. Ann N Y Acad Sci 933:57–67

Sparks PJ, Simon GE, Katon WJ, Altman LC, Ayars GH, Johnson RL (1990) An outbreak of illness among aerospace workers. West J Med 153(1):28–33

Su Y, Liang Y (2015) Foliar uptake and translocation of formaldehyde with bracket plants (Chlorophytum comosum). J Hazard Mater 291:120–128

Swenberg JA, Lu K, Moeller BC, Gao L, Upton PB, Nakamura J, Starr TB (2011) Endogenous versus exogenous DNA adducts: their role in carcinogenesis, epidemiology, and risk assessment. Toxicol Sci 120(Suppl 1):S130–S145

Testa A, Giannuzzi R, Sollazzo F, Petrongolo L, Bernardini L, Daini S (2013) Psychiatric emergencies (part I): psychiatric disorders causing organic symptoms. (PDF). Eur Rev Med Pharmacol Sci 17(Suppl 1):55–64

Tjolsen A, Berge OG, Hunskaar S, Rosland JH, Hole K. (1992) The formalin test: an evaluation of the method. Pain 51:5–17

Tong Z, Luo W, Wang Y, Yang F, Han Y, Li H, Luo H, Duan B, Xu T, Maoying Q et al (2010) Tumor tissue-derived formaldehyde and acidic microenvironment synergistically induce bone cancer pain. PLoS One 5:e10234

Tsilis AG, Tsilidis KK, Pelidou SH, Kitsos G (2014) Systematic review of the association between Alzheimer's disease and chronic glaucoma. Clin Ophthalmol 8:2095–2104

Tulpule K, Dringen R (2013) Formaldehyde in brain: an overlooked player in neurodegeneration? J Neurochem 127:7–21

Tulpule K, Schmidt MM, Boecker K, Goldbaum O, Richter-Landsberg C, Dringen R (2012) Formaldehyde induces rapid glutathione export from viable oligodendroglial OLN-93 cells. Neurochem Int 61:1302–1313

von Ehrenstein OS, Aralis H, Cockburn M, Ritz B (2014) In utero exposure to toxic air pollutants and risk of childhood autism. Epidemiology 25:851–858

Weber R, Budmiger H, Siegenthaler W (1988) Chronic formaldehyde exposure--a misunderstood disease? Schweiz Med Wochenschr 118:457–461

Weisskopf MG, Morozova N, O'Reilly EJ, McCullough ML, Calle EE, Thun MJ, Ascherio A (2009) Prospective study of chemical exposures and amyotrophic lateral sclerosis. J Neurol Neurosurg Psychiatry 80:558–561

Williams CW, Lees-Haley PR (1998) Research on chronic, low-level exposure to formaldehyde: implications for neuropsychological assessment. J Clin Psychol 54:851–862

Worek F, Reiter G, Eyer P, Szinicz L (2002) Reactivation kinetics of acetylcholinesterase from different species inhibited by highly toxic organophosphates. Arch Toxicol 76:523–529

World Health Organization (2009) Pharmacological treatment of mental disorders in primary health care (PDF). Geneva. ISBN 978-92-4-154769-7

Xiong J, Zhu MX (2016) Regulation of lysosomal ion homeostasis by channels and transporters. Sci China Life Sci 59(8):777–791

Zaman A, Khan MSS, Akter L, Syeed SH, Akter J, Al Mamun A, Alam ME, Habib MA, Jalil MA (2015) Exploring new pharmacology and toxicological screening and safety evaluation of one widely used formulation of Nidrakar Bati from South Asia region. Bmc Complem Altern M 15

Zendehdel R, Fazli Z, Mazinani M (2016) Neurotoxicity effect of formaldehyde on occupational exposure and influence of individual susceptibility to some metabolism parameters. Environ Monit Assess 188:648

Zhang Q, Yan W, Bai Y, Zhu Y, Ma J (2014) Repeated formaldehyde inhalation impaired olfactory function and changed SNAP25 proteins in olfactory bulb. Int J Occup Environ Health 20:308–312

Zhou JN, Liu YJ. (2012) Circadian rhythm of melatonin and Alzheimer's disease. Prog Biochem Biophys 39:796–803

Zhou QY, Palmiter RD (1995) Dopamine-deficient mice are severely hypoactive, adiposic, and aphagic. Cell 83:1197e1209

Zhuang XX, Gross C, Santarelli L, Compan V, Trillat AC, Hen R (1999) Altered emotional states in knockout mice lacking 5-HT1A or 5-HT1B receptors. Neuropsychopharmacology 21:S52–S60

Chapter 11
Formaldehyde Affecting Lifespan and Stress Resistance in *Drosophila*

Yining Li and Rongqiao He

Abstract Longevity is a fundamental and fascinating topic in the contemporary biological studies. A lot of genes, signaling pathways, and transcription factors were reported to affect lifespan from yeast to human. Formaldehyde participates in the epigenetic regulating roles on methylation and demethylation of DNA, RNA, and histone and exerts various toxic effects from animals to human at higher concentrations, while the effects of lower concentrations were rarely reported. We used *Drosophila* as a model organism to study the effects of formaldehyde at various concentrations on longevity and found that different concentrations of formaldehyde played different roles on lifespan and stress resistance in *Drosophila*. Since knocking down the expression level of formaldehyde dehydrogenase (FDH) in the neural system of *Drosophila* could also extend lifespan and increase stress resistance, it is a hopeful way to screen for the genes in the *Drosophila* neural system to study the probable molecular mechanisms of longevity. In this chapter, we reviewed the progress of lifespan study in *Drosophila* and its relationship with formaldehyde and formaldehyde dehydrogenase.

Keywords Formaldehyde • Lifespan • Stress resistance • *Drosophila*

1 Introduction

All multicellular organisms such as mammals undergo a lot of physiological, morphological, cellular, and molecular changes, which lead to death as a final end point (Balcombe and Sinclair 2001). Although aging is still regarded as an unsolved problem, it has been already proved that aging process could be regulated by many

---

Y. Li
School of Life Sciences, Peking University, Beijing 100871, China

R. He (✉)
State Key Laboratory of Brain and Cognitive Science, Institute of Biophysics, Chinese Academy of Sciences, Beijing 100101, China

University of Chinese Academy of Sciences, Beijing 100049, China
e-mail: rongqiaohe@163.com

R. He, *Formaldehyde and Cognition*, DOI 10.1007/978-94-024-1177-5_11

genes, signaling pathways, and transcription factors besides environmental effects (Kenyon 2010). Since the short-lived animals might die of novel diseases which are unrelated to normal aging, it is a more hopeful way to search for the mutants and factors which could prolong lifespan in order to study the aging mechanisms (Rose et al. 1992). There have been a lot of ways reported to increase lifespan, including physical stimulus, chemical molecule usage, and genetic mutations (Kenyon 2010).

As described previously (the Environmental Protection Agency, USA "Toxicological Review of Formaldehyde Inhalation Assessment" June 18, 2010), all living organisms are producing and processing formaldehyde (Su et al. 2016). Formaldehyde plays a role in methylation and demethylation of DNA (Yan and Fujimori 2011; Crider et al. 2012), histone (Li et al. 2014; Shi et al. 2004), and RNA (Gerken et al. 2007; Su et al. 2015), suggesting that this compound is involved in gene regulations, in particular of epigenetic events. Furthermore, levels of endogenous formaldehyde increase along with humans' aging older than 75 years (Tong et al. 2013). The best-known phenomenon exemplifying epigenetic drift (the alteration of epigenetic patterns during aging) is the gradual decrease of global DNA methylation (Sierra et al. 2015). DNA methylation appears to be a dynamic tool of transcriptional regulation, with an extra level of complexity due to the recent discovery of the conversion of 5-methylcytosine into 5-hydroxymethylcytosine. This age-related DNA demethylation is associated with changes in histone modification patterns. It has been known that demethylation produces formaldehyde (Lardenoije et al. 2015). Thus, we considered that endogenous formaldehyde might be involved in the lifespan, since it plays a role in epigenetic events (Tong et al. 2011). Here we have observed that different concentrations of formaldehyde could exert different effects on *Drosophila* lifespan. A lower concentration of formaldehyde could prolong the lifespan, while a higher concentration could decrease the lifespan. In this chapter, we would like to discuss the relationship between formaldehyde and lifespan studies in *Drosophila*.

2 Formaldehyde and Lifespan

A lot of studies reported the toxic effects of formaldehyde on animals and human (Kilburn et al. 1987; Qiang et al. 2014). But the differences of effects on longevity among various concentrations of formaldehyde were rarely reported.

As described previously by Su and colleagues, the porcine brain contains 75.5–83.4 μmol/kg formaldehyde in fresh brain tissues measured by HPLC coupled with 2, 4-dinitrophenylhydrazone (Su et al. 2011). After that, Yang and coworkers fed 2-month-old ICR mice with 2% and 3.8% methanol in water (formaldehyde as its metabolite) for 6 months, which resulted in cognitive impairment and tau hyperphosphorylation in their brain (Yang et al. 2014a). The concentrations of blood formaldehyde markedly increased during the administration of methanol within water. Administration with 3% methanol could induce amyloid β deposition, tau hyperphosphorylation, neuron death, and cognitive impairment (Yang et al. 2014b).

So far, it is difficult to feed mammals with formaldehyde because the animals refuse to drink the aldehyde with irritated odor and taste.

It calls for a long period to study longevity with mice and other higher animals. The fruit fly *Drosophila melanogaster* is a powerful genetic model to study aging because of its short life cycle, ease of culture and handling, simple genome, convenient genetic manipulation, and high homology with higher animals (Tower 2011). In order to study the roles of formaldehyde on longevity systematically, we chose *Drosophila* as experimental model, and added different amounts of formaldehyde solution into the fly food, to make up ten kinds of culture medium with different concentrations of formaldehyde as below: 0%, 0.011%, 0.019%, 0.026%, 0.037%, 0.074%, 0.111%, 0.185%, 0.259%, and 0.370%. Afterward, nearly 100 single-gender flies were put into each culture bottle, among which the live ones were counted every 2 or 3 days and transferred into new bottles every week until the last one died.

The average lifespan of each group of female flies under different concentrations of formaldehyde is showed in Fig. 11.1. The statistic results indicated that different concentrations of formaldehyde played different roles on the lifespan of flies, especially in the gender of female. 0.037% formaldehyde could significantly extend the lifespan of female flies to the highest level, whereas the higher concentrations of formaldehyde ($\geq$0.185%) could significantly reduce the lifespan in both male and female flies (Li and He 2016). These results provide a new aspect to study the molecular mechanisms of longevity with the model of *Drosophila melanogaster*.

3 Genes and Lifespan

A lot of genes were reported to be involved in regulating lifespan. The earliest genes to be found were *age-1* and *daf-2* in *C. elegans* (Friedman and Johnson 1987; Kenyon et al. 1993). *Age-1* mutants reduce the propagating fecundity of worms to prolong lifespan, while *daf-2* participants in insulin/IGF-1 pathway to regulate metabolism and then affect lifespan. This pathway is highly conserved in evolution. The mutants of insulin/IGF-1 receptor in *Drosophila* could also extend lifespan (Tatar et al. 2001). In *C. elegans*, this pathway also calls for the participation of *daf-16*. In *Drosophila*, the most similar gene is *dFOXO* (Hwangbo et al. 2004). In mice, insulin and IGF-1 work through different receptors, and both of the two receptors could extend lifespan at different levels (Blüher et al. 2003; Holzenberger et al. 2003).

We are interested in the gene of formaldehyde dehydrogenase (FDH), which is one of the most ancient members of alcohol dehydrogenases (ADH) and belongs to the gene family of medium-chain dehydrogenase/reductase (MDR) (Persson et al. 2008). FDH persists in the organisms from prokaryotes to human, and it is highly conserved among these species (Danielsson and Jörnvall 1992).

FDH in *Drosophila* was separated from other types of alcohol dehydrogenase by a rapid microelectrophoretic technique in 1965 (Ursprung and Leone 1965), and to be localized to the third chromosome, which is different from ADH (Courtright

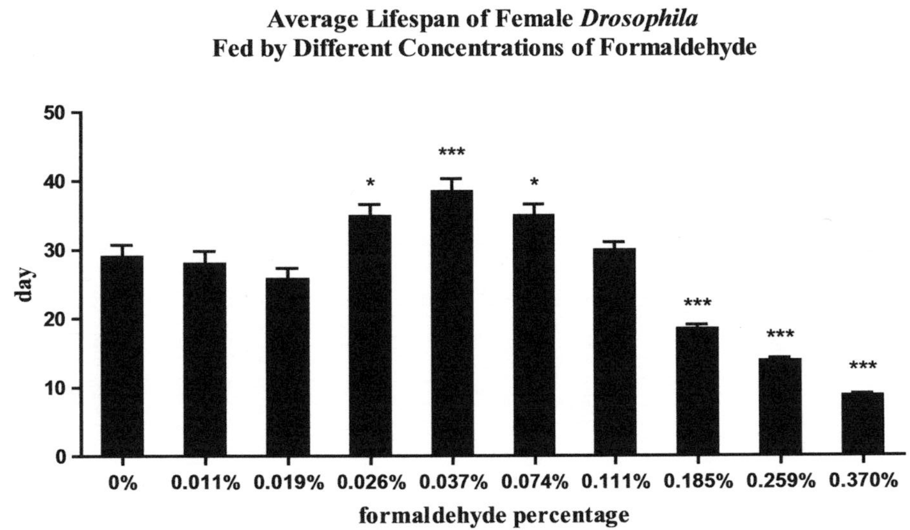

Fig. 11.1 Average lifespan of female *Drosophila* fed by different concentrations of formaldehyde. Compared to the 0% group, female flies fed by 0.026%, 0.037%, and 0.074% formaldehyde in food significantly increased lifespan, whereas 0.185% and higher concentrations significantly decreased lifespan ($n = 100$ for each group, $^{*}p < 0.05$, $^{***}p < 0.001$)

et al. 1966). Besides of the catalytic activity on formaldehyde, FDH could also catalyze octanol efficiently (Sieber et al. 1972; Danielsson et al. 1994).

In mammals, type III alcohol dehydrogenase (ADH III) is also called S-nitrosoglutathione reductase (GSNOR), which could catalyze the reduction reaction of S-nitrosoglutathione (GSNO) (Jensen et al. 1998). Since GSNO is a key protein on the metabolic pathway of NO, ADH III participates in the metabolism of NO through the degradation of GSNO. *Drosophila* FDH is most similar to GSNOR. Overexpression of FDH in *Drosophila* fan-shaped body could decrease the visual learning ability of the flies, and this phenotype could be rescued by overexpression of cGMP, which is necessary for the function of learning and memory (Hou et al. 2011).

Pecsenye and colleagues restricted the sugar input of *Drosophila* and found that the catalytic activity of FDH increased in fly larvae, but reduced in adult flies (Pecsenye et al. 1994). This phenomenon could also be affected by temperature. In larvae, the increasing level was more significant at 18 °C than 26 °C (Pecsenye et al. 1996). Antosh and colleagues performed a screen in flies and mice whose lifespan were increased by food restriction and by adding resveratrol, respectively. *Adh5* (a homolog of *Drosophila fdh*) was one of the genes which were upregulated by both of the two methods in flies and mice (Antosh et al. 2011). This result indicated that FDH might be closely related to longevity in both insects and mammals.

In order to study the roles of *fdh* gene more clearly, we knocked down the expression level of *fdh* gene in the *Drosophila* neural system by RNAi technique.

The preliminary results showed that the lifespan was significantly extended both in the laboratorial culturing condition and starvation condition (Li and He, unpublished data). Going on to investigate the relationship among *fdh* gene, formaldehyde, and lifespan will provide a new way to cut into the mechanism of longevity.

4 Formaldehyde and Stress Resistance

The mutants with lengthened lifespan are always accompanied by strengthened stress resistance, which includes starvation, heat shock, and reactive oxygen species stress (Martin et al. 1996). Some of the long-lived mutants increase resistibility to only one type of these stresses, while other mutants could resist two or more kinds of stresses simultaneously (Wang et al. 2004). The *age-1* and *daf-2* mutants in worms could resist reactive oxygen species, ultraviolet radiation, and heat shock. Meanwhile, the *mth* and *EcR* mutants in flies increased resistibility to paraquat, starvation, and heat shock (Lin et al. 1998; Simon et al. 2003). A study in 2004 screened for the genes which were upregulated or downregulated by three of these stresses and finally found 13 genes responding to all of them (Wang et al. 2004).

We further tested the effect of 0.037% formaldehyde on stress resistance, including starvation, heat shock, and paraquat, and found that 0.037% formaldehyde significantly increased starvation resistance and heat shock resistance, but reduced

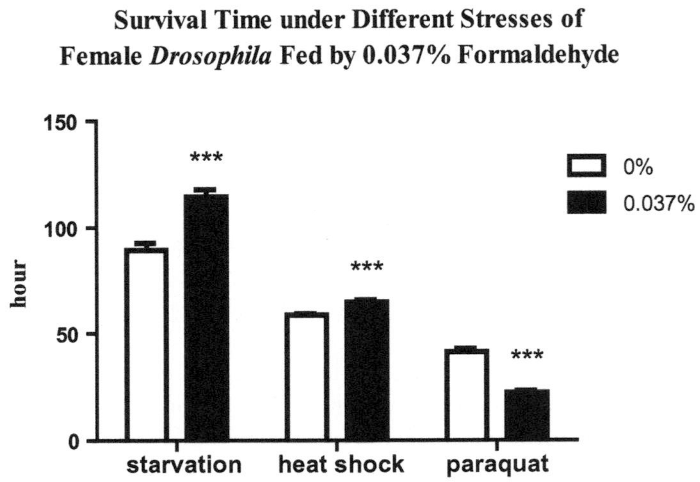

Fig. 11.2 Survival time under different stresses of female *Drosophila* fed by 0.037% formaldehyde. Compared to the 0% group, female flies fed by 0.037% formaldehyde in food significantly increased starvation resistance and heat shock resistance, but reduced oxidative resistance ($n = 75–100$ for each group, $***p < 0.001$)

oxidative resistance in both male and female flies. Figure 11.2 showed the results in female flies as an example.

5 Hormesis and Lifespan

The earliest way to prolong lifespan is hormesis, which is a phenomenon that a mild exposure to an originally lethal stress factor becomes beneficial in many organisms (Sørensen et al. 2007). The most common ways of hormesis include radiation, heat shock stress, food limitation, oxidative stress, and so on (Rattan 2008). The most traditional way to prolong lifespan is dietary restriction (DR). Since food provides most of the calorie in animals, restricting food input is to restrict the calorie input. So dietary restriction is also called caloric restriction (CR). DR was found in rodents in the 1930s, and later its existence was also proved in yeast, worms, flies, and human (Kenyon 2011; Rogina and Helfand 2004).

Since formaldehyde is harmful and even lethal at higher concentrations, but becomes beneficial at lower concentrations, we have regarded the effect of formaldehyde on lifespan as a kind of hormesis, which might activate the homeodynamic pathways of maintenance and repair and then lead to the beneficial hormetic effects (Rattan 2008). Application of the hormesis effect of formaldehyde might develop a new aspect of research and therapy for the aging problem.

6 Neural System and Lifespan

There are many factors and molecules in the neural system playing great roles on the lifespan of animals. In *Drosophila*, the prolonging effect of DR is mainly due to the yeast in the food (Mair et al. 2005). If the DR flies are allowed to smell the odor of yeast, then the lifespan elevation effect of DR is dismissed. Mutating the olfactory receptor OR83b could also prolong the lifespan. These results suggested that the olfactory system could perform an important role in the regulation of lifespan (Libert et al. 2007).

Social relationship also plays important roles on lifespan. The lifespan of male *Drosophila* is reduced after copulation with females, and this phenomenon is regarded as a cost of reproduction. Besides, contacting without copulation of males and females could also reduce lifespan. This is intermediated by a gustatory receptor named ppk23 (Gendron et al. 2014).

Besides the sensory systems such as olfactory and gustatory organs, there are still lots of other molecules regulating lifespan and stress resistance in the neural system. The most famous longevity pathway of insulin works in the neurons of *C. elegans*, but not in muscle or intestine (Wolkow et al. 2000). Overexpression of *Sir2* gene in the neural system could extend lifespan through regulating DR effect. Overexpression of the human *SOD1* gene in the motor neuron of *Drosophila* could also prolong the lifespan (Parkes et al. 1998).

7 Low Concentrations of Formaldehyde Enhance Cell Proliferation

As shown in Fig. 11.1, formaldehyde has an optimal concentration to prolong *Drosophila*'s lifespan, but either higher or lower concentration leads to a decrease of the effect. The same phenomena have been also observed in cell culture in the presence of formaldehyde. Decreasing the dose of formaldehyde in the culture medium could improve cell proliferation in many cell lines. The proliferation of HeLa cells and human malignant melanoma SK-MEL-28 cells was promoted in the presence of 10 μM formaldehyde (Ke et al. 2014; Rizzi et al. 2016). Low concentrations of formaldehyde (10–25 μM) stimulate the proliferation in both human low-invasive A375P and high-invasive SK-MEL-28 cells (Rizzi et al. 2014; Rizzi et al. 2016). As described by Tyihak and colleagues, the proper concentration of formaldehyde not only enhances the proliferation but also reduces apoptosis in human colon carcinoma HT-29 cells and human endothelial HUV-EC-C cells (Tyihak et al. 2001). They first found that formaldehyde (0.1 mM) could decrease apoptotic activity of the cultured cells. For the neural system, we would go on to test whether formaldehyde could improve the growth of neural cell lines from various organisms. Of course, whether there is any relation between the enhancement of proliferation and prolonging lifespan in the presence of formaldehyde needs to be further clarified.

Although a lot of molecules were discovered and reported to affect lifespan, it is still far away to illustrate how they are working together and defining the lifespan of the organisms finally. Stepping on to search for the key molecules, which greatly affect lifespan, will firmly promote the study of this area. Since knocking down the expression level of *fdh* gene in the *Drosophila* neural system could prolong lifespan and increase stress resistance, we decided to perform a screen in the *Drosophila* neural system to investigate the probable neuronal mechanisms and pathways of formaldehyde and *fdh* gene in aging.

8 Prospective

Stress resistance is a convenient experimental method since it has a much shorter period than natural lifespan. Using the knockout mutants in *Drosophila* neurotransmitter system, we plan to feed the flies with 0.037% formaldehyde in food and perform the screen under starvation or heat shock condition, searching for the mutants whose lifespan increasing phenotype by 0.037% formaldehyde is reduced to the same level with wild type, or elevated to an even higher level. Then we will go on to watch the lifespan changing of these mutants under normal culturing condition with or without formaldehyde and study the mechanisms of formaldehyde on lifespan. This research might cast new light on the aging problem and help us approach the dream of prolonging lifespan and deferring senescence in the future which might not be quite faraway.

Acknowledgment This project was supported by grants from the Beijing Municipal Science and Technology Project (Z161100000217141; Z161100000216137), the National Key Research and Development Program of China (2016YFC1306300), the National Basic Research Program of China (973 Program) (2012CB911004), the Natural Scientific Foundation of China (NSFC 31270868), the Foundation of Chinese Academy of Sciences (CAS-20140909), and the Queensland-Chinese Academy of Sciences Biotechnology Fund (GJHZ201302).

Competing Financial Interests The authors declare no competing financial interests.

References

Antosh M, Fox D, Helfand SL, Cooper LN, Neretti N (2011) New comparative genomics approach reveals a conserved health span signature across species. Aging 3(6):576–583

Balcombe NR, Sinclair A (2001) Ageing: definitions, mechanisms and the magnitude of the problem. Best Pract Res Clin Gastroenterol 15(6):835–849

Blüher M, Kahn BB, Kahn CR (2003) Extended longevity in mice lacking the insulin receptor in adipose tissue. Science 299(5606):572–574

Courtright JB, Imberski RB, Ursprung H (1966) The genetic control of alcohol dehydrogenase and octanol dehydrogenase isozymes in *Drosophila*. Genetics 54(5):1251–1260
Crider KS, Yang TP, Berry RJ, Bailey LB (2012) Folate and DNA methylation: a review of molecular mechanisms and the evidence for folate's role. Adv Nutr 3(1):21–38
Danielsson O, Jörnvall H (1992) "Enzymogenesis": classical liver alcohol dehydrogenase origin from the glutathione-dependent formaldehyde dehydrogenase line. Proc Natl Acad Sci U S A 89(19):9247–9251
Danielsson O, Atrian S, Luque T, Hjelmqvist L, Gonzàlez-Duarte R, Jörnvall H (1994) Fundamental molecular differences between alcohol dehydrogenase classes. Proc Natl Acad Sci U S A 91(11):4980–4984
Friedman DB, Johnson TE (1987) A mutation in the *age-1* gene in *Caenorhabditis elegans* lengthens life and reduces hermaphrodite fertility. Genetics 118(1):75–86
Gendron CM, Kuo TH, Harvanek ZM, Chung BY, Yew JY, Dierick HA, Pletcher SD (2014) *Drosophila* life span and physiology are modulated by sexual perception and reward. Science 343(6170):544–548
Gerken T, Girard CA, Tung YL, Webby CJ, Saudek V, Hewitson KS, Yeo GSH, McDonough MA, Cunliffe S, McNeill LA, Galvanovskis J, Rorsman P, Robins P, Prieur X, Coll AP, Ma M, Jovanovic Z, Farooqi IS, Sedgwick B, Barroso I, Lindahl T, Ponting CP, Ashcroft FM, O'Rahilly S, Schofield CJ (2007) The obesity-associated *FTO* gene encodes a 2-oxoglutarate-dependent nucleic acid demethylase. Science 318(5855):1469–1472
Holzenberger M, Dupont J, Ducos B, Leneuve P, Géloën A, Even PC, Cervera P, Le Bouc Y (2003) IGF-1 receptor regulates lifespan and resistance to oxidative stress in mice. Nature 421:182–187
Hou Q, Jiang H, Zhang X, Guo C, Huang B, Wang P, Wang T, Wu K, Li J, Gong Z, Du L, Liu Y, Liu L, Chen C (2011) Nitric oxide metabolism controlled by *formaldehyde dehydrogenase* (*fdh*, homolog of mammalian *GSNOR*) plays a crucial role in visual pattern memory in *Drosophila*. Nitric Oxide 24:17–24
Hwangbo DS, Gersham B, Tu MP, Palmer M, Tatar M (2004) *Drosophila* dFOXO controls lifespan and regulates insulin signalling in brain and fat body. Nature 429:562–567
Jensen DE, Belka GK, Du Bois GC (1998) S-Nitrosoglutathione is a substrate for rat alcohol dehydrogenase class III isoenzyme. Biochem J 331(2):659–668
Ke Y, Qin X, Zhang Y, Li H, Li R, Yuan J, Yang X, Ding S (2014) In vitro study on cytotoxicity and intracellular formaldehyde concentration changes after exposure to formaldehyde and its derivatives. Hum Exp Toxicol 33(8):822–830
Kenyon C (2010) The genetics of ageing. Nature 464:504–512
Kenyon C (2011) The first long-lived mutants: discovery of the insulin/IGF-1 pathway for ageing. Philos Trans R Soc Lond Ser B Biol Sci 366(2011):9–16
Kenyon C, Chang J, Gensch E, Rudner A, Tabtiang R (1993) A *C.elegans* mutant that lives twice as long as wild type. Nature 366:461–464
Kilburn KH, Warshaw R, Thornton JC (1987) Formaldehyde impairs memory, equilibrium, and dexterity in histology technicians: effects which persist for days after exposure. Arch Env Heal 42(2):117–120
Lardenoije R, Iatrou A, Kenis G, Kompotis K, Steinbusch HW, Mastroeni D, Coleman P, Lemere CA, Hof PR, van den Hove DL, Rutten BP (2015) The epigenetics of aging and neurodegeneration. Prog Neurobiol 131(2015):21–64
Li Y, He R (2016) The effects of formaldehyde on life span and stress resistance in *Drosophila melanogaster*. Prog Biochem Biophys 43(4):420–428
Li T, Su T, He Y, He R (2014) Formaldehyde, an epigenetic factor for learning and memory participating histone (de)methylation. Acta Neuropharmacologica 4(6):21–29
Libert S, Zwiener J, Chu X, Van Voorhies W, Roman G, Pletcher SD (2007) Regulation of *Drosophila* life span by olfaction and food-derived odors. Science 315(5815):1133–1137
Lin YJ, Seroude L, Benzer S (1998) Extended life-span and stress resistance in the *Drosophila* mutant *methuselah*. Science 282(5390):943–946

Mair W, Piper MDW, Partridge L (2005) Calories do not explain extension of life span by dietary restriction in *Drosophila*. PLoS Biol 3(7):e223

Martin GM, Austad SN, Johnson TE (1996) Genetic analysis of ageing: role of oxidative damage and environmental stresses. Nat Genet 13(1996):25–34

Parkes TL, Elia AJ, Dickinson D, Hilliker AJ, Phillips JP, Boulianne GL (1998) Extension of *Drosophila* lifespan by overexpression of human *SOD1* in motorneurons. Nat Genet 19:171–174

Pecsenye K, Lefkovitch LP, Giles BE, Saura A (1994) Does *Drosophila melanogaster* use ethanol as an energy source during starvation? Hereditas 121(3):225–236

Pecsenye K, Lefkovitch LP, Giles BE, Saura A (1996) Differences in environmental temperature, ethanol and sucrose associated with enzyme activity and weight changes in *Drosophila melanogaster*. Insect Biochem Molec Biol 26(2):135–145

Persson B, Hedlund J, Jörnvall H (2008) The MDR superfamily. Cell Mol Life Sci 65(2008):3879–3894

Qiang M, Xiao R, Su T, Wu B, Tong Z, Liu Y, He R (2014) A novel mechanism for endogenous formaldehyde elevation in SAMP8 mouse. J Alzheimers Dis 40(4):1039–1053

Rattan S (2008) Hormesis in aging. Ageing Res Rev 7(2008):63–78

Rizzi M, Cravello B, Reno F (2014) Textile industry manufacturing by-products induce human melanoma cell proliferation via ERK1/2 activation. Cell Prolif 47(6):578–586

Rizzi M, Cravello B, Tonello S, Reno F (2016) Formaldehyde solutions in simulated sweat increase human melanoma but not normal human keratinocyte cells proliferation. Toxicol In Vitro 37(2016):106–112

Rogina B, Helfand SL (2004) Sir2 mediates longevity in the fly through a pathway related to calorie restriction. Proc Natl Acad Sci U S A 101(45):15998–16003

Rose MR, Vu LN, Park SU, Graves JRJL (1992) Selection on stress resistance increases longevity in *Drosophila melanogaster*. Exp Gerontol 27(2):241–250

Shi Y, Lan F, Matson C, Mulligan P, Whetstine JR, Cole PA, Casero RA, Shi Y (2004) Histone demethylation mediated by the nuclear amine oxidase homolog LSD1. Cell 119(7):941–953

Sieber F, Fox D, Ursprung H (1972) Properties of octanol dehydrogenase from *Drosophila*. FEBS Lett 26(1–2):274–276

Sierra MI, Fernández AF, Fraga MF (2015) Epigenetics of aging. Curr Genomics 16(6):435–440

Simon AF, Shih C, Mack A, Benzer S (2003) Steroid control of longevity in *Drosophila melanogaster*. Science 299(5611):1407–1410

Sørensen JG, Kristensen TN, Kristensen KV, Loeschcke V (2007) Sex specific effects of heat induced hormesis in Hsf-deficient *Drosophila melanogaster*. Exp Gerontol 42(12):1123–1129

Su T, Wei Y, He R (2011) Assay of brain endogenous formaldehyde with 2,4-Dinitrophenylhydrazine through UV-HPLC. Prog Biochem Biophys 38(12):1171–1177

Su T, Song D, Li T, Wang X, He R (2015) Nucleic acid methylation/demethylation, endogenous formaldehyde and age-related cognitive impairment. Prog Biochem Biophys 42(3):211–219

Su T, Monte WC, Hu X, He Y, He R (2016) Formaldehyde as a trigger for protein aggregation and potential target for mitigation of age-related, progressive cognitive impairment. Curr Top Med Chem 16(5):472–484

Tatar M, Kopelman A, Epstein D, Tu MP, Yin CM, Garofalo RS (2001) A mutant *Drosophila* insulin receptor homolog that extends life-span and impairs neuroendocrine function. Science 292(5514):107–110

Tong Z, Han C, Miao J, Lu J, He R (2011) Excess endogenous formaldehyde induces memory decline. Prog Biochem Biophys 38(6):575–579

Tong Z, Han C, Luo W, Wang X, Li H, Luo H, Zhou J, Qi J, He R (2013) Accumulated hippocampal formaldehyde induces age dependent memory decline. Age 35(3):583–596

Tower J (2011) Heat shock proteins and *Drosophila* aging. Exp Gerontol 46(2011):355–362

Tyihák E, Bocsi J, Timár F, Rácz G, Szende B (2001) Formaldehyde promotes and inhibits the proliferation of cultured tumour and endothelial cells. Cell Prolif 34(3):135–141

Ursprung H, Leone J (1965) Alcohol dehydrogenases: a polymorphism in *Drosophila melanogaster*. J Exp Zool 160(2):147–154
Wang HD, Kazemi-Esfarjani P, Benzer S (2004) Multiple-stress analysis for isolation of *Drosophila* longevity genes. Proc Natl Acad Sci U S A 101(34):12610–12615
Wolkow CA, Kimura KD, Lee MS, Ruvkun G (2000) Regulation of *C. elegans* life-span by Insulinlike signaling in the nervous system. Science 290(5489):147–150
Yan F, Fujimori DG (2011) RNA methylation by radical SAM enzymes RlmN and Cfr proceeds via methylene transfer and hydride shift. Proc Natl Acad Sci U S A 108(10):3930–3934
Yang M, Lu J, Miao J, Rizak J, Yang J, Zhai R, Zhou J, Qu J, Wang J, Yang S, Ma Y, Hu X, He R (2014a) Alzheimer's disease and methanol toxicity (part 1): chronic methanol feeding led to memory impairments and tau hyperphosphorylation in mice. J Alzheimers Dis 41(4):1117–1129
Yang M, Miao J, Rizak J, Zhai R, Wang Z, Huma T, Li T, Zheng N, Wu S, Zheng Y, Fan X, Yang J, Wang J, Yang S, Ma Y, Lü L, He R, Hu X (2014b) Alzheimer's disease and methanol toxicity (part 2): lessons from four rhesus macaques (macaca mulatta) chronically fed methanol. J Alzheimers Dis 41(4):1131–1147

Chapter 12
Potential Interventions on Formaldehyde-Induced Neuronal Toxicity

Zijian Zhang and Rongqiao He

Abstract For over 100 years, a great deal of research has been focused on the dementia of clinical phase and on single-target drugs to intervene the progression of Alzheimer's disease (AD). However, a real breakthrough in the treatment still needs to be made to stave off AD. So far, many drugs and methods have been tried to treat and intervene AD in basic research and clinical trials. However, the long-term and effective drugs with low side effects on the age-related cognitive impairment have been still investigated. In order to stave off the onset and progression of age-related cognitive impairment, different interventions have been emphasized and applied in Alzheimer's disease treatments, such as developing multi-target drugs, ameliorating lifestyle, regulating diet, providing music care, attending physical exercises, and participating in social activities. The multidomain intervention will be one of the comprehensive therapies for Alzheimer's disease patients, especially for those in preclinical phases. Therefore, we would like to emphasize that intervention of Alzheimer's disease needs personalized intervention and medicine, including individualized diagnosis and treatment. Currently, clarifying early pathological changes of preclinical stage is imperative to understand AD. Early diagnosis and intervention could have better effective treatment results. We can identify the dementia patients with high levels of endogenous formaldehyde and treat them with some Chinese herbs as suggested in this chapter. We would like to recommend the integration of Western medicine and traditional Chinese medicine (TCM), for instance, performing the Western diagnosis coupled with the TCM treatment to make the personalized treatment plan for individuals although the direct relation between the so-called etiological factors and age-related cognitive impairment is still waiting to be clarified.

Z. Zhang
Department of Biomedical Sciences, Paul L. Foster School of Medicine, Texas Tech University Health Sciences Center El Paso, El Paso, TX 79905, USA

R. He (✉)
State Key Laboratory of Brain and Cognitive Science, Institute of Biophysics, Chinese Academy of Sciences, Beijing 100101, China

University of Chinese Academy of Sciences, Beijing 100049, China
e-mail: rongqiaohe@163.com

R. He, *Formaldehyde and Cognition*, DOI 10.1007/978-94-024-1177-5_12

Keywords Formaldehyde • Traditional Chinese medicine • Intervention • Neurotoxicity • Neuroprotection

1 Introduction

The world's population is aging rapidly. By the year 2050, the group aged 60 and over is expected to achieve 22% of the world population, and there will be approximately 434 million people in the age of 80 and above worldwide estimated by the World Health Organization. Unfortunately, >20% of people aged 60 and over suffer from neurological or cognitive disorders. Among them, dementia and depression are the most commonly diagnosed neuropsychiatric disorders according to the 2015 world report on mental health and older adults. Alzheimer's disease (AD), which contributes 70% of the dementia, is widely studied. Neuronal disorders including AD usually occur with unknown causes and are incurable, age-related, and featured by progressive degeneration of the central or peripheral nervous systems and/or death of nerve cells (Gao and Hong 2008).

In recent years, the effect of formaldehyde (FA) on the human brain and cognition has attracted more and more attention. It is one of the major pollutants in indoor air and also widely used in some working conditions. So everyone is living around the formaldehyde. The acute or chronic exposure of FA will cause uncomfortable feelings or even severe damages on the tissue and organs, such as the eyes, upper and lower respiratory tract, skin, and central nervous system. In addition, FA is also produced in the normal metabolic process and balanced in the human body. With age, the formaldehyde accumulates in the neuron and other brain cells. The FA-induced chronic impairments are involved in the development of senile dementia. In view of the daily exposure and endogenous generation, it is essential to take action to antagonize the adverse effect of FA on the central nervous system (CNS). In this chapter, the protective strategies to recognition and memory dysfunction induced by FA will be discussed. They may be in the experimental stage, but we will gain enlightenment from their research to maintain our health.

Formaldehyde is not only an environmental pollutant but also commonly found in the indoor air. Exposure to formaldehyde is possible for everyone at any time. In some occupational sites, formaldehyde levels are even higher. The damages of formaldehyde exposure are extensive. Especially for those with long-term living or working in such conditions, many adverse effects, which are not limited in the central nervous system, will occur. There are lots of guidelines, regulations, and references published by different organizations and institutions, including the World Health Organization, centers for disease control and prevention, and environmental protection agency in each state or country. All these documents are available online and easily accessed. It is a pity that no approved drugs are used to prevent the impairment induced by formaldehyde exposure.

Because of the complexity of AD with unclear causes and many related risk factors, traditional Chinese medicine (TCM) with the characteristics of multi-target and multi-link treatments has advantages over the single-target treatment. Therefore, we would like to review some of TCMs in the treatment of AD, which are probably related to formaldehyde dysmetabolism. For decades, drugs from TCM have been researched and developed such as TCM formula *Smart Soup* (Hou et al. 2014), *Shenwu capsule* (Li and Zhang 2012), and *Bushen capsule* (Zhang et al. 2015), as well as single herb extracts such as geniposide (Sun et al. 2013), resveratrol (He et al. 2016), icariin (Song et al. 2015), and DL-3-butylphthalide (Hou et al. 2009). They could act on the complex pathogenesis of AD at multi-targets and multi-pathways, especially that both have neuroprotective and neurotrophic/regenerative effects and protect mitochondria and synapses. Thus, these drugs have significant features and advantages for the treatment of AD. In this chapter, chemicals or compounds which have neuroprotective effect are discussed, including some Chinese medicine herbs, compounds, potential physical methods, and lifestyle interventions. Some of them are related with the dysmetabolism of formaldehyde. So far in clinics, donepezil and rivastigmine are usually used for AD patients with moderate cognitive impairment, and memantine hydrochloride is for severe AD patients. Since no report has been released to clarify the relation between these drugs and metabolism of endogenous formaldehyde, we would not discuss them here.

2 Traditional Chinese Medicine

Traditional Chinese medicine has a long history that spans over thousands of years. It takes a holistic approach to understand normal function and disease processes and includes various forms of herbal medicine, acupuncture, massage, exercise, and dietary therapy. Chinese herbs account for the majority of treatments in TCM. There are several hundreds of herbs from plants, animals, and minerals. Many prescriptions or formula based on the essential theory of TCM have been recorded. During Chinese medical material modernization, lots of chemicals or compounds are separated and purified. They have broad biological activities. There are a few cases to introduce the applications of chemicals from Chinese herbs against the formaldehyde-induced injury on neuronal cells.

As shown in Table 12.1, age-related cognitive impairment is a complex disorder containing different syndromes, which can be divided into five syndromes: the kidney–marrow deficiency, the Qi deficiency and blood stasis, the phlegm-obstructing intelligence, the Yu fire disturbing, and the spleen-Yang impaired by damp, based on clinical differential diagnosis in the traditional Chinese medicine. The purpose of the TCM differential diagnosis of Alzheimer's disease is for the medical doctors to have the personalized diagnosis and personalized treatment on different AD patients (He 2016). According to the individual diagnosis, a personalized TCM prescription

Table 12.1 The syndromes involved in dementia

Syndrome	In Chinese	Characteristics	References
The kidney–marrow deficiency	肾虚髓亏	Memory loss complicated with tinnitus, hearing loss, decline of language competence, etc., which could be related to more AD patients with a high ratio of apoE4/E4	Guo (2011)
The qi deficiency and blood stasis	气虚血瘀	Memory loss complicated with mental fatigue, changes in personality, motive ability decline, hypodynamic behavior, etc., including vascular dementia	Yuan et al. (1994)
The phlegm-obstructing intelligence	痰浊阻窍	Complicated with depression, dizziness headache, less words, abdominal distension, phlegm-spitting saliva, burnout trapped heavy, etc.	Zhu (2010)
The stagnated fire agitating the heart	郁火扰心	Memory loss accompanied with mania, irritability, words and deeds without thinking, disorders of life rhythm, etc.	Chen et al. (2007)
The spleen-yang impaired by dampness	湿困脾阳	Memory loss complicated with dysfunction of the enteric nervous system, etc.	Wang et al. (1998)

(a group of drugs with complementary effects) is made to the patient. Different patients have different prescriptions. So far, many effective natural compounds have been isolated and purified in the world medical market. Here, we would like to suggest that the effective compound should not only be isolated from the plants but also from the group of drugs which have been decocted, because numerous new compounds should be synthesized and produced in the process of boiling. There is an example for this case as the following: Smart Soup, a traditional Chinese medicine formula containing three TCM drugs, can ameliorate amyloid pathology and related cognitive deficits although each of the three components itself has not got the efficacy of the amelioration.

2.1 Smart Soup

The formulae of Smart Soup are prescribed by ancient Chinese medical physicians to patients with aging-related cognitive impairment, and it has been applied in clinics for many centuries. Smart Soup contains three herbs: *rhizoma acori tatarinowii* (RAT) acting in neuroprotection and attenuating learning and memory deficits (Lee et al. 2003), *poria* cum *radix pini* (PRP) having sedative activity (Tong et al. 2010), and *radix polygalae* (RP) ameliorating memory and behavioral deficits (Chen et al. 2004; Lee et al. 2009; Shin et al. 2009) and exhibiting neuroprotective effects (Park et al. 2002; Lin et al. 2012). Each component is frequently used in different TCM formula for their pharmacological efficacies against dysfunctions of the central nervous system. Hou and colleagues have assessed the efficacy of Smart Soup against AD. Oral administration of Smart Soup ameliorated the cognitive impairment of AD transgenic mice, with reduced Aβ levels, retarded Aβ amyloidosis, and reduced Aβ-induced gliosis and neuronal loss in the brains of AD mice. Mechanistic studies showed RP reduced Aβ generation, whereas RAT and PRP exerted neuroprotective effects against Aβ (Hou et al. 2014). Their work provides a practical therapeutic strategy against age-related cognitive impairment, for instance, Smart Soup formulae, which combine drugs with different efficacies.

2.2 *Shenwu Capsule*

Shenwu capsule is different from traditional Chinese decoction. It is developed based on Chinese medicine theory with new preparation formulation. The capsule contains extracts of six herbs including *polygoni multiflori radix praeparata*, *epimedii folium*, *ginseng radix et rhizoma rubra*, *acori tatarinowii rhizoma*, *puerariae lobatae radix*, and *ligusticum chuanxiong hort*. The six herbs are selected from more than 60 classical prescriptions proved in long-term clinical practice. The capsule aims at tonifying the kidney and the spleen and reducing phlegm and removing blood stasis. It has been demonstrated that Shenwu capsule can improve the learning and memory in multiple animal models including APPV717I transgenic mice, Aβ-induced hippocampus injury rats, cholinergic damage-induced AD rats, mitochondrial deficiency-induced AD rats, D-galactose-induced aging mice, cerebral ischemia-induced dementia rats, and diabetes-induced dementia rats (Li and Zhang 2012). Moreover, the clinical trials of Shenwu capsule in China confirmed that it also effectively improved cognitive impairment in patients with mild to moderate AD (Zhong et al. 2007). This may be attributed to the benefits of multiple targets of Chinese herbs.

2.3 *Bushen Capsule*

Bushen capsule is another example of classic TCM prescription. It is based on the explanation of nourishing the kidney and enhancing the brain. It contains two main herbs, *alismatis rhizoma* and *cistanches herba*. The cistanches has neuroprotective action on either middle cognitive impairment (MCI) patients or animal models. In recent study, the effect of Bushen capsule on episodic memory in MCI patient was observed. Episodic memory loss is an important characteristic of early AD symptoms and is considered as one of the core diagnosis criterions. The results of neuropsychological tests and functional magnetic resonance imaging (fMRI) examinations show that Bushen capsule can ameliorate the episodic memory impairment, which is related to the activation or deactivation of the temporal lobe and the putamen (Zhang et al. 2015). This finding not only offers us an fMRI-based early diagnosis marker but also puts forward to a feasible prevention medicine.

2.4 DL-3-Butylphthalide

DL-3-n-butylphthalide (NBP), initially isolated as a pure component from seeds of *Apium graveolens* in 1978 by the researchers of the Institute of Medicine of Chinese Academy of Medical Sciences, is used as an antihypertensive herbal medicine for stroke patients. Afterward, NBP was synthesized (Fig. 12.1), and it was approved by the State Food and Drug Administration of China for clinical use in stroke patients in 2002. Peng and colleagues strongly suggest that NBP has therapeutic potentials for the treatment of vascular dementia (VaD) caused by decreased cerebral blood flow (Peng et al. 2007). In 2009, Hou and coworkers investigated the effects and mechanisms of DL-3-butylphthalide on the treatment of vascular dementia. They recruited 72 vascular dementia patients with MCI. After the treatment of DL-3-butylphthalide for 8 weeks, participants' cognitive scores were significantly elevated by using mini-mental state examination, activity of daily living scale, and clinical dementia rating (Hou et al. 2009). Decreases in malondialdehyde (MDA) and superoxide dismutase (SOD) in the patients' serum were also observed. Their data provided a positive result of DL-3-butylphthalide in the treatment of VaD and improvement of the cognitive abilities and suggest that antioxidation be the mechanism for the efficacy of DL-3-butylphthalide. Jia and colleagues studied the effects of DL-3-n-butylphthalide in patients with vascular cognitive impairment without dementia caused by subcortical ischemic small vessel disease. Over the 6-month treatment period, their results further demonstrated that NBP is effective to improve cognitive and global functions in patients recruited (Jia et al. 2016). The recent advances pertaining to the neuroprotective mechanisms of NBP-derived compounds provide the possibility of their clinical implementation in the management of various neurological conditions (Abdoulaye and Guo 2016).

Fig. 12.1 The structure of DL-3-butylphthalide

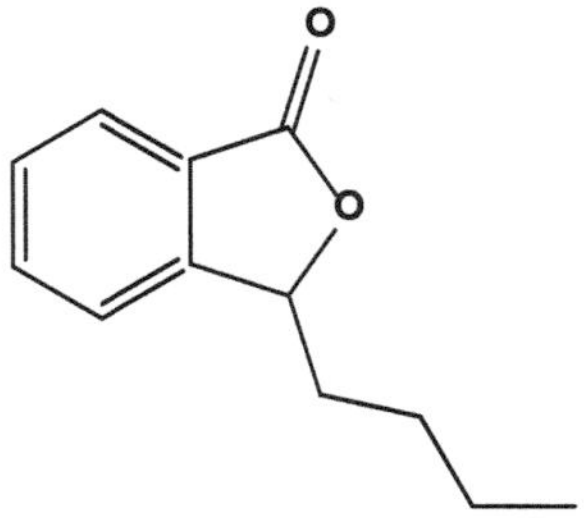

DL-3-butylphthalide

2.5 Epimedium

Epimedium (Yin yang huo, Fig. 12.2a) is a genus of flowering plants in the family Berberidaceae. It is used to enhance functions of the liver, kidney, bone, and cardiovascular system. Icariin (Fig. 12.2b) is an active ingredient of *Epimedium*. It has a wide range of pharmacological activities including an inhibitor of phosphodiesterase type 5, enhancing nitric oxide production and testosterone-like effects. These can explain its application in the treatment of erectile dysfunction and impotence (Li et al. 2015). In addition, it is a weak antioxidant and has neuroprotective and antidepressant-like effects (Shindel et al. 2010). The combination of icariin and *Panax notoginseng* saponins ameliorates learning and memory deficit and reduces the blood viscosity by protecting neurons from oxidative stress in the ischemia–reperfusion brain (Zheng et al. 2008).

Song and colleagues found that formaldehyde can induce the apoptosis of SH-SY5Y, a human neuroblastoma cell line. Icariin attenuates formaldehyde-induced cell death and decreases Tau phosphorylation through the inhibition of glycogen synthase kinase-3β (GSK-3β) (Song et al. 2015). Tau protein hyperphosphorylation is one of the important pathological features of Alzheimer's disease. It suggests that icariin is a promising candidate chemical to antagonize formaldehyde-induced learning and memory loss.

2.6 *Geniposide*

Another important compound from Chinese herbs is geniposide. It is isolated from gardenia fruit (Fig. 12.2c) extracts and is the main active ingredient. Geniposide (Fig. 12.2d) has broad and diverse bioactivities including antioxidative, anti-inflammatory, antithrombotic, and neuroprotective effects (Suzuki et al. 2001; Yin et al. 2010; Song et al. 2014; Lv et al. 2015). It is proposed to be a great promise as a novel treatment of Alzheimer's disease because of its multi-molecular mechanisms to protect the brain from pathologic damages (Liu et al. 2015).

Recently, it is confirmed that geniposide also has protective effects on formaldehyde-induced neuronal cell death. Formaldehyde at the concentration of 0.12 mM causes SH-SY5Y cell apoptosis and morphological changes. Geniposide can restore these effects by regulating the expression of Bcl-2, p53, caspase 3, and caspase 9 and increasing the activities of superoxide dismutase and glutathione peroxidase. Geniposide has anti-apoptotic and antioxidant actions to keep neuronal cell survival (Sun et al. 2013). In another neuronal cell line neuro 2a (N2a), geniposide also can prevent formaldehyde-induced apoptosis, and the abnormal expression of related genes Akt, Bcl-2, FOXO3, and p53 can be rescued (Sun et al. 2013; Chen et al. 2014). The joint use of geniposide and ginsenoside, also called TongLuoJiuNao, has very good efficacy in the treatment of ischemic cerebral stroke and dementia (Chinese SFDA: 2004 L01620). So we speculate that it may have good preventive effect on the formaldehyde-induced neuronal damage.

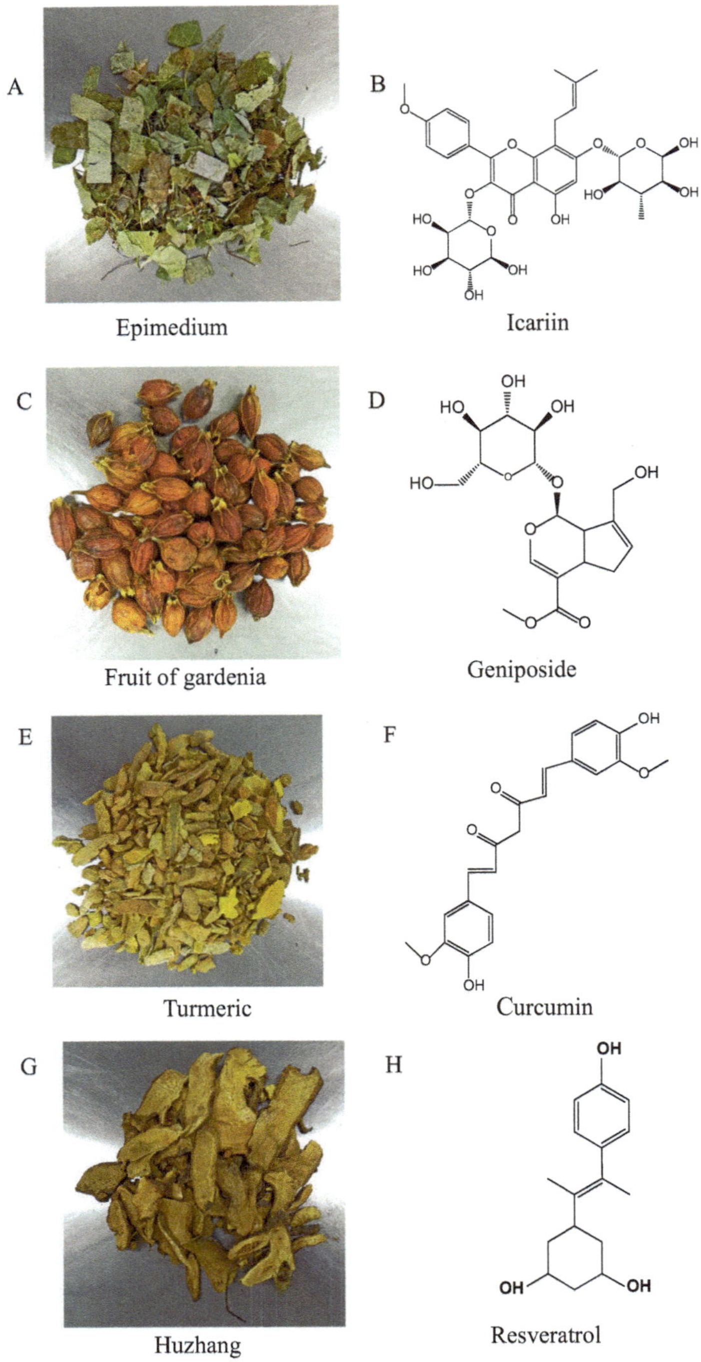

Fig. 12.2 The traditional Chinese herbs and their active compounds to protect against formaldehyde-induced neuronal damage. The herbs are listed on the left column and ingredients on the right. (**a**) Epimedium. (**b**) Icariin, the major active chemical of epimedium. (**c**) and (**d**) are the fruit of gardenia and geniposide, respectively. (**e**) and (**f**) are turmeric and curcumin. (**g**) and (**h**) are huzhang and resveratrol

2.7 Curcuminoids

Various curcuminoids are purified from turmeric (Fig. 12.2e). Curcumin (Fig. 12.2f) is generally considered as the most active constituent. Its potential to prevent and treat cancer, inflammation, dementia, and other diseases attracts much scientific interests (Sharma et al. 2005). The formaldehyde-induced genotoxicity has been studied a lot. Formaldehyde exposure caused DNA–protein cross-links (DPCs). Curcumin is an important antioxidant. Formaldehyde significantly increased MDA levels and decreased SOD and glutathione peroxidase (GSH-Px) activity. In addition, the activation of NF-κB and AP-1 was induced by formaldehyde treatment. Pretreatment with curcumin ameliorated the formation of DNA–protein cross-links caused by formaldehyde exposure. At the same time, curcumin also counteracted formaldehyde-induced oxidative stress (Zhang et al. 2013; Ciftci et al. 2015). The result was obtained from human lung cancer cell line A549. Although the evidence is little and not direct, the rich sources of food and supplements make it a good candidate for research and drug development.

2.8 Resveratrol

In 1940, Takaoka isolated resveratrol (Res) (3,5,4′-trihydroxystilbene) from the roots of white hellebore (*Veratrum grandiflorum O. Loes*) (Takaoka 1940). Later, Nonomura and colleagues separated it from the roots of *Polygonum cuspidatum* (Fig. 12.2g), a plant used in traditional Chinese and Japanese medicine in 1963 (Nonomura et al. 1963). As a polyphenol antioxidant, Res is considered to have therapeutic potential for the treatment of AD (Tredici et al. 1999; Ranney and Petro 2009). Res (10–100 μM, Fig. 12.2h) exerts its neuroprotective effects on AD pathological markers through different mechanisms, such as by reducing oxidative stress, promoting clearance of Aβ, and activating the anti-amyloidogenic cleavage of Aβ precursor protein (Marambaud et al. 2005; Huang et al. 2011; Richard et al. 2011).

Res is also identified as a natural formaldehyde capturer (Tyihák et al. 1998), which decreases the cytotoxicity of formaldehyde depending on cell line and point of time (Marcsek et al. 2007). Resveratrol plays a role in formaldehyde metabolism and also in the estrogen receptor positivity of MCF-7 cells. He and colleagues investigated the effects of Res on the viability of N2a cells in FA exposure, because N2a, a mouse neuroblastoma cell line, is commonly used to study neurotoxicity (Lepage et al. 2005). They observed that Res markedly decreases cell apoptosis rates by resisting the formaldehyde-induced cytotoxicity and Tau hyperphosphorylation (He et al. 2016).

Further study demonstrated that Res attenuates Tau phosphorylation at Thr181 in a dose-dependent manner through suppression of GSK-3-β and calmodulin-dependent protein kinase II (CaMKII) activities. These two kinases are commonly observed to be involved in Tau hyperphosphorylation in the AD brain (Wang et al.

2013). Furthermore, Res is able to increase the activity of phosphoseryl/phosphothreonyl proteinphosphatase-2A (PP2A) in the presence of formaldehyde. These possible mechanisms underlying the neuroprotective effects of Res to FA-induced impairment give an insight into the AD treatment via prevention of Tau from hyperphosphorylation by inhibition of GSK-3β and CaMKII as well as enhancement of dephosphorylation by the activation of PP2A.

2.9 Tetrahydroxystilbene Glucoside

2,3,5,4′-Tetrahydroxystilbene-2-O-β-D-glucoside (TSG) is the major active ingredient of Chinese herb, the tuberous root of *Polygonum multiflorum*. This compound has been tested in multiple dementia or AD-like animal models. The TSG treatment has significantly improved the learning and memory capacity of different chemical-induced AD models. In the APPV717I transgenic mice, the number of amyloid plaques, A-β content, and presenilin 1 expression was reduced in the presence of TSG (Zhang et al. 2006). TSG also ameliorated learning and memory deficits in Aβ40-induced (Chu et al. 2006), aluminum-induced (Luo et al. 2009), and sodium azide-induced (Zhang 2010) AD-like models. All these suggested that TSG may have multiple targets involved in the complicated pathogenesis of AD. Enhancing cholinergic function, inhibiting the A-β cascade, and protecting synapses and mitochondria may play their roles together to prevent the dementia loss and neuronal injury (Zhou et al. 2012 and Hou et al. 2011).

2.10 Cornel Iridoid Glycoside

The mainly water-soluble component purified from the sarcocarp of *Cornus officinalis* is a cornel iridoid glycoside (CIG). More than ten compounds are obtained, which are categorized into CIG. This extract has broad biological activities, especially on the central neuronal system. In the hippocampus cholinergic neuron damage in AD animal model, the learning and memory ability and neuronal cell loss were recovered after CIG treatment. It significantly elevated the expression of nerve growth factor, TrkA, brain-derived neurotropic factor (BDNF), synaptophysin, and GAP-43, which are beneficial to the neuron survival and axon growth (Zhao et al. 2010 and Ding et al. 2010). In human neuroblastoma SK-N-SH cells, in the presence of wortmannin, a phosphatidylinositol-3 kinase (PI3K) inhibitor, and GF-109203X, a protein kinase C inhibitor, CIG treatment attenuated Tau hyperphosphorylation, improved cell morphology, and reduced microtubule cytoskeleton damage (Yang et al. 2013).

2.11 Cassia obtusifolia

The seeds of *Cassia obtusifolia* (CO) have been traditionally used as an herbal remedy for acute inflammatory diseases. Ten years ago, the ethanolic fraction of the seeds (COE) has been shown to attenuate memory impairments in mice (Kim et al. 2007). Drever and collaborators studied the effects of COE with calcium dysregulation and cell death models in mouse primary hippocampal cultures and showed mechanisms for the therapeutic use of COE in the treatment of neurodegenerative disorders (Drever et al. 2008). Yi and colleagues found that COE has neuroprotective activities against Aβ-induced brain disorders (Yi et al. 2016). Jung and co-workers demonstrated that cassiae semen has neuroprotective effects, attributable to its anti-inflammatory actions, in ischemic stroke and Alzheimer's disease (AD) models (Jung et al. 2016). Levels of endogenous formaldehyde were elevated in cells during oxidative stress and also in the brain of mice with ischemic stroke (Chen et al. 2017). It appears that CO could have the efficacy to decrease formaldehyde in the oxidative stress and ischemic stroke because of its inhibition of inflammation.

3 Candidate Compounds Intervening with Formaldehyde

Antioxidants are molecules which can interact with free radicals to prevent or slow cell damage. During the normal metabolism, free radicals resulting in the oxidative stress are balanced by the body's endogenous antioxidant systems. In the above section, it is mentioned that formaldehyde itself is a quite reactive chemical and can cause oxidative damage to cells or tissues. Therefore, the application of antioxidants may have the potential to reduce the toxicity of formaldehyde.

3.1 *N-Acetyl Cysteine*

N-Acetyl cysteine (NAC, Fig. 12.3a) is one common antioxidant. It is a precursor of glutathione, an endogenous chemical with thiol group and antioxidant activity. It can also be used as medication for acetaminophen overdose, mucolytic therapy, psychiatry, and other diseases. In animal experiments, formaldehyde was intraperitoneally injected into ICR male mice for 1 week, and the performances of water maze task in both the learning and memory periods were poor. It was indicated that formaldehyde accumulation would impair the learning and memory capability. When NAC was supplemented, it significantly improved the performance of formaldehyde-treated mice (Feng et al. 2008). In addition, NAC shows beneficial effects on damages in the human lung epithelial cells and osteoblastic cells induced by formaldehyde (Huang et al. 2005; Wang et al. 2014).

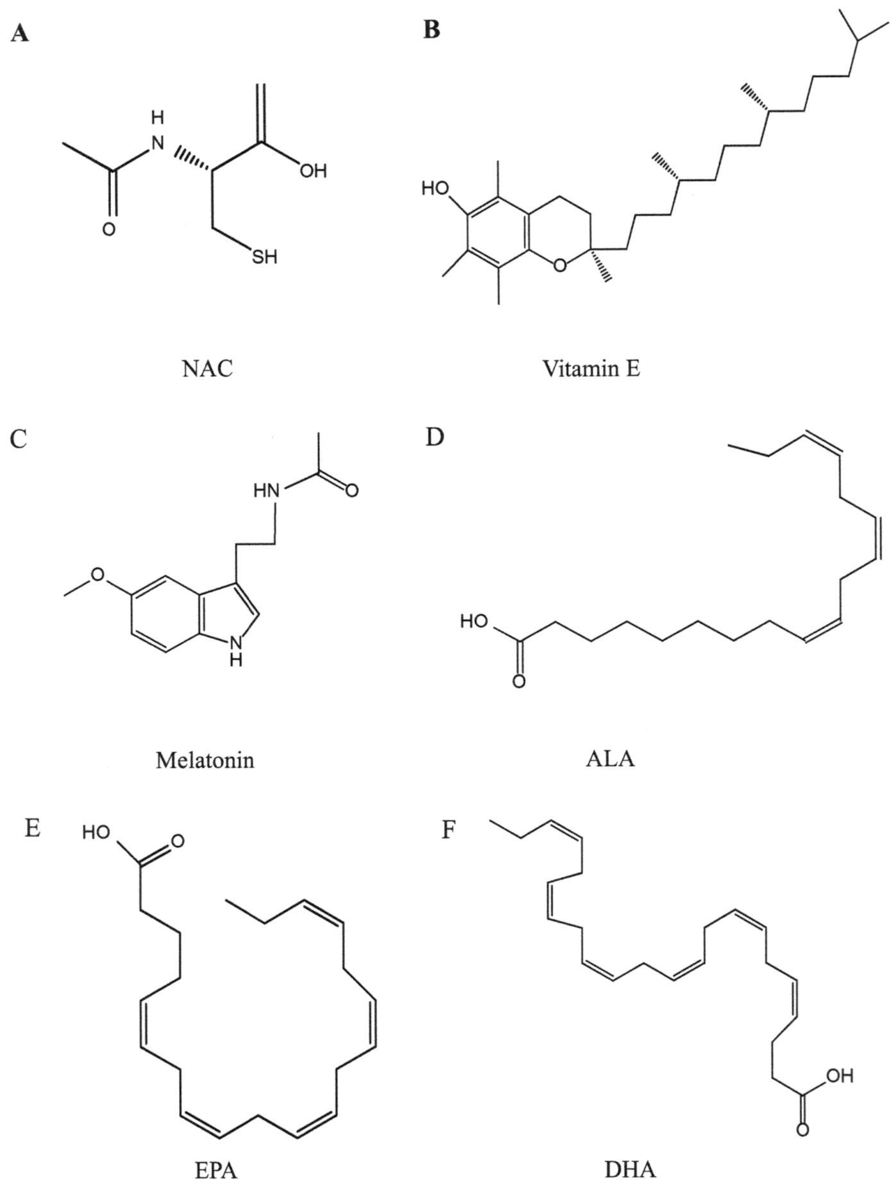

Fig. 12.3 The molecular structures of antioxidants to protect against formaldehyde-induced neuronal damage. (**a**) N-acetyl cysteine (NAC), (**b**) vitamin E, (**c**) melatonin, (**d–f**) omega-3 polyunsaturated fatty acid (ALA, EPA, and DHA, respectively)

3.2 Vitamin E

Vitamin E (VE, Fig. 12.3b) is a well-known antioxidant and acts as a peroxyl radical scavenger. It is a group of compounds including both tocopherols and tocotrienols. They are fat insoluble and easily incorporated into cell membranes to protect them from oxidative damage. The VE treatment improved the performance of formaldehyde-poisoned mice in the water maze task (Zhao et al. 2008). According to the behavior assay, it can be concluded that the antioxidant agency can restore the reduction of formaldehyde-induced learning and memory ability.

3.3 Melatonin

Melatonin (Fig. 12.3c) is a hormone secreted by the pineal gland in the brain. It maintains the body's circadian rhythm and regulates the release of female reproductive hormones. However, it also exists in bacteria, plants, and unicellular organisms, which do not have the pineal gland. Numerical studies have demonstrated that melatonin functions as an antioxidant (Reiter et al. 2005). Prefrontal cortex like the hippocampus plays a fundamental role in short-term and working memory (Preston and Eichenbaum 2013). Melatonin has the protective effects against formaldehyde-induced neurotoxicity in prefrontal cortex (Zararsiz et al. 2007). In the meantime, the activities of SOD and GSH-Px increased, and MDA levels decreased after melatonin treatment. Moreover, melatonin also prevents the rat testes from formaldehyde-induced oxidative damage and apoptosis (Ozen et al. 2008). In the experiments simulating occupational formaldehyde exposure, Mei and colleagues observed a decrease of melatonin in the mouse brain with cognitive impairment. Their work suggested that addition of melatonin may be beneficial to individuals who are usually exposed to formaldehyde in the workplace (Mei et al. 2016).

3.4 Selenium

Selenium is an essential trace element and component of some selenoproteins and enzymes, such as GSH-Px, thioredoxin reductase, and other peroxidases. That is why selenium has the antioxidant activity. Shi and his colleagues have found that pretreatment with selenium attenuated the formaldehyde-induced DPC production in A549 cell line and counteracted the formaldehyde-induced oxidative stress (Shi et al. 2014). In fact, the selenium content in our bodies is quite low and becomes toxic if taken in excess. For this reason, the selenium supplementation requires serious consideration.

3.5 Omega-3 Fatty Acids

Omega-3 fatty acids are polyunsaturated and include α-linolenic acid (ALA, Fig. 12.3d), found in plant foods, and eicosapentaenoic acid (EPA, Fig. 12.3e) and docosahexaenoic acid (DHA, Fig. 12.3f), found in fish. Omega-3s are beneficial in treating heart diseases, rheumatoid arthritis, asthma, and Crohn's disease (Defilippis et al. 2010; Gil et al. 2012). One human study shows that taking omega-3 supplements enhances the antioxidant defense system of patients with type 2 diabetes and probably prevents diabetes complications and related disease progress (Hajianfar et al. 2013). It is confirmed that after 10% formaldehyde intraperitoneal injection for 14 days, the activities of catalase (CAT) and SOD of cerebellum decreased, while the administration of omega-3 fatty acids alleviates the excessive free radicals (Zararsiz et al. 2011). Because mammals are unable to synthesize omega-3 fatty acids, the omega-3s from food have great benefits to human bodies. It may serve as potential food supplements to prevent the adverse effects of formaldehyde exposure.

4 Hydrogen Sulfide

Accumulating evidence shows that hydrogen sulfide (H_2S) can be produced enzymatically in all mammalian species including human (Li et al. 2011). It may function as a third gaseous signaling molecule participating with a wide range of physiological process (Modis et al. 2013). Abnormally increased or decreased endogenous H_2S production is associated with various diseases involving the cardiovascular, endocrine, gastrointestinal, and nervous systems (Wallace and Wang 2015). Endogenous formaldehyde accumulation in the brain will result in memory-deteriorating diseases such as age-related dementia. It has been demonstrated that endogenous FA inhibits the expression of cystathionine beta-synthase (CBS), the predominant H_2S-generating enzyme, and decreases the production of endogenous H_2S rat in hippocampus (Tang et al. 2013b). Excess of nitric oxide inhibits the CBS activity in the central nervous system. However, the specific inhibitor of nitric oxide synthase significantly attenuates FA-induced decreases in endogenous H_2S generation, neurotoxicity, and intracellular ROS accumulation in PC12 cells (Tang et al. 2013a). It is suggested that controls of endogenous H_2S may be a suitable novel therapeutic avenue for FA-induced neurotoxicity.

It has been reported that hydrogen sulfide also has cytoprotective effects on the neurons and cardiac muscle from oxidative stress and ischemia–reperfusion injury, respectively (Kimura et al. 2012). Tang and colleagues first reported that H_2S protects PC12 cells from formaldehyde-induced cytotoxicity and apoptosis, and this effect is related to attenuation of ROS accumulation, preserving mitochondrial function and pro-apoptotic gene expression (Tang et al. 2012). Endoplasmic reticulum (ER) stress is related with formaldehyde-induced neurotoxicity. To study the mechanism of hydrogen sulfide protection, the ER stress and SIRT-1 expression were assessed (Li et al. 2014). It was indicated that protection action of H_2S is due to overcome ER stress via upregulation of SIRT-1. BDNF signaling has been studied widely for its neuroprotective properties. Jiang and colleagues confirmed that BDNF–TrkB pathway also involved the neuroprotection of H_2S against FA-induced cytotoxicity (Jiang et al. 2015). Recently, drug research and development based on H_2S attract much attention as it can exert crucial effects on a wide range of cellular signaling processes. Some are currently in clinical trials. But more evidence on animals and human is required when the people use H_2S to treat against formaldehyde-induced CNS disorder.

5 Alcohol Treatments

Formaldehyde has been demonstrated to cause microtubule dissociation and Tau protein hyperphosphorylation, while methanol and formic acid cannot produce such effects. It is indicated that formaldehyde-induced neurodegeneration is in close relationship with Alzheimer's disease. However, in two back-to-back papers in 2014, Yang and colleagues found that chronic methanol exposure to rats and rhesus macaques induced partial AD-like symptoms by behavioral experiments and immunohistochemical analysis (Yang et al. 2014a, b). Therefore, formaldehyde may be involved in methanol-induced memory loss.

It is known that methanol acute poisoning can be treated with ethanol. In our body, methanol can be converted into formaldehyde by alcohol dehydrogenase. The application of ethanol can reduce the action of alcohol dehydrogenase on methanol by competitive inhibition, so methanol is excreted by the kidneys rather than being transformed into toxic metabolite. Here we suggest that alcohol treatment may be a feasible method to alleviate the memory loss caused by methanol and its metabolite formaldehyde.

6 Water Intake or Drinking Tea or Coffee

Chronic dehydration is a common symptom of patients with dementia. This may result the declined thirst perception and memory loss, and the dehydration can induce formaldehyde accumulation and hyperosmotic stress, both of which will lead to cognitive impairment. In mice, the frequency and quantity of drinking decreased with aging. When mice were water-fast but fed with food for 72 h, the brain formaldehyde level increased (Li et al. 2012). The accumulation of endogenous formaldehyde is related with aging and some memory-deteriorating disease. It is reasonable that drinking decreases the concentration of endogenous formaldehyde. Regular water intake relieves chronic dehydration and decreases formaldehyde level. It may protect the brain and prevent other diseases. It is recommended to intake 350–450 ml water for the three periods morning, afternoon, and evening, respectively. Of course, it depends upon one's water intake habit. We have a brief discussion of regular drinking tea or coffee in Chap. 4 (Eskelinen et al. 2009). Now that AD belongs to life-style disease, regulatioin of one's daily habit may be useful and beneficial to our cognitive abilities (He 2016).

7 Renal Dialysis

According to the physical property, formaldehyde is a small gaseous molecule and can permeate and diffuse easily. The formaldehyde level in the blood increased with aging-related dementia. Blood dialysis is a possible way to decrease the concentration of formaldehyde. But there is one concern. The setup of dialysis is usually disinfected with formalin. The residual formaldehyde may have the opposite action (Klein 1986). More evidences and practices are required to assess the effectiveness and safety of renal dialysis on the formaldehyde removal.

8 Others

There are several compounds contained in food or some clinical drugs. They have been investigated to counteract the formaldehyde-induced memory loss and oxidative damage. We will introduce them simply to start discussions. Triphlorethol-A (Fig. 12.4a), an open-chain trimer of phloroglucinol, is a phlorotannin component isolated from *Ecklonia cava*. It was reported to exert a cytoprotective effect against oxidative stress induced by ultraviolet, γ-ray radiation, and H_2O_2 (Tierney et al. 2010). Zhang and colleagues have also found that it can prevent lung fibroblast cells from formaldehyde-induced cell damage and apoptosis via mitochondria-mediated pathway (Zhang et al. 2010). It also protects cells from formaldehyde-induced DNA damage through the PI3K/Akt pathway (Zhang et al. 2011).

Caffeic acid phenethyl ester (CAPE, Fig. 12.4b) is a natural phenolic chemical compound of propolis extract. It has antimicrobial, antioxidant, anti-inflammatory, and cytotoxic properties (Murtaza et al. 2014). The formaldehyde-induced neurotoxicity is also prevented by CAPE (Turkoglu et al. 2007).

Theophylline (Fig. 12.4c) is a phytonutrient found in large quantities in green and black tea. It is widely used in the treatment of respiratory diseases. In the Morris water maze test, it showed that theophylline attenuates memory loss induced by formaldehyde in mice and increases the activities of SOD in the brain (Hao et al. 2008). Another clinical drug for antidepressant, phenelzine (Fig. 12.4d), was approved to protect rat primary cortical neurons and astrocytes from formaldehyde-induced toxicity (Song et al. 2010). It is a monoamine oxidase inhibitor and semicarbazide-sensitive amine oxidase that coverts methylamine to formaldehyde. The protective effect of phenelzine is thought to be due, at least in part, to its ability to sequester reactive formaldehyde.

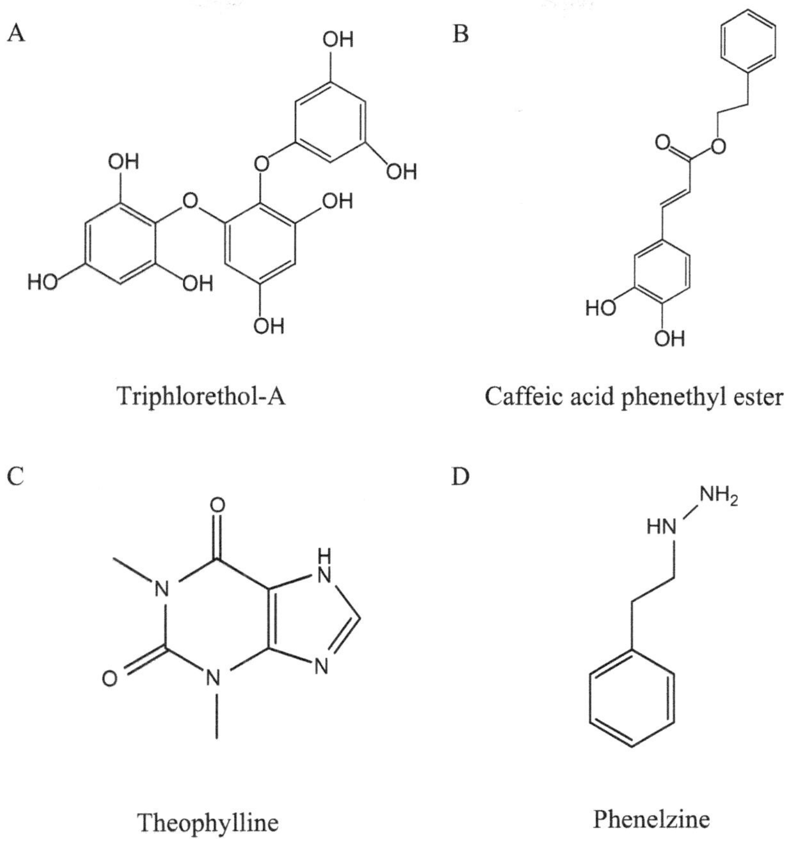

Fig. 12.4 The potential chemicals to antagonize the adverse effects caused by formaldehyde exposure. (**a**) Triphlorethol-A, (**b**) caffeic acid phenethyl ester, (**c**) theophylline, (**d**) phenelzine

HSP70 is a member of conserved ubiquitously expressed heat shock protein family and can act to protect cells from oxidative stress. When it was overexpressed in human bronchial epithelium cells, the damage of formaldehyde exposure was reduced and the DPC content decreased (Duan 2011). It was indicated that targeting HSP70 may be a potential tragedy against formaldehyde exposure injury.

9 Conclusion

Increasing people are suffering from age-related cognitive function impairment including Alzheimer's disease and vascular dementia. At present, no effective drugs can control the process of AD. Formaldehyde is recognized as one of the key risk factors involved in the senile dementia disease. In this chapter, several intervening methods including traditional Chinese medicine, antioxidant chemical, diet regulation, and physical therapy are discussed, and more details are introduced. Herein, we proposed one comprehensive program to prevent and treat age-related dementia from altering diet behavior to the chemical intervention and from Western medical diagnostics to Chinese medicine prescriptions or in other style collaborations. This systemic action and invention will hopefully benefit the aged people to delay the onset of dementia and slow the development process.

Acknowledgment This project was supported by grants from the National Key Research and Development Program of China (2016YFC1306300), the Beijing Municipal Science and Technology Project (Z161100000217141; Z161100000216137), the National Basic Research Program of China (973 Program) (2012CB911004), the National Natural Science Foundation of China (NSFC 31270868), the Foundation of Chinese Academy of Sciences (CAS-20140909), and the Queensland-Chinese Academy of Sciences Biotechnology Fund (GJHZ201302).

Competing Financial Interests The authors declare no competing financial interests.

References

Abdoulaye IA, Guo YJ (2016) A review of recent advances in neuroprotective potential of 3-N-butylphthalide and its derivatives. Biomed Res Int 2016:5012341. doi:10.1155/2016/5012341

Chen YL, Hsieh CL, Wu PH, Lin JG (2004) Effect of Polygala Tenuifolia root on behavioral disorders by lesioning nucleus basalis magnocellularis in rat. J Ethnopharmacol 95:47–55

Chen GH, Shan P, Qiu X (2007) The clinical study on Huanglianjiedu decoction on for patients with senile dementia, the type hyperactivity of heart- fire and liver in TCM. J Emerg Trad Chin Med 16(4):386–387

Chen J, Sun M, Wang X, Lu J, Wei Y, Tan Y, Liu Y, Götz J, He R, Hua Q (2014) The herbal compound geniposide rescues formaldehydeinduced apoptosis in N2a neuroblastoma cells. Sci China Life Sci 57(4):412–421

Chen XX, Su T, He YG He RQ (2017) Spatial cognition decline caused by excess formaldehyde in the lysosome. Prog Biochem Biophys 44. doi:10.16476/j.Pibb.2017.0050

Chu J, Ye C, Li L, Zhang L (2006) Tetrahydroxystilbene glucoside improves behavior changes and cholinergic function in dementia model mice induced by β-amyloid. Basic Clin Med 26(2):197–198

Ciftci G, Aksoy A, Cenesiz S, Sogut MU, Yarim GF, Nisbet C, Guvenc D, Ertekin A (2015) Therapeutic role of curcumin in oxidative DNA damage caused by formaldehyde. Microsc Res Tech 78(5):391–395

Defilippis AP, Blaha MJ, Jacobson TA (2010) Omega-3 fatty acids for cardiovascular disease prevention. Curr Treat Options Cardiovasc Med 12(4):365–380

Ding YX, Zhang L, Ye CF, Wang W, Li L (2010) Effects of cornel iridoid glycoside on learning-memory ability and synaptophysin in fimbria-fornix transacted rats. Chin J New Drugs 19(2):133–138

Drever BD, Anderson WG, Riedel G, Kim DH, Ryu JH, Choi DY, Platt B (2008) The seed extract of Cassia obtusifolia offers neuroprotection to mouse hippocampal cultures. J Pharmacol Sci 107(4):380–392

Duan YY (2011) Effects of overexpression of heat shock protein 70 on the damage induced by formaldehyde in vitro. Zhonghua Lao Dong Wei Sheng Zhi Ye Bing Za Zhi 29(5):349–352

Eskelinen MH, Ngandu T, Tuomilehto J, Soininen H, Kivipelto M (2009) Midlife coffee and tea drinking and the risk of late-life dementia: a population-based CAIDE study. J Alzheimers Dis 16(1):85–91

Feng Y, Ding S, Zhai J (2008) Learning and memory damage induced by formaldehyde and protective effect of N-acetyl-cysteine in mice. J Environ Health 08:680–683

Gao HM, Hong JS (2008) Why neurodegenerative diseases are progressive: uncontrolled inflammation drives disease progression. Trends Immunol 29(8):357–365

Gil A, Serra-Majem L, Calder PC, Uauy R (2012) Systematic reviews of the role of omega-3 fatty acids in the prevention and treatment of disease. Br J Nutr 107(Suppl 2):S1–S2

Guo HM (2011) Relation between ApoE genes and traditional Chinese medicinal syndrome of Alzheimer's. J Nanjing TCM Univ 27(2):124–127

Hajianfar H, Paknahad Z, Bahonar A (2013) The effect of omega-3 supplements on antioxidant capacity in patients with type 2 diabetes. Int J Prev Med 4(Suppl 2):S234–S238

Hao D, Zhang L, Wang S, Zhang Y, Wu C, Yang H, Gu R (2008) Theophylline attenuates amnesia induced by formaldehyde in mice. J Jilin Med Coll 29(2):75–77

He RQ (2016) Early multidomain intervention to stave off Alzheimer's disease. Chin Sci Bull 61(32):3420–3427

He X, Li Z, Rizak JD, Wu S, Wang Z, He R, Qin D, Hu X (2016) Resveratrol attenuates formaldehyde induced hyperphosphorylation of tau protein and cytotoxicity in N2a cells. Front Neurosci 10:598. doi:10.3389/fnins.2016.00598

Hou DR, Chen K, Wang Y, Tian Y, Wan S, Tang JC, Song Z (2009) The effect and mechanism of DL-3-butylphthalide on the treatment of vascular dementia. J South Medical Univ 29(3):574–578

Hou Y, Yang Q, Zhou L, Du X, Li M, Yuan M, Zhou Z, Li Z (2011) Tetrahydroxystilbene glucoside improves learning and (or) memory ability of aged rats and maybe connected to the APP pathway. Can J Physiol Pharmacol 89(11):801–809

Hou Y, Wang Y, Zhao J, Li X, Cui J, Ding J, Wang Y, Zeng X, Ling Y, Shen X, Chen S, Huang C, Pei G (2014) Smart soup, a traditional Chinese medicine formula, ameliorates amyloid pathology and related cognitive deficits. PLoS One 9(11), e111215

Huang FM, Chou LS, Chou MY, Chang YC (2005) Protective effect of NAC on formaldehyde-containing-ZOE-based root-canal-sealers-induced cyclooxygenase-2 expression and cytotoxicity in human osteoblastic cells. J Biomed Mater Res B Appl Biomater 74(2):768–773

Huang TC, Lu KT, Wo YY, Wu YJ, Yang YL (2011) Resveratrol protects rats from Aβ -induced neurotoxicity by the reduction of iNOS expression and lipid peroxidation. PLoS One 6(12), e29102

Jia J, Wei C, Liang J, Zhou A, Zuo X, Song H, Wu L, Chen X, Chen S, Zhang J, Wu J, Wang K, Chu L, Peng D, Lv P, Guo H, Niu X, Chen Y, Dong W, Han X, Fang B, Peng M, Li D, Jia Q, Huang L (2016) The effects of DL-3-n-butylphthalide in patients with vascular cognitive impairment without dementia caused by subcortical ischemic small vessel disease: a multicentre, randomized, double-blind, placebo-controlled trial. Alzheimers Dement 12(2):89–99

Jiang JM, Zhou CF, Gao SL, Tian Y, Wang CY, Wang L, Gu HF, Tang XQ (2015) BDNF-TrkB pathway mediates neuroprotection of hydrogen sulfide against formaldehyde-induced toxicity to PC12 cells. PLoS One 10(3):e0119478

Jung HA, Ali MY, Jung HJ, Jeong HO, Chung HY, Choi JS (2016) Inhibitory activities of major anthraquinones and other constituents from Cassia obtusifolia against β-secretase and cholinesterases. J Ethnopharmacol 191:152–160

Kim DH, Yoon BH, Kim YM, Lee S, Shin BY, Jung JW, Kim HJ, Lee YS, Choi JS, Kim SY, Lee KT, Ryu JH (2007) The seed extract of cassia obtusifolia ameliorates learning and memory impairments induced by scopolamine or transient cerebral hypoperfusion in mice. J Pharmacol Sci 105(1):82–93

Kimura H, Shibuya N, Kimura Y (2012) Hydrogen sulfide is a signaling molecule and a cryoprotectant. Antioxid Redox Signal 17(1):45–57

Klein E (1986) Effects of disinfectants in renal dialysis patients. Environ Health Perspect 69:45–47

Lee B, Choi Y, Kim H, Kim SY, Hahm DH, Lee HJ, Shim I (2003) Protective effects of methanol extract of Acori graminei rhizoma and Uncariae Ramulus et uncus on ischemia-induced neuronal death and cognitive impairments in the rat. Life Sci 74(4):435–450

Lee JY, Kim KY, Shin KY, Won BY, Jung HY, Suh YH (2009) Effects of BT-11 on memory in healthy humans. Neurosci Lett 454(2):111–114

Lepage KT, Dickey RW, Gerwick WH, Jester EL, Murray TF (2005) On the use of neuro-2a neuroblastoma cells versus intact neurons in primary culture for neurotoxicity studies. Crit Rev Neurobiol 17(1):27–50

Li L, Zhang L (2012) Action characteristics of traditional Chinese medicine in treatment of Alzheimer's disease. Prog Biochem Biophys 39(8):816–828

Li L, Rose P, Moore PK (2011) Hydrogen sulfide and cell signaling. Annu Rev Pharmacol Toxicol 51:169–187

Li T, Qiang M, He R (2012) Chronic dehydration and regularly drinking water to mitigate age-related cognitive impairment. Acta Neuropharmacologica 2(03):43–51

Li X, Zhang KY, Zhang P, Chen LX, Wang L, Xie M, Wang CY, Tang XQ (2014) Hydrogen sulfide inhibits formaldehyde-induced endoplasmic reticulum stress in PC12 cells by upregulation of SIRT-1. PLoS One 9(2):e89856

Li C, Li Q, Mei Q, Lu T (2015) Pharmacological effects and pharmacokinetic properties of icariin, the major bioactive component in Herba Epimedii. Life Sci 126:57–68

Lin Z, Gu J, Xiu J, Mi T, Dong J, Tiwari JK (2012) Traditional Chinese medicine for senile dementia. Evid Based Complement Alternat Med 2012:692621

Liu W, Li G, Holscher C, Li L (2015) Neuroprotective effects of geniposide on Alzheimer's disease pathology. Rev Neurosci 26(4):371–383

Luo HB, Yang JS, Shi XQ, Fu XF, Yang QD (2009) Tetrahydroxystilbene glucoside reduces the cognitive impairment and overexpression of amyloid precursor protein induced by aluminum exposure. Neurosci Bull 25(6):391–396

Lv C, Wang L, Liu X, Yan S, Yan SS, Wang Y, Zhang W (2015) Multi-faced neuroprotective effects of geniposide depending on the RAGE-mediated signaling in an Alzheimer mouse model. Neuropharmacology 89:175–184

Marambaud P, Zhao H, Davies P (2005) Resveratrol promotes clearance of Alzheimer's disease amyloid-β peptides. J Biol Chem 280(45):37377–37382

Marcsek ZL, Kocsis Z, Szende B, Tompa A (2007) Effect of formaldehyde and resveratrol on the viability of Vero, HepG2 and MCF-7 cells. Cell Biol Int 31(10):1214–1219

Mei Y, Duan C, Li X, Zhao Y, Cao F, Shang S, Ding S, Yue X, Gao G, Yang H, Shen L, Feng X, Jia J, Tong Z, Yang X (2016) Reduction of endogenous melatonin accelerates cognitive decline in mice in a simulated occupational formaldehyde exposure environment. Int J Environ Res Public Health 13(3): pii:E258

Módis K, Wolanska K, Vozdek R (2013) Hydrogen sulfide in cell signaling, signal transduction, cellular bioenergetics and physiology in C. elegans. Gen Physiol Biophys 32(1):1–22

Murtaza G, Karim S, Akram MR, Khan SA, Azhar S, Mumtaz A, Bin Asad MH (2014) Caffeic acid phenethyl ester and therapeutic potentials. Biomed Res Int 2014:145342

Nonomura S, Kanagawa H, Makimoto A (1963) Chemical constituents of Polygonaceous plants. I. Studies on the components of Ko-jo-kon. (Polygonum cuspidatum Sieb. Et Zucc.) Yakugaku Zasshi 83:988–990

Turkoglu AO, Sarsilmaz M, Kus I, Songur A, Ozyurt H, Akpolat N, Ogeturk M (2007) Caffeic acid phenethyl ester (CAPE) prevents formaldehydeinduced neuronal damage in hippocampus of rats. Neuroanatomy 6:66–71

Ozen OA, Kus MA, Kus I, Alkoc OA, Songur A (2008) Protective effects of melatonin against formaldehyde-induced oxidative damage and apoptosis in rat testes: an immunohistochemical and biochemical study. Syst Biol Reprod Med 54(4–5):169–176

Park CH, Choi SH, Koo JW, Seo JH, Kim HS, Jeong SJ, Suh YH (2002) Novel cognitive improving and neuroprotective activities of Polygala Tenuifolia Willdenow extract, BT-11. J Neurosci Res 70(3):484–492

Feng Y, Xu S, Chen G, Wang L, Feng Y, Wang X (2007) DL-3-n-butylphthalide improves cognitive impairment induced by chronic cerebral hypoperfusion in rats. J Pharmacol Exp Ther 321(3):902–910
Preston AR, Eichenbaum H (2013) Interplay of hippocampus and prefrontal cortex in memory. Curr Biol 23(17):R764–R773
Ranney A, Petro MS (2009) Resveratrol protects spatial learning in middle aged C57BL/6 mice from effects of ethanol. Behav Pharmacol 20(4):330–336
Reiter RJ, Tan DX, Maldonado MD (2005) Melatonin as an antioxidant: physiology versus pharmacology. J Pineal Res 39(2):215–216
Richard T, Pawlus AD, Iglesias ML, Pedrot E, Waffo-Teguo P, Merillon JM, Monti JP (2011) Neuroprotective properties of resveratrol and derivatives. Ann N Y Acad Sci 1215:103–108
Sharma RA, Gescher AJ, Steward WP (2005) Curcumin: the story so far. Eur J Cancer 41(13):1955–1968
Shi YQ, Chen X, Dai J, Jiang ZF, Li N, Zhang BY, Zhang ZB (2014) Selenium pretreatment attenuates formaldehyde-induced genotoxicity in A549 cell lines. Toxicol Ind Health 30(10):901–909
Shin KY, Lee JY, Won BY, Jung HY, Chang KA, Koppula S, Suh YH (2009) BT-11 is effective for enhancing cognitive functions in the elderly humans. Neurosci Lett 465(2):157–159
Shindel AW, Xin ZC, Lin G, Fandel TM, Huang YC, Banie L, Breyer BN, Garcia MM, Lin CS, Lue TF (2010) Erectogenic and neurotrophic effects of icariin, a purified extract of horny goat weed (Epimedium spp.) in vitro and in vivo. J Sex Med 7(4 Pt 1):1518–1528
Song MS, Baker GB, Dursun SM, Todd KG (2010) The antidepressant phenelzine protects neurons and astrocytes against formaldehyde-induced toxicity. J Neurochem 114(5):1405–1413
Song X, Zhang W, Wang T, Jiang H, Zhang Z, Fu Y, Yang Z, Cao Y Zhang N (2014) Geniposide plays an anti-inflammatory role via regulating TLR4 and downstream signaling pathways in lipopolysaccharide-induced mastitis in mice. Inflammation 37(5):1588–1598
Song YX, Miao JY, Qiang M, He RQ, Wang XM, Li WW (2015) Icariin protects SH-SY5Y cells from formaldehyde-induced injury through suppression of tau phosphorylation. Chin J Integr Med 22(6):430–437
Sun P, Chen JY, Li J, Sun MR, Mo WC, Liu KL, Meng YY, Liu Y, Wang F, He RQ, Hua Q (2013) The protective effect of geniposide on human neuroblastoma cells in the presence of formaldehyde. BMC Complement Altern Med 13:152
Suzuki Y, Kondo K, Ikeda Y, Umemura K (2001) Antithrombotic effect of geniposide and genipin in the mouse thrombosis model. Planta Med 67(9):807–810
Takaoka M (1940) Of the phenolic substances of white hellebore (Veratrum Grandiflorum Loes. Fil.). J. Fac. Sci. Hokkaido Imp Univ 3:1–16
Tang XQ, Ren YK, Zhou CF, Yang CT, Gu HF, He JQ, Chen RQ, Zhuang YY, Fang HR, Wang CY (2012) Hydrogen sulfide prevents formaldehyde-induced neurotoxicity to PC12 cells by attenuation of mitochondrial dysfunction and pro-apoptotic potential. Neurochem Int 61(1):16–24
Tang XQ, Fang HR, Zhou CF, Zhuang YY, Zhang P, Gu HF, Hu B (2013a) A novel mechanism of formaldehyde neurotoxicity: inhibition of hydrogen sulfide generation by promoting overproduction of nitric oxide. PLoS One 8(1):e54829
Tang XQ, Zhuang YY, Zhang P, Fang HR, Zhou CF, Gu HF, Zhang H, Wang CY (2013b) Formaldehyde impairs learning and memory involving the disturbance of hydrogen sulfide generation in the hippocampus of rats. J Mol Neurosci 49(1):140–149
Tierney MS, Croft AK, Hayes M (2010) A review of antihypertensive and antioxidant activities in macroalgae. Bot Mar 53(5):387–408
Tong XG, Liu JL, Cheng YX (2010) A new pregnane steroid from Poria cum radix pini. J Asian Nat Prod Res 12(5):419–423
Tredici G, Miloso M, Nicolini G, Galbiati S, Cavaletti G, Bertelli A (1999) Resveratrol, map kinases and neuronal cells: might wine be a neuroprotectant? Drugs Exp Clin Res 25(2–3):99–103
Tyihák E, Albert L, Németh ZI, Kátay G, Király-Véghely Z, Szende B (1998) Formaldehyde cycle and the natural formaldehyde generators and capturers. Acta Biol Hung 49(2–4):225–238

Wallace JL, Wang R (2015) Hydrogen sulfide-based therapeutics: exploiting a unique but ubiquitous gasotransmitter. Nat Rev Drug Discov 14(5):329–345
Wang CX, Lin SR, Xia SJ, Li DX (1998) Experimental study on dementia rats in Alzheimer type and the models. Liaoning J TCM 3:141–142
Wang JZ, Xia YY, Grundke-Iqbal I, Igbal K (2013) Abnormal hyperphosphorylation of tau: sites, regulation, and molecular mechanism of neurofibrillary degeneration. J Alzheimer's Dis 33(Suppl 1):S123–S139
Wang M, Chen S, Ba J, Pan H, Tao Y (2014) N-acetylcysteine: a promising drug against formaldehyde-induced damage in lung epithelial cells. Med Hypotheses 83(5):633–635
Yang CC, Kuai XX, Li Y, Zhang L, Yu JC, Li L, Zhang L (2013) Cornel iridoid glycoside attenuates tau hyperphosphorylation by inhibition of PP2A demethylation. Evid Based Complement Alternat Med 2013:108486
Yang M, Lu J, Miao J, Rizak J, Yang J, Zhai R, Zhou J, Qu J, Wang J, Yang S, Ma Y, Hu X, He R (2014a) Alzheimer's disease and methanol toxicity (part 1): chronic methanol feeding led to memory impairments and tau hyperphosphorylation in mice. J Alzheimers Dis 41(4):1117–1129
Yang M, Miao J, Rizak J, Zhai R, Wang Z, Huma T, Li T, Zheng N, Wu S, Zheng Y, Fan X, Yang J, Wang J, Yang S, Ma Y, Lu L, He R, Hu X (2014b) Alzheimer's disease and methanol toxicity (part 2): lessons from four rhesus macaques (Macaca mulatta) chronically fed methanol. J Alzheimers Dis 41(4):1131–1147
Yi JH, Park HJ, Lee S, Jung JW, Kim BC, Lee YC, Ryu JH Kim DH (2016) Cassia obtusifolia seed ameliorates amyloid β-induced synaptic dysfunction through anti-inflammatory and Akt/GSK-3β pathways. J Ethnopharmacol 178:50–57
Yin F, Liu J, Zheng X, Guo L, Xiao H (2010) Geniposide induces the expression of heme oxygenase-1 via PI3K/Nrf2-signaling to enhance the antioxidant capacity in primary hippocampal neurons. Biol Pharm Bull 33(11):1841–1846
Yuan PJ, Zhao FC, Guo AZ (1994) Treatment of 30 cases of Alzheimer's patients with qi deficiency and blood stasis using decoction invigorating yang for recuperation. TCM Res 7(4):26–27
Zararsiz I, Kus I, Ogeturk M, Akpolat N, Kose E, Meydan S, Sarsilmaz M (2007) Melatonin prevents formaldehyde-induced neurotoxicity in prefrontal cortex of rats: an immunohistochemical and biochemical study. Cell Biochem Funct 25(4):413–418
Zararsiz I, Meydan S, Sarsilmaz M, Songur A, Ozen OA, Sogut S (2011) Protective effects of omega-3 essential fatty acids against formaldehyde-induced cerebellar damage in rats. Toxicol Ind Health 27(6):489–495
Zhang R (2010) Pharmacological effects and mechanisms of tetrahydroxystilbene glucoside and icariin on Alzheimer's disease model induced by mitochondrial complex IV deficiency. PhD thesis, Capital Medical University, Beijing
Zhang L, Xing Y, Ye CF, Ai HX, Wei HF, Li L (2006) Learning-memory deficit with aging in APP transgenic mice of Alzheimer's disease and intervention by using tetrahydroxystilbene glucoside. Bahev Brain Res 173(2):246–254
Zhang R, Lee IK, Kang KA, Piao MJ, Kim KC, Kim BJ, Lee NH, Choi JY, Choi J, Hyun JW (2010) Cytoprotective effects of triphlorethol-a against formaldehyde-induced oxidative damage and apoptosis: role of mitochondria-mediated caspase-dependent pathway. J Toxicol Environ Health A 73(21–22):1477–1489
Zhang R, Kang KA, Piao MJ, Kim KC, Lee NH, You HJ, Hyun JW (2011) Triphlorethol-a improves the non-homologous end joining and base-excision repair capacity impaired by formaldehyde. J Toxicol Environ Health A 74(12):811–821
Zhang BY, Shi YQ, Chen X, Dai J, Jiang ZF, Li N, Zhang ZB (2013) Protective effect of curcumin against formaldehyde-induced genotoxicity in A549 cell lines. J Appl Toxicol 33(12):1468–1473
Zhang J, Xu K, Wei D, Guo R, Li H, Wang Y, Zhang Z (2015) The effects of Bushen capsule on episodic memory in amnestic mild cognitive impairment patients: a pilot placebo controlled fMRI study. J Alzheim Dis 46(3):665–676

Zhao J, Feng Y, Ding S (2008) Effects of vitamin E and N-acetyl-cysteine on learning and memory ability impairment induced by formaldehyde in mice. Chin Occup Med 35(6):471–474

Zhao LH, Ding YX, Zhang L, Li L (2010) Cornel iridoid glycoside improves memory ability and promotes neuronal survival in fimbria-fornix transected rats. Eur J Pharmacol 647(1–3):68–74

Zheng M, Qu L, Lou Y (2008) Effects of icariin combined with Panax notoginseng saponins on ischemia reperfusion-induced cognitive impairments related with oxidative stress and CA1 of hippocampal neurons in rat. Phytother Res 22(5):597–604

Zhong J, Zhu A, Yang C, Tian J (2007) A clinical controlled trial of Shenwu capsule in the treatment of mild cognitive impairment. China J Chin Mat Medica 32(17):1800–1803

Zhou L, Hou Y, Yang Q, Du X, Li M, Yuan M, Zhou Z (2012) Tetrahydroxystilbene glucoside improves the learning and memory of amyloid-beta(1-42)-injected rats and may be connected to synaptic changes in the hippocampus. Can J Physiol Pharmacol 90(11):1446–1455

Zhu JM (2010) Clinical effect of Xunyunning tablet in treating senile patients MCI of turbid phlegm blocking orifice syndrome. Chin J Med Guide 12(12):2107–2109

Chapter 13
Molecular, Cellular, and Animal Experiments in Formaldehyde Study

Min Qiang, Tao Su, Beibei Wu, and Rongqiao He

Abstract Alzheimer's disease (AD), which is involved in many molecular pathways in the human central nervous system, is a multifactorial complex neurodegenerative disease. The failure of the recent development of AD therapeutic agents in clinical trials is mainly due to their single-target or single-pathogenic pathway effect or late time window for treatments. Anabolism and metabolism of formaldehyde are also involved in many pathways and genes. According to the recent studies, endogenous formaldehyde is positively related with the severity of age-related cognitive impairment, showing approximately 40% AD patients suffering from dysmetabolism of formaldehyde. Furthermore, administration of low concentration of methanol in monkeys, a precursor compound for formaldehyde, induces cognitive impairment accompanied with Aβ deposition and formation of senile plaque, Tau hyperphosphorylation, and aggregation. These evidences suggest that formaldehyde plays a role in the progression of AD. To clarify the relationship between formalde-

M. Qiang
State Key Laboratory of Brain and Cognitive Science, Institute of Biophysics, Chinese Academy of Sciences, Beijing 100101, China

Shanghai Institute for Advanced Immunochemical Studies, ShanghaiTech University, Shanghai 201210, China

T. Su
State Key Laboratory of Brain and Cognitive Science, Institute of Biophysics, Chinese Academy of Sciences, Beijing 100101, China

B. Wu
State Key Laboratory of Brain and Cognitive Science, Institute of Biophysics, Chinese Academy of Sciences, Beijing 100101, China

Department of Physiology, David Geffen School of Medicine, University of California, Los Angeles (UCLA), Los Angeles, CA 90095, USA

R. He (✉)
State Key Laboratory of Brain and Cognitive Science, Institute of Biophysics, Chinese Academy of Sciences, Beijing 100101, China

University of Chinese Academy of Sciences, Beijing 100049, China
e-mail: rongqiaohe@163.com

R. He, *Formaldehyde and Cognition*, DOI 10.1007/978-94-024-1177-5_13

hyde and age-related cognitive impairment, useful models are urgently required. Establishment of proper models is beneficial to enlighten the understanding of the role of formaldehyde in the neurodegenerative disease, especially age-related cognitive impairment. In this chapter, we review some useful protocol examples in the experiments including molecular, cellular, and animal models involved in formaldehyde study from or for our colleagues.

Keywords Formaldehyde • Cell culture • Compounds • Animal • Models • Intraperitoneal injection • Intracerebroventricular injection

1 Introduction

Many models have been used in the study of effects that formaldehyde (FA) plays on cells and organisms and the related mechanisms. This chapter will show the formaldehyde-related molecular, cellular, and animal models which are commonly used in recent years. We show the production of formaldehyde from the reaction of biochemical compounds such as oxidation of lipid (Li et al. 2008) and sugar decomposition (Wang et al. 2017b). Concentrations of endogenous formaldehyde increase in animals with aging. We injected rodents with formaldehyde or administrated rodents and primates including rhesus monkeys with methanol to establish formaldehyde-induced models with cognitive impairment. Mice with high concentrations of formaldehyde were induced by exposing them to limited water intake. Transgenic rodents and *Drosophila* are also introduced. Establishment of these animal models is beneficial to study the relationship between formaldehyde and cognitive impairment. The animal experiments and the animal models, which were established in the authors' laboratory, are performed in accordance with the National Institutes of Health's Guide for the Care and Use of Laboratory Animals and were approved by the Biological Research Ethics Committee, Institute of Biophysics, Chinese Academy of Sciences, Beijing, China.

According to many research reports (Su et al. 2016), the accumulation of formaldehyde may be one of the key causes of memory loss in the brain especially in the hippocampi of animals. Formaldehyde is involved in triggering abnormal phosphorylation of Tau protein and promoting Aβ deposits. Excess brain FA resulting from either exogenous or endogenous pathways can induce memory impairment and cognitive decline in different kinds of animal models such as aging animals, transgenic mice, and formaldehyde- or methanol-administrated animals.

2 Preparation of Formaldehyde with Chemical Compounds

2.1 *Oxidation of Myelin Yielding Formaldehyde*

Lipid oxidation plays an important role in the progression of Alzheimer's disease (Ward et al. 2014; Giudetti et al. 2016). Oxidation of lipid such as myelin oxidation produces aldehydes including formaldehyde (Li et al. 2008). The experiment was performed as follows: N-Acyl-4-sphingosine-1-phosphocholine (1 mg/ml) was incubated with 1 mM H_2O_2 in 50 mM phosphate buffer (pH 7.2) at 37 °C for 24 h to induce lipid oxidation. Then formaldehyde level in the reaction mixture was analyzed with HPLC coupled with DNPH as described (Su et al. 2011).

2.2 *D-Ribose Decomposition Producing Formaldehyde*

D-ribose, present in all living cells, comes from food that contains high amounts of riboflavin and nucleic acids, such as eggs, meat, wheat bran, and RNA. Recent studies suggest that type 2 diabetes mellitus (T2DM) is not only a disorder in D-glucose metabolism but also in D-ribose metabolism (Su et al. 2013; Chen et al. 2016). Administration of mouse with D-ribose by intraperitoneal injection (once daily for at least 1 month) or gavage (once daily for 6 months) leads to decline of learning and memory ability, followed by glia activation, Aβ deposition, and Tau hyperphosphorylation (Han et al. 2014; Wei et al. 2015; Wu et al. 2015).

All reaction mixtures were filtered with 0.22 μm membranes (Millipore, USA) before incubation at 37 °C for different durations. D-Ribose was purchased from Amresco, USA. Other standard reagents and amino acid were obtained from Sigma-Aldrich, USA. Wang and her coworkers incubated D-ribose (100 mM final concentration) dissolved in 20 mM Tris-HCl buffer (pH 8.0) at 37 °C for up to 3 days and took aliquots for the measurement of formaldehyde concentration with 1.0 g/l 2, 4-dinitrophenylhydrazine (DNPH) at 60 °C for 30 min (Wang et al. unpublished data).

2.3 *Formaldehyde from Protein Reacted with Malondialdehyde*

Malondialdehyde is an active modifying agent of proteins both in vitro and in vivo (Foettinger et al. 2006; Weismann et al. 2011), causing protein aggregation and amyloid deposition (Allen et al. 1984). MDA-protein aggregates continuously enter into cells within our body, especially under conditions of oxidative stress, which can be induced in various pathological conditions such as cardiovascular or neurodegenerative diseases.

Bovine serum albumin (0.05 mM) was incubated with malondialdehyde (10 mM) in 10 mM phosphate buffer (pH 7.2) at 37 °C for 24 h or longer. 10 ng/μl formaldehyde (33 μM) alone was used as control. DNPH was added to reaction mixtures as described before. Aliquots were analyzed on an HPLC and a GC-mass spectrometry (HP6890 GC, USA) for formaldehyde quantification (Su et al. 2011).

2.4 Formaldehyde Resulted from O-Demethylation Catalyzed by Dioxygenase

As described by Farrow and Facchini, the production of formaldehyde increases in response to the amount of codeine *O*-demethylase used in the assays containing codeine, which undergoes *O*-demethylation. Similarly, the yield of formaldehyde increases in response to the amount of protopine *O*-dealkylase rising in the assay in the presence of cryptopine (Farrow and Facchini 2013).

3 Production of Formaldehyde in Cell Culture

3.1 Formaldehyde in Daily Cell Culture

Generation of formaldehyde is resulted from different metabolic pathways, such as demethylation of DNA, RNA and histone, protein denaturation, lipid oxidation, reduced sugar decomposition, and oxidative stress (Chap. 2). Of course, every cell produces, processes, and uses formaldehyde at each moment. Daily cell culture generates formaldehyde and releases it into the medium. Thus, formaldehyde concentrations can be measured in the culture medium.

Murine neuroblastoma neuro-2a cells (N2a cell line; China Cell Resource Confederation, Beijing, China) were maintained in DMEM-F12 (Corning, Steuben county, NY, USA) supplemented with 10% (v/v) fetal bovine serum (FBS; PAA Laboratories, Pasching, Austria) as monolayer in petri dishes (Corning, USA), at 37 ° C and 5% CO_2 with >95% relative humidity. Aliquots of medium were taken for measurements of formaldehyde concentration. In general, the average concentration of formaldehyde in the medium released by the cells is 8-11 μM (Wang et al. 2017a).

3.2 *Yield of Formaldehyde in the Culture of MRMT-1 Cells In Vitro*

Formaldehyde concentration in cultured rat breast carcinoma cells MRMT-1 increased significantly with time from 24 h to 72 h. Concentrations of formaldehyde increased significantly in tumors of the MRMT-1 subcutaneous vaccination model 7 days after inoculation. Concentrations of formaldehyde increased significantly in bone marrows and sera of rats with bone cancer pain from day 7 to day 21 (Liu et al. 2013). Formaldehyde levels were determined on an HPLC coupled with fluorescence according to Luo et al. (2001).

3.3 *Cells Producing Formaldehyde in the Presence of H_2O_2*

Cells (blood endothelial cell line, bEnd.3) were plated (1×10^5 cell/dish) and cultured in DMEM containing 10% fetal calf serum in petri dish (35 mm) in an incubator (37 ° C, 5% CO_2) for 12 hours or longer. The DMEM culture was replaced by BBS buffer (136.9 mM NaCl, 5.37 mM KCl, 1.26 mM $CaCl_2$, 0.81 mM $MgSO_4$, 0.44 mM KH_2PO_4, 0.335 mM Na_2HPO_4, 10 mM PIPES; pH = 7.2) containing formaldehyde fluorescence probe (FAP-1, final concentration 10 μM) at 37 ° C, for 30 min. The cells were then washed with fresh BBS buffer without the FAP-1 probe. After that, cells were observed under a confocal laser scanning microscope (ZEISS LSM 700 confocal microscope). At the excitation wavelength of 633 nm, the fluorescence of FAP-1 was observed (emission wavelength, 688 nm). At the same time, MitoTracker Green FM (Cell Signaling, USA) (λ_{ex} = 490 nm; λ_{em} = 516 nm) and LysoTracker Green DND-26 (Cell Signaling, USA) (λ_{ex} = 504 nm; λ_{em} = 511 nm) could be employed for staining of mitochondrion and lysosome, respectively. Lysosome contributes to the degradation of not only biomacromolecules but also regulation of channel transporters (Jian and Zhu 2016). The staining signals would show the location of formaldehyde in lysosomes, but not much in mitochondria. Afterwards, H_2O_2 was added to bEnd.3 cells when they were cultured in DEME medium. The signals of formaldehyde markedly increased, following the addition of H_2O_2 (Chen et al. 2017). In other words, H_2O_2 promotes the cells to produce formaldehyde significantly (Fig. 13.1a).

3.4 *ROS Production from Cells in the Presence of Formaldehyde*

Chen and her colleagues found that intracellular ROS levels were increased along with increased FA concentration (Chen et al. 2017). FA-induced ROS was generated from eosinophils recruited to the inflammatory sites of the airways (Jung et al.

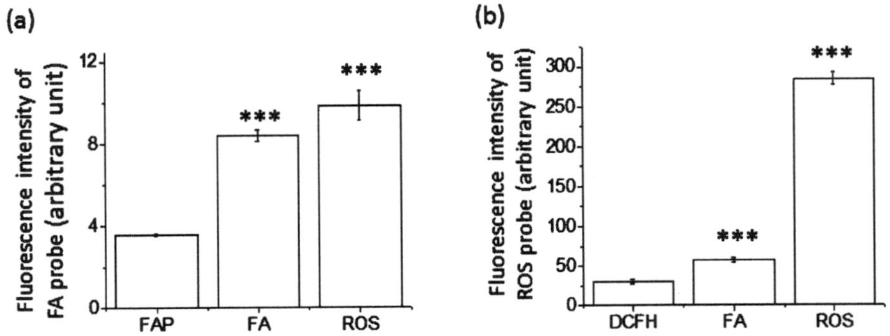

Fig. 13.1 Relation of formaldehyde and ROS in bEnd.3 cells detected by flow cytometry. (**a**) Changes in the fluorescence intensity of formaldehyde probe after addition of ROS to culture medium of bEnd.3 cells; (**b**) Changes in the fluorescence intensity of ROS probe after addition of formaldehyde. The data were from at least three independent experiments. All values are expressed as the mean ± s.d.; *FA*, formaldehyde. ***$p<0.001$

2007). In fact, formaldehyde is able to trigger oxidative stress of cells and produces ROS during the culture. Cells such as bEnd.3 and N2a cells (1×10^6) were plated in the 60 mm tissue culture plates. After 12 h of incubation, the cells were treated with 0.2 mM formaldehyde, followed by staining with DCFDA dye, and ROS content was analyzed with flow cytometry as shown in Fig. 13.1b (Chen et al. 2017).

3.5 *Culture of Cells in the Presence of Formaldehyde at the Pathological Concentration*

In many cases (Petro and Schengrund 2009; Lu et al. 2013), concentrations of formaldehyde used in cell experiments were much higher than those detected in the AD patients and normal people. To evaluate the effect of exposure to a pathological dose of formaldehyde (Li et al. 2016b), N2a cells were maintained in the cell culture medium with the presence of a low concentration of FA (Sigma-Aldrich, St. Louis, MO, USA) by using a serial passage strategy. To simulate the chronic exposure to the pathological dose of formaldehyde, cells were seeded into 100 mm petri dishes at 5×10^5 cells/milliliter (ml) in 8 ml of cell culture medium (4×10^6 cells), which was supplemented with formaldehyde (10 μM as final concentration) in the medium. Inoculated and fresh cell culture media were collected. Cells were passaged every day for the following 6 days or longer. The medium containing formaldehyde was collected before the passaging. After trypsinization, aliquots of cell samples were harvested, and 4×10^6 cells were seeded into a new 100 mm petri dish with 8 ml of medium containing formaldehyde (Fig. 13.2). The control cells were serially passaged in cell culture medium without the addition of FA (Wang et al. 2017a).

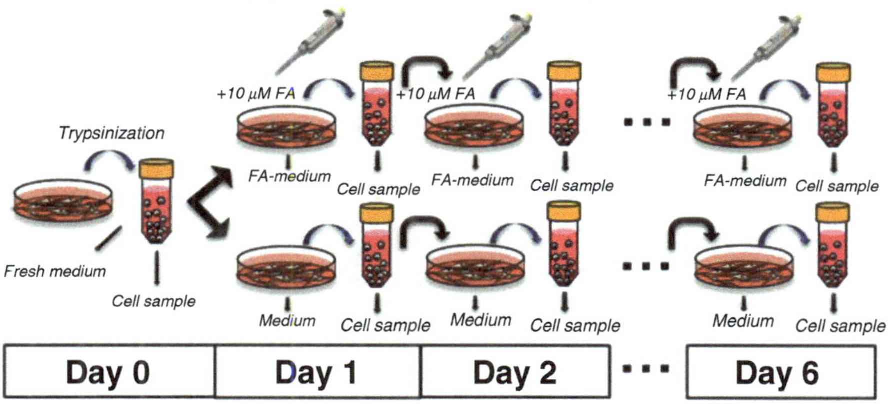

Fig. 13.2 Cell culture in the presence and absence of formaldehyde. Neuro-2a (N2a) cells were inoculated at a seeding density (5 × 10^5 cells/ml) and cultured in the presence of formaldehyde (final concentration, 10 μM) for 24 hours, after which aliquots were taken for another inoculation at the same seeding density and in the presence of the same concentration of formaldehyde. Then, this culture procedure was continued until the end of day 6. Aliquots were taken each day for the analysis as indicated

3.6 *Formaldehyde from Intestinal Microbiota*

Microbe models are used for production of formaldehyde, such as the *Porphyromonas gingivalis vimA* mutant during hydrogen peroxide-induced oxidative stress (McKenzie et al. 2015; Hu et al. 2016) and formaldehyde dismutase in *Pseudomonas putida* J3 (Blaschke et al. 2017). In this chapter, literature for bacteria and fungi producing formaldehyde was not described in details (Liu et al. 2017). The highest yield of formaldehyde was detected in the cecum of APP/PS1 transgenic mice (Chap. 6).

4 Production of Formaldehyde from Animal Tissues

4.1 *Formaldehyde from Frozen Fish*

The storage of fish meat under low temperature is often accompanied with the production of formaldehyde which resulted from trimethylamine oxidation. Formaldehyde was produced during frozen storage of gadoid (Atlantic cod, *Gadus morhua*; pollock, *Pollachius virens*; cusk, *Brosme brosme*; and silver hake, *Merluccius bilinearis*) fillets at −5 °C, bringing about similar reductions in the extractable proteins. This is to say that more attention should be paid to the fact that the storage of fish produces formaldehyde though under low temperatures (Bernheim 1973; Tokunaga 1973).

4.2 *Formaldehyde in Brain Tissue*

Neural cell death or loss of neuron is one of the pathological characteristics for AD. Cytotoxicity resulting from Aβ deposition and Tau hyperphosphorylation leads to neurodegeneration and the death of cells by means of a process called apoptosis (Gandia et al. 2006). The loss of synapses in the affected brain regions correlates best with cognitive impairment in AD patients and has been considered as the early mechanism that precedes neuronal loss (Tonnies and Trushina 2017). The concentrations of formaldehyde in the parietal lobe, frontal lobe, temporal lobe, occipital lobe, hippocampus, cerebellum, and brainstem of porcine brain were measured under room temperature (25 °C). The concentration for formaldehyde was in μmol per kilogram of fresh brain tissues by using 2,4-dinitrophenylhydrazine (DNPH) on a HPLC as described previously (Su et al. 2011).

4.3 *Decomposed Brain Tissues Producing Formaldehyde*

Brain tissues maintained at room temperature produce formaldehyde. Fresh porcine brain from a meat processing plant in Beijing was allowed to stand at room temperature (25 °C). Aliquots of the brain tissue were taken at different time intervals for measurements of the concentration of formaldehyde (Fig. 13.3a) by 2,4-dinitrophenylhydrazine (DNPH) on an HPLC as described previously (Su et al. 2011), compared with the standard curve of formaldehyde concentration (Fig. 13.3b).

5 Animal Models

5.1 *Aged Rodents*

5.1.1 Production of Formaldehyde in Aging Mice

Aging is the greatest risk factor for age-related memory decline. Therefore, an aged C57BL/6 mouse *(wild type)* can be directly used as an elevated formaldehyde model. As described by this laboratory, the level of brain formaldehyde in 10-month mice significantly increased, compared with those of 3-month mice (Li et al. 2016c). C57 BL/6 mice were maintained under pathogen-free conditions (22 ± 2 °C, humidity 50%) and provided a regular diet and sterile water. Their brain formaldehyde was determined at different time intervals (3 months, 6 months, and 10 months) by using HPLC coupled with a chromogenic agent DNPH as described (Su et al. 2011). The average concentration of brain formaldehyde of 10-month-old mice was significantly higher than that of 3-month-old ones.

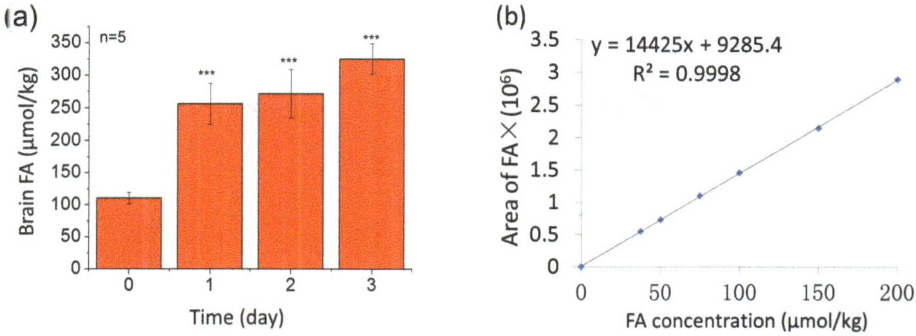

Fig. 13.3 Changes in concentrations of formaldehyde in porcine brain at room temperature. A piece of fresh porcine brain (10 g) from a meat processing plant in Beijing was allowed to stand at room temperature (25 °C), and aliquots (1 g) of the brain tissue were taken at different time intervals for measurements of formaldehyde (panel **a**). Formaldehyde was determined by 2,4-dinitrophenylhydrazine (DNPH) on an HPLC as described (Su et al. 2011), compared to the standard curve of concentration of formaldehyde solution (panel **b**). FA concentration was in μmol per kilogram of fresh brain tissues. Data were from five separate experiments. ***$p < 0.001$

5.1.2 Aged Rats

The concentration of endogenous formaldehyde of Sprague-Dawley (SD) rats was measured at different ages. SD rats which were reared for 1 year had a significantly higher concentration of FA in the hippocampus than those for 1 month (Tong et al. 2013). This is to say, the endogenous level of formaldehyde of rodents increased with aging (after 8-month old). Mice and rats that were reared under normal conditions can be used to study dysmetabolism of endogenous formaldehyde when they are old enough.

5.1.3 Senescence-Accelerated Mice (SAMP8)

The senescence-accelerated mouse (SAM) is a model of accelerated senescence that was established through phenotypic selection from a common genetic pool of AKR/J strain of mice (Takeda et al. 1981; Takeda 1999). The SAM strain includes nine senescence-prone strains (SAMP) and four senescence-resistant strains (SAMR). The SAMP8 strain has been proposed as a model for the study of human AD (Butterfield and Poon 2005), as it shows age-related learning and memory deficiency in behavioral tests (Miyamoto et al. 1986; Carp et al. 2000). Hyperphosphorylation of Tau protein was observed in SAMP8 (Morley et al. 2012). In comparison, the SAMR1 strain, which is commonly used as a control, has a normal life span and displays no distinct cognitive disorders.

As reported previously, SAMP8 displays impaired spatial learning abilities as early as 3 months of age (Chen et al. 2004). Interestingly, Qiang and her coworkers found the imbalance of FA metabolism was related to cognitive impairments in 3-month-old SAMP8 mice (Fig. 13.4) (Qiang et al. 2014). The capacity for spatial

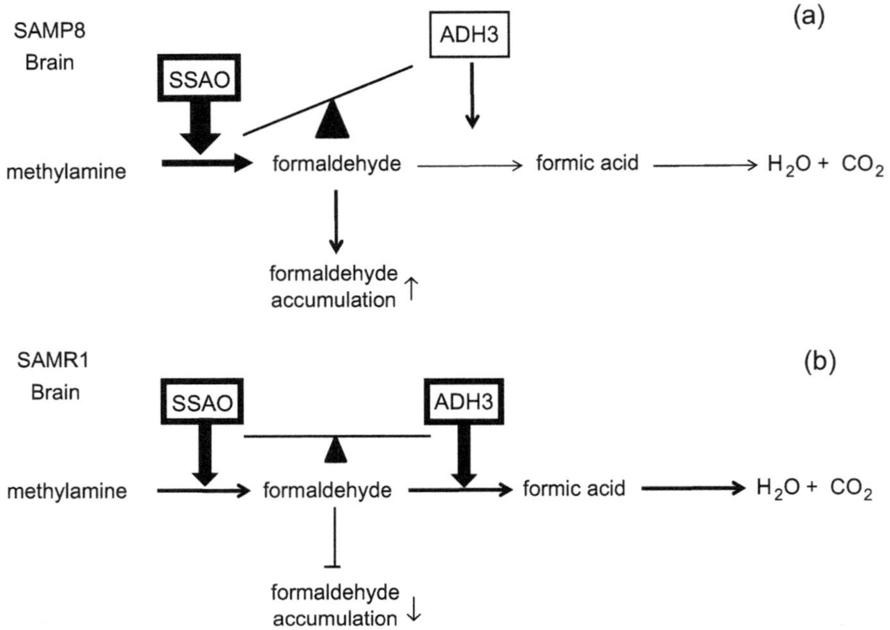

Fig. 13.4 A hypothetical mechanism of formaldehyde accumulation as observed in SAMP8. A hypothetical mechanism for formaldehyde accumulation in 3-month-old SAMP8 (panel **a**) compared to age-matched SAMR1 (panel **b**). Three-month-old SAMP8 showed spatial cognitive decline but 3-month-old SAMR1 did not. Imbalance in the cellular activities/expression levels of the enzymes SSAO and ADH3 results in formaldehyde accumulation in the brain of SAMP8, but the imbalance was not observed in SAMR1 control mice

learning and memory of 3-month-old SAMP8 mice was lower, while brain formaldehyde levels were higher, compared with age-matched SAMR1 mice. To clarify the underlying reasons of formaldehyde elevation in neurodegenerative diseases, Qiang and her colleagues detected the expression level of (anabolic) semicarbazide-sensitive amine oxidase (SSAO) and (catabolic) alcohol dehydrogenase III (ADH3), which were involved in formaldehyde metabolism. As a result, they found that SSAO expression levels were increased, whereas ADH3 exhibited reduced expression levels of mRNA, protein, and enzymatic activity in 3-month-old SAMP8 mice. The imbalance of these metabolic enzymes may represent an explanation for the formaldehyde elevation in the SAMP8 brain. Such an increase may be related to the Tau hyperphosphorylation in the brain, ultimately leading to cognitive impairment. SAMP8, as a classical AD mouse model, is helpful to reveal the underlying mechanisms of age-related accumulation of formaldehyde.

5.2 *Animals Exposed with Formaldehyde*

5.2.1 Formaldehyde Steaming Contacted Rodents

Exposure to exogenous gaseous formaldehyde led to its accumulation, and the number of hippocampal neurons decreased in brain of rats, which further caused memory loss and cognitive decline (Cui 1996). Formaldehyde, as the smallest aldehyde molecule, could pass through the blood-brain barrier after dissolving in blood (Shcherbakova et al. 1986). Results from rat experiments showed that formaldehyde (i.p. for 10 days) induces oxidative frontal cortex and hippocampal tissue damage (Gurel et al. 2005). Different concentrations of gaseous formaldehyde exposure resulted in different effects on cognitive ability, anxiety, and depression-like behaviors which may be associated with alterations in hippocampal glucocorticoid receptors and brain tyrosine hydroxylase levels (Li et al. 2016a).

As described by Li and her colleagues, different concentrations of gaseous formaldehyde were generated by evaporation of formalin solution (37%) in static wood toxication chambers (54 cm × 54 cm × 54 cm). The cage contained five mice during each inhaled FA treatment. The concentration of evaporated formaldehyde was monitored by a formaldehyde digital electrochemical analyzer (DM100-CH2O) to be around 1 ± 0.04 ppm and 2 ± 0.12 ppm during the procedure. The mice were affected as gas dissolved into their body fluids when they were exposed to gaseous formaldehyde similar to the situations for humans. The inhaled groups were exposed to different concentrations of gaseous formaldehyde (2 h once daily for 7 days). The animals were not allowed to drink or eat during exposure. The concentration of gaseous formaldehyde generated from chamber emissions were measured three times daily. The experimental results (0, 1 ± 0.04 and 2 ± 0.12 ppm) were very close to designed concentrations (0, 1 and 2 ppm). Gaseous FA levels from the chamber were steady and reliable (Li et al. 2016a).

5.2.2 Intracerebroventricular Injection to Study Memory Impairment in the Presence of Formaldehyde

Intracerebroventricular injection (i.c.v.) is much more precise and requires much less quantity of formaldehyde to deliver it into the animal brain than intraperitoneal injection. First, anesthesia should be performed. The rats were anesthetized with sodium pentobarbital (45 mg/kg, i.p.) and placed into a stereotaxic apparatus for operation. After that, 2.5 μl PBS or formaldehyde (0.1, 1, or 10 μmol) was injected into the bilateral ventricle at the following coordinates: dorsal/ventral, 3.0 mm; medial/lateral, 1.8 mm; and anterior/posterior, 1.4 mm from bregma, with an injection rate of 0.5 μl/min under the control of micropump. To ensure that the entire injection had been delivered, the injection cannula was allowed to remain in place for an additional minute before being removed.

Hydrogen sulfide (H_2S), the third gasotransmitter, is an endogenous neuromodulator, which facilitates the induction of hippocampal long-term potentiation, involving the functions of learning and memory. Tang and colleagues performed injection (i.c.v.) to observe the effect of formaldehyde on learning and memory and endogenous H_2S generation in the rat hippocampus. Formaldehyde disturbs the generation of endogenous H_2S in the hippocampus, leading to the oxidative stress-mediated neuron damage and impairment of learning and memory (Tang et al. 2013).

5.2.3 Intracerebroventricular Injection to Induce LTP and Spatial Memory Decline

To study the effect of formaldehyde on LTP and LTD, the SD rats were injected (i.c.v.) with different concentrations of formaldehyde (dissolved in saline, 0.3 and 0.5 mM, 5 μl for each injection, i.c.v. in 5 min) before high-frequency electrical stimulation (HFS) for 30 min and were then subjected to electrophysiological experiments or hippocampal formaldehyde detection (Tong et al. 2013). The injection with formaldehyde interferes the formation of LTP and spatial memory process. In the male SD rats which were injected (i.c.v.) with FA (0.3 and 0.5 mM, 5 μl, i.c.v. in 5 min), the hippocampal formaldehyde level rose 30 min after injection, and the fEPSP amplitude was markedly reduced at this concentration of formaldehyde, respectively, demonstrating that excess hippocampal formaldehyde inhibited the LTP formation in SD rats.

5.2.4 Intraperitoneal Injection

5.2.4.1 Formaldehyde Injection for Seven Consecutive Days to See Acute Effect

The male wild-type SD rats (250–300 g) were intraperitoneally (i.p.) administered with formaldehyde once a day at 0.5 mM (60 mg/kg) prior to Morris water maze training and were then subjected to learning and memory behavioral tests, as well as formaldehyde level detection. In the following work, the male rats treated with formaldehyde (0.5 mM, i.p.) for 7 days showed significantly spatial learning function impairments in the Morris water maze test (Tong et al. 2013). At the same time, memory retrieval ability was markedly reduced along with markedly increased hippocampal formaldehyde levels. This result demonstrated that excess hippocampal formaldehyde could inhibit spatial memory formation in SD rats.

5.2.4.2 Effect of Formaldehyde Accumulation by Injection for 30 Days

The male wild-type SD rats were intraperitoneally injected with 0.5 mM formaldehyde (i.p. 60 mg/kg) over a period of 30 consecutive days, because age-related dementia involves chronic gradual memory decline (Gurel et al. 2005; Tong et al. 2011). Hippocampal formaldehyde concentrations were then detected by fluo-HPLC after completion of Morris water maze testing. The excess formaldehyde suppressed hippocampal LTP formation in vivo, and decline in learning and memory recovery ability was observed in the FA-treated group (Tong et al. 2013). The mechanism of suppressing the hippocampal LTP formation might be that the excess FA blocked the NMDA receptor, which is a critical molecule for LTP and memory formation.

5.2.5 Tail Intravenous Injection

C57BL/6 N mice were maintained in animal facilities under pathogen-free conditions. Mice were injected with formaldehyde (3.7 μg per gram of body weight, as described) (Erkrath et al. 1981) via the tail vein and sacrificed at the desired time points. For Western blotting, mouse brain tissues were homogenized in lysis buffer and proteins were extracted. For immunofluorescence detection, mouse brains were fixed with 4% paraformaldehyde and 10-μm-thick sagittal frozen sections were prepared for immunohistochemistry. Using tail intravenous injection, Lu and her colleagues observed Tau hyperphosphorylation in the hippocampus of mice (Lu et al. 2013).

5.2.6 *Drosophila* on Formaldehyde Diet

Different proportions of formaldehyde were added into the *Drosophila melanogaster*'s food (Whittinghill and Lewis 1961), and the effects of formaldehyde on *Drosophila*'s life span and stress resistance were observed. The life span of female flies depended on the formaldehyde concentrations. 0.037% formaldehyde could significantly extend the life span of female flies, whereas the higher concentrations of formaldehyde (≥0.185%) could significantly reduce the life span of male and female flies. These *Drosophila* results provided a new way to study the molecular mechanisms of longevity and stress resistance in *Drosophila* (Li and He 2016).

5.3 *Methanol Feeding*

5.3.1 Mouse Drinking Water with Methanol

Methanol is reliably found in just one of the major food groups, fruits and vegetables, where it is chemically locked safely to pectin, which can pass digestion without absorption by the gut (Dorokhov et al. 2015). As an FA precursor, methanol ingested in large quantities is metabolized first to formaldehyde and then to formic acid or formate salts, which is poisonous to the central nervous system and may cause blindness, coma, and death (https://en.wikipedia.org/wiki/Methanol). Methanol is converted to FA in liver and brain vascular cells, which caused some concerns to the chronic exposure to methanol, in respect to long-term and several age-related pathologies. Although FA was more effective in causing tau protein phosphorylation and changes to microtubule than methanol in in vitro experiments, the research of relationships between FA, methanol, and the progression of AD attracted more and more attentions.

Previous research of Bhatt and coworkers indicated that memory-related behavioral changes in rats were connected to chronic oral and systemic formaldehyde exposure (Bhatt and Panchal 1992). In Yang's work (Yang et al. 2014b), 8-week-old male ICR mice were fed with methanol solution (concentrations of 2% or 3.8%) for more than 6 weeks. The results of the following experiments, such as Y-maze spatial recognition memory test, olfactory memory test, and immunohistochemical analysis, showed that methanol feeding in mice lead to some AD-like symptoms, including cognitive impairments, hippocampal neuron loss, and tau protein hyperphosphorylation. This methanol-fed mice model, as a comparatively simple and easily acquiring animal model, might be helpful to researches related to pathological features of AD as a stand-alone tool or in combination with other animal models.

However, administration of formaldehyde or methanol either by oral intake or injection did not induce the formation of senile plaque in the brain of wild-type mice. It may be because of the different gene sequence of APP between mouse and human. Under recently, formation of senile plaque (amyloid-beta deposits) has not been clearly observed in the brains of wild-type mice except in the transgenic mice which were transferred with human APP genes.

5.3.2 Monkey Drinking Water with Methanol

In order to further study the important pathological features in AD, Yang and her colleagues (Yang et al. 2014a) expanded their research to the nonhuman primate, rhesus monkeys (*Macaca mulatta*), ranging in age from 3 to 4 years which were at the beginning of the study. In this experiment, the young male rhesus monkeys were chronically fed with 3% methanol ad libitum for 1 to 2 years (Yang et al. 2014a). Then, behavioral tests including working memory and biochemical tests were evaluated on their cognitive ability and on levels of age-related impairment markers. In the variable spatial delay response tasks, the monkeys showed memory loss after methanol feeding. Simultaneously, Tau hyperphosphorylation at residues T181 and S396 was detected in cerebrospinal fluid during feeding. Amyloid β deposition (the formation of senile plaque) is distinctly observed in the hippocampus and cerebral cortex, accompanied with significant increase of Tau phosphorylated aggregates in monkey brains (Yang et al. 2014a).

Tau hyperphosphorylation in cerebrospinal fluid is showed in methanol feeding dependent manner; however, Tau phosphorylated aggregates in the brain were persistent 6 months after the feeding ended, indicating that methanol feeding causes a long-time pathological change. The use of methanol feeding in monkeys may provide an animal model to study the pathology and evaluate the efficacy of drug treatment for AD (Su et al. 2016).

5.4 Transgenic Animal Models

5.4.1 Transgenic with APP/PS1

In transgenic AD animal models, increases in brain formaldehyde level and the development of abnormal LTP concurred in APP/PS1 transgenic and APP transgenic mice (Trinchese et al. 2004). APP/PS1 transgenic mice were known as they started to deposit Aβ in the brain at the age of 2 months and show abnormal long-term potentiation (LTP) as early as the age of 3 months. Tong and his colleagues (Tong et al. 2011) found that brain FA level of APP/PS1 transgenic mice was significantly higher at 3 months than that of their respective age-matched controls during the initiation of Aβ deposits. It appeared that the increase of brain FA accompanies the initiation of Aβ deposits and the onset of abnormal memory in these transgenic mice. In APP transgenic mice, Aβ deposits begin at 6–7 months of age (Kawarabayashi et al. 2001). Interestingly, their brain formaldehyde level significantly increased at 6 months and became similar to that of the controls at 12 months. This suggests that the onset of Aβ deposits is accompanied with an increase in brain formaldehyde. Moreover, in both APP/PS1 (6 months old) and APP (12 months old) transgenic mice, brain formaldehyde levels were not significantly different from those of their respective age-matched controls, but more and more typical senile plaques were detected in the brain of these two transgenic types of AD mice (Trinchese et al. 2004; Karuppagounder et al. 2008). Therefore, whether excessive endogenous formaldehyde plays a role in the initiation of Aβ deposition in these transgenic mice needs to be further investigated.

5.4.2 ADH3 Knockout in *Drosophila*

Nitric oxide (NO) plays a very important role in learning and memory, and alcohol dehydrogenase III (ADH III) which is also named S-nitrosoglutathione reductase (GSNOR) is a key protein in the regulation of NO metabolism (Liu et al. 2001). To study the relationship between NO metabolism and learning and memory, the expression of gene (formaldehyde dehydrogenase) *fdh*, a homolog to mammalian GSNOR, was modulated in *Drosophila*. Hou and coworkers observed that *fdh* overexpression resulted in denitrozation of multiple proteins functionally enriched in vesicle-mediated transport, which was important for learning and memory (Hou et al. 2011). They also found that ADH3-KO *Drosophila*, with its considerably decreased FA, exhibited a decline in the ability to learn and memory, suggesting that overdose of endogenous FA could affect animal cognition (Su et al. 2016).

5.4.3 ADH3 Knockin Mouse

The transgenic C57/BL6 mouse with ADH3 gene becomes a slow learner in Morris water maze, compared with the wild-type mouse. We will describe this transgenic mouse in detail after Chen and her colleagues' data are published.

5.4.4 Aldehyde Dehydrogenase and Related Enzyme Knockout Mice

Aldehyde dehydrogenase (ALDH) detoxifies acetaldehyde into acetate, which has been deeply studied because alcohol use is widespread and related to numerous diseases, including oral, colorectal, liver, pancreatic, aerodigestive, breast, and colon cancers (Matsumoto et al. 1990). ALDH also converts formaldehyde to formic acid (Hedberg et al. 2001). Therefore, loss-of-function mutations in ALDH2 also lead to formaldehyde toxicosis. Chaudhry and colleagues have evaluated the effect of ALDH2 deficiency with transgenic mouse on EtOH-induced disruption of intestinal epithelial tight junctions and adherens junctions, gut barrier dysfunction, and liver injury. Their data demonstrated that ALDH2 deficiency enhances EtOH-induced disruption of intestinal epithelial tight junctions, barrier dysfunction, and liver damage (Chaudhry et al. 2015).

Heit and coworkers have briefly described the transgenic mice with different types of ALDH (Heit et al. 2015). The knockout mouse models for enzymes metabolizing ethanol (ADH1, CAT, and CYP2E1), acetaldehyde (ALDH2, ALDH1A1, and ALDH1B1), and enzymes involved in GSH synthesis (GCLC and GCLM) were also possibly related to the metabolism of formaldehyde, such as the *Adh1*$^{-/-}$ mouse (Molotkov et al. 2002), the *Aldh2*$^{-/-}$ strain (Isse et al. 2002; Yu et al. 2009) and the *Aldh1a1*$^{-/-}$ mouse (Fan et al. 2003), the *Aldh2* conditional knockout (Skarnes et al. 2011), and the *Cyp2e1*$^{-/-}$ mouse (Lee et al. 1996). These models represent highly relevant animal models for alcohol as well as formaldehyde research. These mice models are beneficial to understand methanol and formaldehyde metabolism and resultant oxidative stress (Evans et al. 2016), which are primary pathogenic events mediating alcohol- or formaldehyde-induced organ damage and neurobehavioral changes. Utilization of these models will deliver valuable information about the fundamental mechanisms underlying not only ethanol but also formaldehyde toxicity (Heit et al. 2015). Such knowledge should accelerate the development of more effective, targeted therapies to both prevent and treat health issues associated with excessive alcohol consumption and dysmetabolism of endogenous formaldehyde.

5.4.5 Semicarbazide-Sensitive Amine Oxidase Knockout

Semicarbazide-sensitive amine oxidase (SSAO) (E.C.1.4.3.6.) is a family of enzymes that is expressed in prokaryotes as well as in eukaryotes. In mammals, large amounts of SSAO are located in smooth muscle cells of blood vessels (Lewinsohn 1984). SSAO catalyzes the reaction of methylamine and polyamine to generate formaldehyde. Yu and his coworkers (Yu et al. 2004) proposed that SSAO-mediated deamination might contribute to protein deposition, formation of senile plaque, and inflammation, and thus might be involved in the pathophysiology of chronic vascular and neurological disorders, such as diabetic complications, atherosclerosis, and Alzheimer's disease. As reported earlier by Hernandez and colleagues (Hernandez et al. 2006), an increase of SSAO was observed in moderate-severe and severe AD patients, which contributes to the vascular damage and age-related dementia. Su and his colleagues found that SSAO-KO rats showed an enhancement in the ability to learn and in memory capacity with decreased levels of FA in both the brain and urine (unpublished data from Su et al.). As described by Gokturk and colleagues, the transgenic mice have their SSAO activity and mRNA expression increased. Histological studies revealed a specific aorta phenotype with a condensed and rigid vessel wall in some of the transgenic mice, but wild-type animals did not display this phenotype (Gokturk et al. 2003). This transgenic mouse model may be of great value for increasing the knowledge about to what extent human SSAO contributes to vascular complications in diabetes and also to which extent inhibition of SSAO can prevent the development of such complications. Those changes suggest a high degree of rigidity of the vessel wall (Gokturk et al. 2003). In the lab, SSAO-KO rats and the transgenic rats showed a slower learning rate than the wild-type rats.

5.4.6 Methylase Knockout and Knockin

Some studies have revealed that experience-mediated DNA methylation, which is regulated by DNA methyltransferase (DNMT) activity, is required for the formation of recent memory as well as the maintenance of remote memory. Studies showed that simultaneous knockout of DNMT1 and DNMT3a results in a reduction in global DNA methylation, as well as deficits in both memory acquisition and retrieval in mice (Miller et al. 2008; Feng et al. 2010). Furthermore, overexpression of

DNMT3a in the hippocampus can reverse spatial memory deficits in aged mice (Oliveira et al. 2012; Su and Tsai 2012). As hypothesized recently, endogenous FA is involved in learning and memory process, in particular learning ability, because FA acts as a donor for the methylation of DNA, RNA, and histones. Homeostasis of endogenous FA metabolism may benefit from these methylations, which are involved in cognitive behaviors in learning and memory (Tong et al. 2013).

5.5 Increase of Formaldehyde in Animal Administrated with Small Compounds

5.5.1 Intraperitoneal Injection of D-Ribose

C57BL/6 J mice were intraperitoneally injected (i.p.) with D-ribose (rib, 3.2 g/kg/d), D-glucose (glc, 3.82 g/kg/d), or saline (0.9% NaCl in double-distilled water) as the control for 10 days. Blood formaldehyde levels were measured with 1.0 g/l 2,4-dinitrophenylhydrazine by HPLC. The results showed that D-ribose-treated mice had a significantly ($p<0.05$) higher level of blood formaldehyde than the control group. However, the increase of serum formaldehyde in D-glucose-treated mice was not significant ($p>0.05$). The typical pathological characteristics of AD could be observed such as cognitive impairment, Aβ deposition, Tau hyperphosphorylation, and glia activation after mice had gavage of D-ribose for 6 months (Han et al. 2014; Wu et al. 2015).

Male C57BL/6 J mice (8 weeks) were obtained from Vital River Laboratory Animal Technology Co. Ltd. (Beijing, China). After 1 week of acclimatization to the cages, mice were randomly divided into three groups and received intraperitoneal injections each day for 10 days with D-ribose at doses of 3.2 g/kg, D-glucose at doses of 3.82 g/kg, or 0.9% saline as control. All mice were maintained in animal facilities under pathogen-free conditions. The mice showed cognitive impairment accompanied with deposition of advanced glycation end products (AGEs) in their brains (Han et al. 2011; Han et al. 2014).

5.5.2 Administration with Folic Acid

Both formaldehyde and folic acid are important components of one-carbon cycle, while folic acid is also one of the vital nutrients in neural system development. To find out whether folic acid dysregulation is related to formaldehyde metabolism, we studied the effect of folic acid feeding on formaldehyde accumulation. According to the Picolab Mouse Diet 20 composition, normal mouse chow contains 2.9 mg/kg of folic acid (http://www.labdiet.com). C57 BL/6 mice were randomly divided to three groups: those fed with regular food supplemented with 0 mg/kg (blank control group), 2 mg/kg (normal dose group), and 10 mg/kg folic acid (excess dose group) for consecutive 30 days, respectively. After that, mice were subjected to Morris

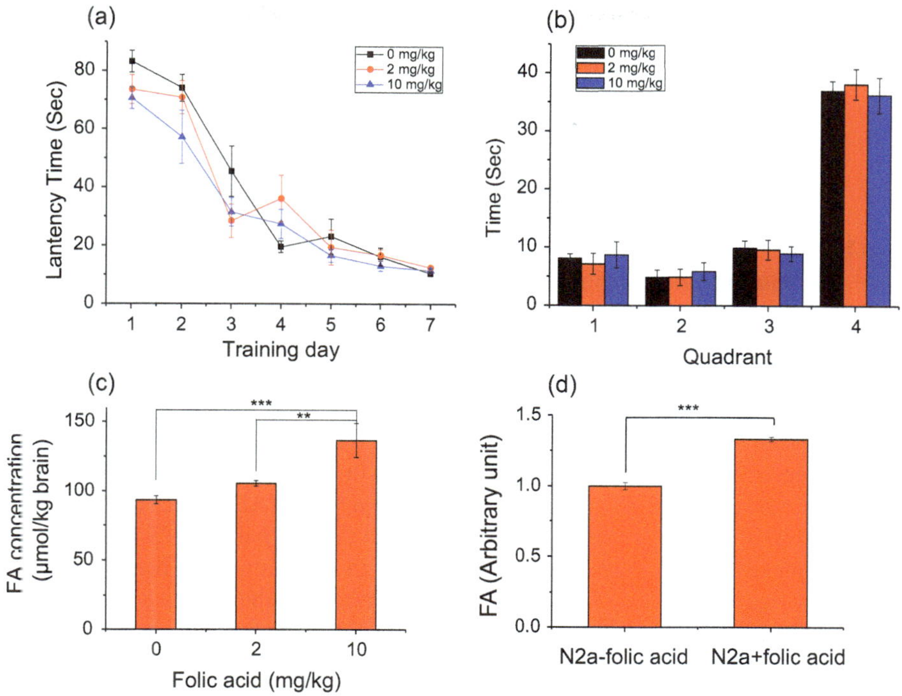

Fig. 13.5 Folic acid contained chow feeding induces brain formaldehyde elevation. C57 BL/6 J mice ($n = 9$) were divided randomly into three groups and were fed with chows (AIN93G) containing different concentration of folic acids (0, 2, 10 mg/kg) for 30 consecutive days. Morris water maze was used to detect mice learning ability (panel **a**) and memory capacity (panel **b**) after 1 month of feeding. Formaldehyde concentrations of mice brain were measured after the behavioral test (panel **c**). N2a cells were incubated in a medium with or without folic acid for 24 hours (panel **d**); then media were collected. Formaldehyde concentrations were measured as soon as possible. Data were shown as mean ± S.E.M., ** indicates $p < 0.01$, ***indicates $p < 0.001$

water maze test before formaldehyde concentration test. Mice fed with different concentration of folic acid showed no difference in spatial learning and memory abilities in the behavioral test. In other words, supplementation of folic acid could not improve learning and memory for the wild-type mouse (Fig. 13.5a, b). On the contrary, the brain formaldehyde concentration significantly ($p<0.01$) increased in mice fed with 10 mg/kg folic acid than that of control mice. However, 2 mg/kg folic feeding did not show marked ($p>0.05$) increase of formaldehyde although the increased formaldehyde level depends on the feeding amount of folic acid (Fig. 13.5c). To demonstrate the contribution of folic acid to formaldehyde production, mouse neuro-2a cells were cultured in a medium with or without folic acid for 24 hours; the formaldehyde released into the media were measured. As shown in Fig. 13.5d, formaldehyde level is markedly lower in the media without folic acid than that supplemented with folic acid. Folic metabolism is involved in formaldehyde formation. Excess folic acid intake resulted in higher concentration of formaldehyde formation in mice brain and neuronal cells. These results suggest that a long-term observation (longer than 6 months) of animal behaviors by feeding excess folic acid should be performed in the future.

5.6 *Dehydration Producing Formaldehyde*

5.6.1 Rodents Exposed of Less Water Intake Suffer from High Level of Endogenous Formaldehyde

Chronic dehydration is regarded as a common symptom of patients with age-related cognitive impairment, particularly those with AD. Li and her colleagues established an animal model for a chronic dehydration model with C57 BL/6 mouse that was administrated with water containing 4% NaCl for 3 months. Brain formaldehyde was increased with an imbalance of the activities between semicarbazide-sensitive amine oxidase (SSAO) and formaldehyde dehydrogenase 3 (ADH3) (Li et al. 2016b).

The level of brain 5-hydroxytryptamine (5-HT) of mice fed with 4% NaCl markedly decreased and showed slower learning rate in the shuttle box behavioral test than those fed with water as control. To demonstrate the relationship between dysmetabolism of formaldehyde and the level of 5-HT leading to the slow learning, Li and her coworkers intraperitoneally injected the mice with formaldehyde (once daily) for 7 days. Formaldehyde injection resulted in a significant decrease of 5-HT level in the mice brain and slow learning rate in shuttle box, compared with those injected with 4% NaCl, which showed neither significant differences in 5-HT levels nor impaired cognitive behavior (Li et al. 2016b).

Note that water deprivation of mice (wild-type C57BL/6) for 3 days leads to a significant increase of the brain formaldehyde concentration (Li et al. 2016b).This is to say that the acute dehydration also induces high levels of endogenous formaldehyde, especially the brain formaldehyde.

5.6.2 Determination of Endogenous Formaldehyde in Humans Subjected to Water Deprivation

As described by Li and her colleagues, the volunteer participants' ($n = 20$) activities were prescribed as indicated in the recruiting requirements (Li et al. 2016c). Urine samples (middle stream) were collected in the morning ahead of breakfast (8:00 a.m.), prior to lunch (12:30), and before going to sleep (11:00 p.m.) to measure the uric formaldehyde (FA/Urc ratios). The participants were divided into three groups: those who were forbidden from drinking water, those who drank water according to the prescribed quantities, and those who drank water based on their own daily habits as control. Measurements were carried out to compare the formaldehyde concentrations between the water-deprived group and the water-prescribed group. The results showed that lack of water intake could lead to formaldehyde accumulation in vivo. Thus, we must pay attention to water intake in daily life.

6 Estimating Neuropsychiatric Disorders Related to Endogenous Formaldehyde

In clinical study, we measured endogenous formaldehyde and cognitive abilities of the AD patients compared with those of age-matched participants (Tong et al. 2011; Li et al. 2016c). For estimation of their cognitive abilities, the two most frequently used screening tests are the mini-mental status examination (MMSE) and the Montreal cognitive assessment (MoCA). The second method to assess a patient's mental status is by using specific neuropsychological tests that focus on specific domains of cognition, such as frontal executive functions, attention, episodic verbal and visuospatial memory, declarative knowledge such as language (speech, reading, and writing) and arithmetical, as well as visuospatial and perceptual abilities (Finney et al. 2016). Data analysis uses stepwise linear regression (Kilburn et al. 1998), and nerve behavior tests included balance, reaction time, strength, hearing, visual performance and cognitive recall, and perceptual motor and memory functions. Regression equations modeled the performance of each test and the influences of demographic factors (Kilburn et al. 1998). Symptoms include excessive fatigue, somnolence, headache, difficulty remembering, irritability, and instability of mood; choice reaction time was prolonged; blink latency was delayed; balance was abnormal; visual fields were constricted; delayed verbal recall and visual reproduction were impaired; errors on fingertip number; writing was abnormal; choice reaction time was prolonged; long-term memory was decreased; and so on (Kilburn 1994). FA-exposed animals showed cognitive and memory impairment by testing open field, water mirror maze (Pitten et al. 2000; Malek et al. 2003), new object recognition, etc. (Lu et al. 2008; Li et al. 2016a).

7 Conclusion

Models for the effect of high or low concentration of formaldehyde on organisms should be diversely and variously established because anabolism and metabolism of formaldehyde are related with multiple pathways (Chap. 2). Along with the study of the relationship between formaldehyde and neurodegeneration, more and more molecular, cellular, and animal models will be established in different patterns and methods. For the sake of homeostasis of metabolism of endogenous formaldehyde, excess or lack of formaldehyde could lead to disorders of related gene regulation, protein expression, and neural functions. Since rodents have much stronger metabolic abilities of formaldehyde than humans, the data for cognitive abilities resulted from mouse and rat should not be directly referred to humans. The mouse or rat did not show Aβ deposition in the brain by administration of formaldehyde. However, the situation is different in monkey trials. Administration of low concentration (3%) of methanol for at least 6 months or longer in young monkey induces decline of working memory ability and Tau hyperphosphorylation at residues T181 and S396 in cerebrospinal fluid and phosphorylated Tau aggregation in the hippocampus and cerebral cortex, and β-amyloid deposition were observed. These models including molecular, cellular, and animal cases could support the investigation of the pathomechanism and evaluate the efficacy of drug and physical treatment for AD (Su et al. 2016).

Acknowledgment We would like to thank Jian Yang who had attended the analysis of brain formaldehyde using HPLC in this laboratory. This project was supported by grants from the National Key Research and Development Program of China (2016YFC1306300), the Beijing Municipal Science and Technology Project (Z161100000217141; Z161100000216137), the National Basic Research Program of China (973 Program) (2012CB911004), the Natural Scientific Foundation of China (NSFC 31270868), the Foundation of Chinese Academy of Sciences CAS-20140909, and the Queensland-Chinese Academy of Sciences Biotechnology Fund (GJHZ201302).

Competing Financial Interests The authors declare no competing financial interests.

References

Allen DW, Burgoyne CF, Groat JD, Smith CM 2nd, White JG (1984) Comparison of hemoglobin Koln erythrocyte membranes with malondialdehyde-reacted normal erythrocyte membranes. Blood 64(6):1263–1269

Bernheim F (1973) The effect of formaldehyde and acetaldehyde on the swelling of cells of Pseudomonas Aeruginosa in salt solutions and their interactions with amines and carboxylic acids. Proc Soc Exp Biol Med 142(2):675–679

Bhatt HV, Panchal GM (1992) Behavioural change in rats due to chronic oral and systemic formaldehyde. Indian J Physiol Pharmacol 36(4):270–272

Blaschke L, Wagner W, Werkmeister C, Wild M, Gihring A, Rupp S, Zibek S (2017) Development of a simplified purification method for a novel formaldehyde dismutase variant from Pseudomonas putida J3. J Biotechnol 241:69-75

Butterfield DA, Poon HF (2005) The senescence-accelerated prone mouse (SAMP8): a model of age-related cognitive decline with relevance to alterations of the gene expression and protein abnormalities in Alzheimer's disease. Exp Gerontol 40(10):774–783

Carp RI, Meeker HC, Kozlowski I, Sersen EA (2000) An endogenous retrovirus and exogenous scrapie in a mouse model of aging. Trends Microbiol 8(1):39–42

Chaudhry KK, Samak G, Shukla PK, Mir H, Gangwar R, Manda B, Isse T, Kawamoto T, Salaspuro M, Kaihovaara P, Dietrich P, Dragatsis I, Nagy LE, Rao RK (2015) ALDH2 deficiency promotes ethanol-induced gut barrier dysfunction and fatty liver in mice. Alcohol Clin Exp Res 39(8):1465–1475

Chen GH, Wang YJ, Wang XM, Zhou JN (2004) Accelerated senescence prone mouse-8 shows early onset of deficits in spatial learning and memory in the radial six-arm water maze. Physiol Behav 82(5):883–890

Chen L, Cheng CY, Choi H, Ikram MK, Sabanayagam C, Tan GS, Tian D, Zhang L, Venkatesan G, Tai ES, Wang JJ, Mitchell P, Cheung CM, Beuerman RW, Zhou L, Chan EC, Wong TY (2016) Plasma Metabonomic profiling of diabetic retinopathy. Diabetes 65(4):1099–1108

Chen XX, Su T, He YG, He, RQ (2017) Spatial cognition decline caused by excess formaldehyde in the lysosome. Prog Biochem Biophys44(6):486–494

Cui X (1996) Inhaled formaldehyde on the effects of GSH level and distribution of formaldehyde. China J Prevent Med 30(186)

Dorokhov YL, Shindyapina AV, Sheshukova EV, Komarova TV (2015) Metabolic methanol: molecular pathways and physiological roles. Physiol Rev 95(2):603–644

Erkrath KD, Adebahr G, Kloppel A (1981) Lethal intoxication by formalin during dialysis (author's transl). Z Rechtsmed 87(3):233–236
Evans AM, Fameli N, Ogunbayo OA, Duan J, Navarro-Dorado J (2016) From contraction to gene expression: nanojunctions of the sarco/endoplasmic reticulum deliver site- and function-specific calcium signals. Sci China Life Sci 59(8):749–763
Fan X, Molotkov A, Manabe S, Donmoyer CM, Deltour L, Foglio MH, Cuenca AE, Blaner WS, Lipton SA, Duester G (2003) Targeted disruption of Aldh1a1 (Raldh1) provides evidence for a complex mechanism of retinoic acid synthesis in the developing retina. Mol Cell Biol 23(13):4637–4648
Farrow SC, Facchini PJ (2013) Dioxygenases catalyze O-demethylation and O,O-demethylenation with widespread roles in benzylisoquinoline alkaloid metabolism in opium poppy. J Biol Chem 288(40):28997–29012
Feng J, Zhou Y, Campbell SL, Le T, Li E, Sweatt JD, Silva AJ, Fan G (2010) Dnmt1 and Dnmt3a maintain DNA methylation and regulate synaptic function in adult forebrain neurons. Nat Neurosci 13(4):423–430
Finney GR, Minagar A, Heilman KM (2016) Assessment of mental status. Neurol Clin 34(1):1–16
Foettinger A, Leitner A, Lindner W (2006) Derivatisation of arginine residues with malondialdehyde for the analysis of peptides and protein digests by LC-ESI-MS/MS. J Mass Spectrom 41(5):623–632
Gandia L, Alvarez RM, Hernandez-Guijo JM, Gonzalez-Rubio JM, de Pascual R, Rojo J, Tapia L (2006) Anticholinesterases in the treatment of Alzheimer's disease. Rev Neurol 42(8):471–477
Giudetti AM, Romano A, Lavecchia AM, Gaetani S (2016) The role of brain cholesterol and its oxidized products in Alzheimer's disease. Curr Alzheimer Res 13(2):198–205
Gokturk C, Garpenstrand H, Nilsson J, Nordquist J, Oreland L, Forsberg-Nilsson K (2003) Studies on semicarbazide-sensitive amine oxidase in patients with diabetes mellitus and in transgenic mice. Biochim Biophys Acta 1647(1-2):88–91
Gurel A, Coskun O, Armutcu F, Kanter M, Ozen OA (2005) Vitamin E against oxidative damage caused by formaldehyde in frontal cortex and hippocampus: biochemical and histological studies. J Chem Neuroanat 29(3):173–178
Han C, Lu Y, Wei Y, Liu Y, He R (2011) D-ribose induces cellular protein glycation and impairs mouse spatial cognition. PLoS One 6(9):e24623
Han C, Lu Y, Wei Y, Wu B, Liu Y, He R (2014) D-ribosylation induces cognitive impairment through RAGE-dependent astrocytic inflammation. Cell Death Dis 5:e1117
Hedberg JJ, Grafstrom RC, Vondracek M, Sarang Z, Warngard L, Hoog JO (2001) Micro-array chip analysis of carbonyl-metabolising enzymes in normal, immortalised and malignant human oral keratinocytes. Cell Mol Life Sci 58(11):1719–1726
Heit C, Dong H, Chen Y, Shah YM, Thompson DC, Vasiliou V (2015) Transgenic mouse models for alcohol metabolism, toxicity, and cancer. Adv Exp Med Biol 815:375–387
Hernandez M, Sole M, Boada M, Unzeta M (2006) Soluble semicarbazide sensitive amine oxidase (SSAO) catalysis induces apoptosis in vascular smooth muscle cells. Biochim Biophys Acta 1763(2):164–173
Hou QL, Jiang HQ, Zhang X, Guo C, Huang B, Wang P, Wang TP, Wu KY, Li JA, Gong ZF, Du LB, Liu Y, Liu L, Chen C (2011) Nitric oxide metabolism controlled by formaldehyde dehydrogenase (fdh, homolog of mammalian GSNOR) plays a crucial role in visual pattern memory in Drosophila. Nitric Oxide-Biol Chem 24(1):17–24
Hu X, Wang T, Jin F (2016) Alzheimer's disease and gut microbiota. Sci China Life Sci 59(10):1006–1023
Isse T, Oyama T, Kitagawa K, Matsuno K, Matsumoto A, Yoshida A, Nakayama K, Nakayama K, Kawamoto T (2002) Diminished alcohol preference in transgenic mice lacking aldehyde dehydrogenase activity. Pharmacogenetics 12(8):621–626
Jian X, Zhu MX (2016) Regulation of lysosomal ion homeostasis by channels and transporters. Sci China Life Sci 59(8):777–791

Jung WW, Kim EM, Lee EH, Yun HJ, Ju HR, Jeong MJ, Hwang KW, Sul D, Kang HS (2007) Formaldehyde exposure induces airway inflammation by increasing eosinophil infiltrations through the regulation of reactive oxygen species production. Environ Toxicol Pharmacol 24(2):174–182

Karuppagounder SS, Xu H, Pechman D, Chen LH, DeGiorgio LA, Gibson GE (2008) Translocation of amyloid precursor protein C-terminal fragment(s) to the nucleus precedes neuronal death due to thiamine deficiency-induced mild impairment of oxidative metabolism. Neurochem Res 33(7):1365–1372

Kawarabayashi T, Younkin LH, Saido TC, Shoji M, Ashe KH, Younkin SG (2001) Age-dependent changes in brain, CSF, and plasma amyloid beta protein in the Tg2576 transgenic mouse model of Alzheimer's disease. J Neurosci 21(2):372–381

Kilburn KH (1994) Neurobehavioral impairment and seizures from formaldehyde. Arch Environ Health 49(1):37–44

Kilburn KH, Thornton JC, Hanscom B (1998) Population-based prediction equations for neurobehavioral tests. Arch Environ Health 53(4):257–263

Lee SS, Buters JT, Pineau T, Fernandez-Salguero P, Gonzalez FJ (1996) Role of CYP2E1 in the hepatotoxicity of acetaminophen. J Biol Chem 271(20):12063–12067

Lewinsohn R (1984) Mammalian monoamine-oxidizing enzymes, with special reference to benzylamine oxidase in human tissues. Braz J Med Biol Res 17(3-4):223–256

Li YN, He RQ (2016) The effects of formaldehyde on life span and stress resistance in Drosophila melanogaster. Prog Biochem Biophys 43(4):420–428

Li FX, Lu J, Xu YJ, Tong ZQ, Nie CL, He RQ (2008) Formaldehyde-mediated chronic damage may be related to sporadic neurodegeneration. Prog Biochem Biophys 35(4):393–400

Li Y, Song Z, Ding Y, Xin Y, Wu T, Su T, He R, Tai F, Lian Z (2016a) Effects of formaldehyde exposure on anxiety-like and depression-like behavior, cognition, central levels of glucocorticoid receptor and tyrosine hydroxylase in mice. Chemosphere 144:2004–2012

Li T, Su T, He YG, He RQ (2016b) Chronic-dehydrated Dysmetabolism of formaldehyde in mouse brain and decline of learning in the shuttle box. Prog Biochem Biophys 43(4):429–438

Li T, Su T, He Y, Lu J, Mo W, Wei Y, He R (2016c) Brain formaldehyde is related to water intake behavior. Aging Dis 7(5):561–584

Liu LM, Hausladen A, Zeng M, Que L, Heitman J, Stamler JS (2001) A metabolic enzyme for S-nitrosothiol conserved from bacteria to humans. Nature 410(6827):490–494

Liu J, Liu FY, Tong ZQ, Li ZH, Chen W, Luo WH, Li H, Luo HJ, Tang Y, Tang JM, Cai J, Liao FF, Wan Y (2013) Lysine-specific demethylase 1 in breast cancer cells contributes to the production of endogenous formaldehyde in the metastatic bone cancer pain model of rats. PLoS One 8(3):e58957

Liu KL, He YG, Yu LX, He RQ (2017) Elevated formaldehyde in the cecum of APP/PS1 mouse. Microbiol China 44(8):1761–1766

Lu Z, Li CM, Qiao Y, Yan Y, Yang X (2008) Effect of inhaled formaldehyde on learning and memory of mice. Indoor Air 18(2):77–83

Lu J, Miao J, Su T, Liu Y, He R (2013) Formaldehyde induces hyperphosphorylation and polymerization of tau protein both in vitro and in vivo. Biochim Biophys Acta 1830(8):4102–4116

Luo W, Li H, Zhang Y, Ang CY (2001) Determination of formaldehyde in blood plasma by high-performance liquid chromatography with fluorescence detection. J Chromatogr B Biomed Sci Appl 753(2):253–257

Malek FA, Moritz KU, Fanghanel J (2003) A study on the effect of inhalative formaldehyde exposure on water labyrinth test performance in rats. Ann Anat 185(3):277–285

Matsumoto K, Moriya F, Nanikawa R (1990) The movement of blood formaldehyde in methanol intoxication. II. The movement of blood formaldehyde and its metabolism in the rabbit. Nihon Hoigaku Zasshi 44(3):205–211

McKenzie RM, Aruni W, Johnson NA, Robles A, Dou Y, Henry L, Boskovic DS, Fletcher HM (2015) Metabolome variations in the Porphyromonas gingivalis vimA mutant during hydrogen peroxide-induced oxidative stress. Mol Oral Microbiol 30(2):111–127

Miller CA, Campbell SL, Sweatt JD (2008) DNA methylation and histone acetylation work in concert to regulate memory formation and synaptic plasticity. Neurobiol Learn Mem 89(4):599–603

Miyamoto M, Kiyota Y, Yamazaki N, Nagaoka A, Matsuo T, Nagawa Y, Takeda T (1986) Age-related changes in learning and memory in the senescence-accelerated mouse (SAM). Physiol Behav 38(3):399–406

Molotkov A, Deltour L, Foglio MH, Cuenca AE, Duester G (2002) Distinct retinoid metabolic functions for alcohol dehydrogenase genes Adh1 and Adh4 in protection against vitamin a toxicity or deficiency revealed in double null mutant mice. J Biol Chem 277(16):13804–13811

Morley JE, Farr SA, Kumar VB, Armbrecht HJ (2012) The SAMP8 mouse: a model to develop therapeutic interventions for Alzheimer's disease. Curr Pharm Des 18(8):1123–1130

Oliveira AM, Hemstedt TJ, Bading H (2012) Rescue of aging-associated decline in Dnmt3a2 expression restores cognitive abilities. Nat Neurosci 15(8):1111–1113

Petro KA, Schengrund CL (2009) Membrane raft disruption promotes axonogenesis in n2a neuroblastoma cells. Neurochem Res 34(1):29–37

Pitten FA, Kramer A, Herrmann K, Bremer J, Koch S (2000) Formaldehyde neurotoxicity in animal experiments. Pathol Res Pract 196(3):193–198

Qiang M, Xiao R, Su T, Wu BB, Tong ZQ, Liu Y, He RQ (2014) A novel mechanism for endogenous formaldehyde elevation in SAMP8 mouse. J Alzheimers Dis 40(4):1039–1053

Shcherbakova LN, Tel'pukhov VI, Trenin SO, Bashilov IA, Lapkina TI (1986) Permeability of the blood-brain barrier to intra-arterial formaldehyde. Biull Eksp Biol Med 102(11):573–575

Skarnes WC, Rosen B, West AP, Koutsourakis M, Bushell W, Iyer V, Mujica AO, Thomas M, Harrow J, Cox T, Jackson D, Severin J, Biggs P, Fu J, Nefedov M, de Jong PJ, Stewart AF, Bradley A (2011) A conditional knockout resource for the genome-wide study of mouse gene function. Nature 474(7351):337–342

Su SC, Tsai LH (2012) DNA methylation in cognition comes of age. Nat Neurosci 15(8):1061–1062

Su T, Wei Y, He RQ (2011) Assay of brain endogenous formaldehyde with 2,4-Dinitrophenylhydrazine through UV-HPLC. Prog Biochem Biophys 38(12):1171–1177

Su T, Xin L, He YG, Wei Y, Song YX, Li WW, Wang XM, He RQ (2013) The abnormally high level of uric D-ribose for type-2 diabetics. Prog Biochem Biophys 40(9):816–825

Su T, Monte WC, Hu XT, He YG, He RQ (2016) Formaldehyde as a trigger for protein aggregation and potential target for mitigation of age-related, progressive cognitive impairment. Curr Top Med Chem 15(6):472–484

Takeda T (1999) Senescence-accelerated mouse (SAM): a biogerontological resource in aging research. Neurobiol Aging 20(2):105–110

Takeda T, Hosokawa M, Takeshita S, Irino M, Higuchi K, Matsushita T, Tomita Y, Yasuhira K, Hamamoto H, Shimizu K, Ishii M, Yamamuro T (1981) A new murine model of accelerated senescence. Mech Ageing Dev 17(2):183–194

Tang XQ, Zhuang YY, Zhang P, Fang HR, Zhou CF, Gu HF, Zhang H, Wang CY (2013) Formaldehyde impairs learning and memory involving the disturbance of hydrogen sulfide generation in the hippocampus of rats. J Mol Neurosci 49(1):140–149

Tokunaga T (1973) The effect of decomposed products of trimethylamine oxide on quality of frozen Alaska pollack fillet. Nippon Suisan Gakkaishi 40:167–174

Tong Z, Zhang J, Luo W, Wang W, Li F, Li H, Luo H, Lu J, Zhou J, Wan Y, He R (2011) Urine formaldehyde level is inversely correlated to mini mental state examination scores in senile dementia. Neurobiol Aging 32(1):31–41

Tong ZQ, Han CS, Luo WH, Wang XH, Li H, Luo HJ, Zhou JN, Qi JS, He RQ (2013) Accumulated hippocampal formaldehyde induces age-dependent memory decline. Age 35(3):583–596

Tonnies, E., and Trushina, E. (2017) Oxidative stress, synaptic dysfunction, and Alzheimer's disease. J Alzheimers Dis,

Trinchese F, Liu S, Battaglia F, Walter S, Mathews PM, Arancio O (2004) Progressive age-related development of Alzheimer-like pathology in APP/PS1 mice. Ann Neurol 55(6):801–814

Wang J, Zhou J, Mo WC, He YG, Wei Y, He RQ, Yi FP (2017) Accumulation of simulated pathological level of formaldehyde decreases cell viability and adhesive morphology in neuronal cells. Prog Biochem Biophys 44(7):601–614

Wang YJ, Su T, Yang J, Wei Y, He RQ (2017b) Decomposition of D-ribose producing reactive carbonyl species including formaldehyde. Prog Biochem Biophys (Submited)

Ward RJ, Zucca FA, Duyn JH, Crichton RR, Zecca L (2014) The role of iron in brain ageing and neurodegenerative disorders. Lancet Neurol 13(10):1045–1060

Wei Y, Han CS, Wang YJ, Wu BB, Su T, Liu Y, He RQ (2015) Ribosylation triggering Alzheimer's disease-like tau hyperphosphorylation via activation of CaMKII. Aging Cell 14(5):754–763

Weismann D, Hartvigsen K, Lauer N, Bennett KL, Scholl HP, Charbel Issa P, Cano M, Brandstatter H, Tsimikas S, Skerka C, Superti-Furga G, Handa JT, Zipfel PF, Witztum JL, Binder CJ (2011) Complement factor H binds malondialdehyde epitopes and protects from oxidative stress. Nature 478(7367):76–81

Whittinghill M, Lewis BM (1961) Clustered crossovers from male Drosophila raised on formaldehyde media. Genetics 46:459–462

Wu BB, Wei Y, Wang YJ, Su T, Zhou L, Liu Y, He RQ (2015) Gavage of D-ribose induces a beta-like deposits, tau hyperphosphorylation as well as memory loss and anxiety-like behavior in mice. Oncotarget 6(33):34128–34142

Yang MF, Miao JY, Rizak J, Zhai RW, Wang ZB, Huma T, Li T, Zheng N, Wu SH, Zheng YW, Fan XN, Yang JZ, Wang JH, Yang SC, Ma YY, Lu LB, He RQ, Hu XT (2014a) Alzheimer's disease and methanol toxicity (part 2): lessons from four rhesus macaques (Macaca mulatta) chronically fed methanol. J Alzheimers Dis 41(4):1131–1147

Yang M, Lu J, Miao J, Rizak J, Yang J, Zhai R, Zhou J, Qu J, Wang J, Yang S, Ma Y, Hu X, He R (2014b) Alzheimer's disease and methanol toxicity (part 1): chronic methanol feeding led to memory impairments and tau hyperphosphorylation in mice. J Alzheimers Dis 41(4):1117–1129

Yu PH, Wang M, Fan H, Deng Y, Gubisne-Haberle D (2004) Involvement of SSAO-mediated deamination in adipose glucose transport and weight gain in obese diabetic KKAy mice. Am J Physiol Endocrinol Metab 286(4):E634–E641

Yu HS, Oyama T, Isse T, Kitakawa K, Ogawa M, Pham TT, Kawamoto T (2009) Characteristics of aldehyde dehydrogenase 2 (Aldh2) knockout mice. Toxicol Mech Methods 19(9):535–540

Chapter 14
Methods in Determination of Formaldehyde

Tao Su and Rongqiao He

Abstract In order to analyze formaldehyde, a simplest aldehyde closely related to human health from both exogenous and endogenous environments, many methods have been developed and employed with different techniques and instruments. According to the universal principle, measurements by different methods result in different values for the concentration of exogenous and endogenous formaldehyde. Among the methods, high-performance liquid chromatography (HPLC) is one of the most commonly used approaches in the determination of formaldehyde in the liquid and body tissues such as urine, blood, the brain, and liver. Of course, HPLC and gas chromatography (GC) coupled with mass spectrometry (MS) or fluorescent techniques show their high sensitivity and other advantages. 2,4-Dinitrophenylhydrazine (DNPH) is one of the most frequently used reagents applied with spectrophotometer, HPLC, and gas chromatography (GC). HPLC coupled with DNPH or ampicillin has been used to determine formaldehyde levels of the morning urine from Alzheimer's disease (AD) patients and age-matched participants. The results showed that the levels of urine formaldehyde are positively correlated with the severity of cognitive impairment in the AD patients. Approximately 40% of the AD patients have got high levels of endogenous formaldehyde, suggesting that endogenous formaldehyde could be developed as a biomarker for progression and intervention of AD. Currently, the specific fluorescent probes such as formaldehyde probe-1 (FAP-1) and formaldehyde-induced 2-aza-Cope reaction have been synthesized to label formaldehyde molecules in cells. Using these fluorescent probes, the endogenous formaldehyde is found to be localized in lysosomes. Exogenous formaldehyde is also localized and restricted in lysosome inside a living cell monitored by confocal microscopy. Furthermore, nano-techniques have been also introduced to analyze formaldehyde. Quartz crystal microbalance (QCM) can measure nanogram-scale changes of formaldehyde. As the technique advances, current efforts are underway to develop greater sensitivity, lower cost, improved portability, and more straightforward detection methods. The operational principles

T. Su
State Key Laboratory of Brain and Cognitive Science, Institute of Biophysics, Chinese Academy of Sciences, Beijing 100101, China

R. He (✉)
State Key Laboratory of Brain and Cognitive Science, Institute of Biophysics, Chinese Academy of Sciences, Beijing 100101, China

University of Chinese Academy of Sciences, Beijing 100049, China
e-mail: rongqiaohe@163.com

R. He, *Formaldehyde and Cognition*, DOI 10.1007/978-94-024-1177-5_14

and sensing performance of sensor have been created as well as the convenient and high-performance formaldehyde-determining applications will appear more in the near future.

Keywords Formaldehyde • Generation • Metabolism • Formaldehyde dehydrogenase

1 Introduction

Abundant methods in analysis of formaldehyde have been established in the recent decades (Table 14.1). More attention is paid to formaldehyde, with which many human diseases are found be related, especially the brain impairment and disorder. Formaldehyde, as a common environment pollutant, is widely used in industry to manufacture building materials and numerous household products, such as some plywood adhesives, abrasive materials, insulation insecticides, and embalming fluids. Formaldehyde is found everywhere from the air to water, from soil to natural food, from lotion to food preservatives, from skin care products to bacteria, and from digestion contents to intestinal microbiota (Liu et al. 2017) (see Chap. 1). In other words, every organism including human beings is naturally producing formaldehyde at each moment through material metabolisms. In recent years, publics and government of many countries have paid more attention to formaldehyde as pollution becomes worse. Therefore, determination or analysis of formaldehyde becomes a process that has to be performed in our daily life. Different methods in analysis of formaldehyde have been developed, such as the fluorescent molecular probe (Tang et al. 2016a, b), the nanoporous silica aerogel sorbent (Barkhordari et al. 2017), the TiO_2 nanotube array (Tang et al. 2017), and the derivatization procedure adapted to HPLC-MS/MS (Backe 2017). In this chapter, we would like to briefly introduce the methods in analysis of formaldehyde in vivo and in vitro.

2 A Variety of Methods for Analysis of Formaldehyde

Environmental Protection Agency (EPA) of the United States suggested that the acceptable daily intake (ADI) of formaldehyde is 0.2 mg/kg body weight and warned that formaldehyde exposure higher than ADI may potentially harm health (EPA 1999). Overload of exogenous formaldehyde leads to many health-related problems (Tang et al. 2009), including cognitive impairment (Marceaux et al. 2008) such as learning, memory, and equilibrium and even damage to the brain (Flyvholm and Menne 1992; Perna et al. 2001). Furthermore, according to International Agency for Research on Cancer (IARC) (IARC 2004), formaldehyde is considered as a dangerous human carcinogen. Thus, detection and determination of both exogenous and endogenous formaldehyde are imperative to human health and daily life.

Table 14.1 The formaldehyde detection methods

Authors	Year	Methods	Reagents	Samples (value)	Limit of detection
Sawicki et al.	1961	Colorimetry	3-Methyl-2-benzothiazolone hydrazone	Aqueous solution (90 μg/100 ml)	0.1 μg/ml
Lorrain et al.	1981	Ion chromatography	–	–	0.5 nmol
Septon et al.	1982	Polarography	Analyzed by differential pulse polarography at a dropping mercury electrode	Workplace air range of 5.8–17.7 mg/m^3	–
Szarvas et al.	1986	Dimedone ^{14}C radiometric method	5,5-Dimethyl-cyclohexane-1, 3-dione	Blood (0.4–0.6 g/ml); urine (2.5–4.0 g/ml)	–
Munoz et al.	1989	Spectrophotometry	Pararosaniline	Formaldehyde in air (1.25 mg/m^3)	0.1 μg/ml
Houlgate et al.	1989	HPLC	2,4-Dinitrophenylhydrazine	Mainstream cigarette smoke (5 ± 0.07 μg per cigarette)	–
Gromping et al.	1993	Fluorometric HPLC	DAIH, DNSH, DDL, CTA, PRA, DNPH	Candle smoke (DNPH 1.6 ppm; DNSH 0.5 ppm; DAIH 0.6 ppm; DDL 0.5 ppm; CTA 0.7 ppm; PRA 0.5 ppm)	2 ppbv, 0.5 ppbv, 1 ppbv, 5 ppbv, 50 ppbv, 10 ppbv
Sardi et al.	1994	HPLC	Dimedone, methanol	Fresh animal tissues (e.g., liver, meat) or the fresh parts of plants	–
Shrivastaw et al.	1995	Spectrophotometry	Phenylhydrazine (PH) and potassium ferricyanide (PF)	Diphtheria-tetanus-pertussis vaccine (3.4 μg/ml), diphtheria-tetanus vaccine (3.96 μg/ml), tetanus toxoid vaccine (0.61 μg/ml), Japanese encephalitis vaccine (8.10 μg/ml)	100 ng
Quesenberry et al.	1996	Spectrophotometry	Periodate, purpald	Aqueous solution	1 nmol
Chan et al.	1997	Adsorption voltammetry	Girard's reagent T, EDTA	Air in chem lab (4.8 μg/m^3); air in a newly reconstructed chem lab (24.4 μg/m^3)	0.7 ppbv
Ebeler et al.	1997	GC	–	Tumor-bearing mice (1.43–2.98 μM) and control mice (0.77–1.01 μM)	–

(continued)

Table 14.1 (continued)

Authors	Year	Methods	Reagents	Samples (value)	Limit of detection
Possanzini et al.	1997	HPLC, Fluorometric	2-Diphenylacetyl-l,3-indandione-l-hydrazone	Atmospheric formaldehyde (1–5 ppb)	0.25 ppb
Bechmann et al.	1998	Flow injection analysis (FIA)	Enzymatic oxidation of formaldehyde	Mixture of fish tissue (17.67 ± 3.05 mg/kg)	–
Bolta et al.	1998	Gas chromatography	2,4-Dinitrophenylhydrazine	Spiked atmosphere (0.5 mg/m^3)	0.17 mg/m^3
				Interior of new furniture (1.3 mg/m^3)	
Martos et al.	1998	Gas chromatography, MS	PFBHA	Gaseous formaldehyde (640 ppbv)	4.6 ppbv
Karlberg et al.	1998	Closed container diffusion (CCD) method	2,4-Dinitrophenylhydrazine	Cream base (2400 μg/g) diazolidinyl urea (350 μg/g) in shin-care products	0.06 μg/L
Komazaki et al.	1999	HPLC	2,4-Dinitrophenylhydrazine	Averaged concentrations in air in Kawasaki (7.7 ± 5.5 ppbv)	0.05 ppbv
Spanel et al.	1999	Selected-ion flow-tube mass spectrometry	–	Patients with prostate cancer (0.83 μM) and bladder cancer (2.83 μM)	–
Rozylo et al.	2000	HPLC-OPLC-MS	4,4-Dimethyl-cyclohexane-1,3-dione	Teeth samples (0.05–0.40 μg/g)	–
Chan et al.	2001	Spectrophotometry	3-Methyl-2-benzothiazolinone hydrazone hydrochloride (MBTH), iron(III) chloride-sulfamic acid solution	Indoor air (14.4 ± 1.5, 17.8 ± 2.4, 6.6 ± 0.9, 7.5 ± 1.2 μg/m^3)	3 μg/m^3
Luo et al.	2001	HPLC	Ampicillin	Blood plasma (1.34 μg/ml)	0.46 μg/ml
Kato et al.	2001	Preconcentration-chemical ionization mass spectrometry	–	HeLa-S3 cervical cancer, K562 leukemia, and MCF-7 breast cancer cell lines (1.5–4.0 μM)	–
Ivanyi et al.	2002	HPLC	2,4-Dinitrophenylhydrazine	Shampoo (0.038 g/100 ml)	4×10^{-6}%
Maboudou et al.	2002	GC-MS	Single-ion monitoring (SIM)	Rat brain (10–900 ng/mL)	20 ng/mL
Yu et al.	2003	HPLC/electrochemical	Dopamine, DNPH	Mouse liver, kidney, and brain tissues 63, 51, and 11 nmol/g	–

Li et al.	2003	Flow injection chemiluminescence	Potassium bromate, rhodamine 6G	Formaldehyde in air sample (0.097 (±2.5%), 0.256 (±1.7%), 0.139 (±2.4%), 0.305 (±2.0%) mg/m-3)	0.3 μg/L
Rivero et al.	2004	Combined solid-phase microextraction-isotope dilution mass spectrometry	Pentafluorophenyl hydrazine(PFPH)	Formaldehyde standard solution (4 μg/ml)	0.39 μg/L
Motyka et al.	2004	Fluorescence	2,4-Pentanedione	Indoor air (27.2 ± 0.4 μg/m^{-3})	1.6 μg/m^3
Burini et al.	2004	HPLC and DAD	Ethyl 3-oxobutanoate and ammonia	Spirit sample (ranged between 0.27 and 3.01 mg/l)	0.024 μg/ml
Sritharathikhun et al.	2005	Flow injection technique with spectrophotometric and fluorometric detection	Acetylacetone	Indoor air 5.14 ± 0.08 ppbv for 20 ml of the air sample	Spectrophotometric (0.6 ppb) and the fluorometric (0.2 ppb)
Mariscal et al.	2005	Fluorescence	–	Cotton(1.27 ± 0.20 μg/cm^2)	–
				Filter paper (1.67 ± 0.290 μg/cm^2)	
				Polyvinyl chloride (0.30 ± 0.12 μg/cm^2)	
				Silicone-coated (0.49 ± 0.14 μg/cm2)	
Tomcik et al.	2005	Interdigitated microelectrode array	BrO^-	Textile pieces samples A, B, C, D (207, 240, 353, 400 mg/kg)	4 × 10^{-6} mol/dm^3
Bertoni et al.	2005	Column liquid chromatography, UV detection	1-Methyl-1-(2,4-dinitrophenyl)hydrazine	Sampling rate for HCHO, 15.0 mL/min	0.2 ppbv in 100 L air samples
Liu et al.	2005	Liquid-phase microextraction coupled with HPLC	2,4-Dinitrophenylhydrazine	Shiitake mushroom (355 ± 15 μg/g)	5 μg/L
Largiuni et al.	2005	Spectrofluorometry	Acetylacetone, acetic acid, and acetate	Seawater (mean 15 mg/L)	0.7 mg/L
Li et al.	2005	Spectroscopy	2,4-Pentanedione	Formaldehyde concentrations are higher at lower altitudes 60 m (3.5 ppbv), dropped to near zero at 750 m and 1500 m	Aqueous (16 nM), gaseous (70 pptv)

(continued)

Table 14.1 (continued)

Authors	Year	Methods	Reagents	Samples (value)	Limit of detection
Kawamura et al.	2005	HCHO gas sensor	4-Amino hydrazine-5-mercapto-1,2,4-triazole (AHMT)	Air from living and dining rooms and kitchen(0.017 ppm) and air from bedroom (0.007 ppm)	0.04 ppm
Motyka et al.	2006	Chemiluminescence flow method	Gallic acid, hydrogen peroxide in an alkaline solution	Formaldehyde in air, the calibration graph is linear up to 300 μg/m^3 (244 ppb)	0.60 μg/m^3 (0.49 ppb)
Zhao et al.	2006	Staircase voltammetry	Electrocatalytic oxidation of formaldehyde, Ni(III)	The linear range of detection is from 46.8 to 1640 μg/L (7.80 × 10^{-7} M)	23.4 μg/L
Motyka et al.	2007	Chemiluminescence	Gallic acid, hydrogen peroxide	Rainwater sample (7.61 ± 0.02, 1.58 ± 0.08, 1.92 ± 0.09, 9.83 ± 0.05 0.90 ± 0.08 μmol/L)	4 × 10^{-8} M
Wang et al.	2007	Colorimetry	Acetylacetone	The linear range (0.8–20.0 μg/mL)	0.8 μg/mL
Takigawa et al.	2007	Headspace gas chromatography	O-(2,3,4,5,6-pentafluorobenzyl) hydroxylamine	Urine from healthy individuals ranges (1.89–4.81 μM)	1.08 μg/L
Bianchi et al.	2007	SPEM-GC-MS	Pentafluorobenzyl-hydroxyl-amine hydrochloride	Haddock (1.47 ± 0.27 mg/kg); cod (106 ± 15 mg/kg); hake (293 ± 26 mg/kg); mullet (3.38 ± 0.71 mg/kg)	17 μg/kg
Cui et al.	2007	Spectrophotometry	Rhodamine B, potassium bromate, sulfuric acid	Sea cucumber (12.3 ± 0.1 μg/kg); bamboo shoot (23.6 ± 0.1 μg/kg); sleeve-fish (86.0 ± 0.2 μg/kg); tripe (36.1 ± 0.3 μg/kg); mushroom (83.5 ± 0.2 μg/kg); melon seeds (13.4 ± 0.1 μg/kg); vermicelli (34.8 ± 0.2 μg/kg); bean curd (67.2 ± 0.2 μg/kg); confect (56.9 ± 0.1 μg/kg); shrimp (24.9 ± 0.3 μg/kg)	2.90 μg/L
Abbasi et al.	2007	Spectrophotometry	Janus green, bromate, sulfuric acid	Melamine-formaldehyde resin (6.5 ± 0.71% μg/mL)	0.0015 μg/mL
				Chemical industrial wastewater (0.70 ± 0.01%μg/mL)	

Trenholm et al.	2008	Gas chromatography-mass spectrometry	PFBHA	NOM reagent water (195 μg/L), wastewater matrix (760 μg/L)	10 μg/L
Chen et al.	2008	Chromatography	2,4-Dinitrophenylhydrazine	Cotton (89.6 ± 2.1 mg/kg); wool (45.5 ± 1.6 mg/kg); cashmere (58.4 ± 1.2 mg/kg); terylene (2.8 ± 0.2 mg/kg); nylon (2.9 ± 0.1 mg/kg)	0.06 mg/kg
Soman et al.	2008	HPLC-UV	2,4-Dinitrophenylhydrazine (DNPH)	Three drug substances (ND, 0.4, 0.3 ppm)	0.03 ng (0.1 ppm)
Zhang et al.	2009	Electronic nose analysis	–	Fresh octopus sample (20%, 10%, 5%, 2.5%, 1.25%, 0.625%, 0.313%, 0.156%, 0.0781%, 0.0391%, 0.0195% and 0%, respectively)	–
Iqbal et al.	2009	HPLC	2,4-Dinitrophenylhydrazine, $Ca(OH)_2$, glycolaldehyde	Aqueous solution (170 μmol/L)	–
Hill et al.	2009	Colorimetric solid-phase extraction	Sodium hydroxide, purpald	Solutions with formaldehyde concentrations up to 20 ppm	0.08 ppm
Weng et al.	2009	Polydimethylsiloxane (PDMS) microfluidic chip	Acetylacetone	Shiitake mushroom (187.9 ± 30 mg/kg); frozen catfish (28.2 ± 5.3 mg/kg); pork (5.4 ± 1 mg/kg); pear (47.6 ± 3.6 mg/kg); codfish ball (28.4 ± 4.5 mg/kg); dried shrimp (50.7.2 ± 10.1 mg/kg); dried anchovy (25.5 ± 5.2 mg/kg); ginseng (150.6 ± 20 mg/kg)	Food samples (5.0 mg/kg); standard formaldehyde solution (2.0 μg/ml)
Emeis et al.	2010	^{13}C NMR spectroscopy	–	Benzylhemiformal (0.37%); diazolidinyl urea (0.29%); imidazolidinyl urea (0.054%); N,N′-methylenebis (0.15%); quaternium-15 (0.11%); tris-hexahydrotriazine (0.019%)	0.01%
Wei et al.	2010	Gas chromatography	DNPH	Engine exhaust formaldehyde (32.68 μg/L)	28.2×10^{-9} g/L
Wang et al.	2011	Resonance fluorometry	PY, KBrO3	Blood plasma samples (5.94×10^{-2} μg/mL)	3.80 ng/mL

(continued)

Table 14.1 (continued)

Authors	Year	Methods	Reagents	Samples (value)	Limit of detection
Teshima et al.	2011	Spectrophotometry	Hydroxylamine sulfate, iron(III)-ferrozine complex, hydroxylamine	Formaldehyde in industrial wastewater (157 ± 3.7 mg/L)	1.6 μg/L
Wang et al.	2012b	Electro-spinning/netting nano-fiber/nets	Nylon 6 NFN membranes	Gaseous formaldehyde (0–5 ppm)	50 ppb
Nengsih et al.	2012	Spectrometry	Gold nanoparticles	Aqueous formaldehyde range (0–37%)	3%
Lefebvre et al.	2012	HPLC	2,4-Dinitrophenylhydrazine	Facial moisturizer (2.4 ± 0.5 μg/m^3); body lotion (2.3 ± 1.0 μg/m^3); foundation (1.9 ± 0.4 μg/m^3); shower gel (2.0 ± 0.7 μg/m^3); shampoo (2.2 ± 0.7 μg/m^3); deodorant (2.7 ± 0.6 μg/m^3); hair conditioner (2.6 ± 0.6 μg/m^3); hair styling gel (2.5 ± 1.7 μg/m^3)	0.07 μg/m^3
Wang et al.	2012a	HPLC	2,4-Dinitrophenylhydrazine	Fruit juice samples, tomato (27.1 ng/ml); peach (19.7 ng/ml); orange (24.0 ng/ml)	6.0 ng/ml
Arvand et al.	2012	IL-based DLLME/ spectrophotometry	Methyl acetoacetate, ammonia	Shampoo 1 (75 ± 3.5 ng/ml); shampoo 2 (50 ± 4.1 ng/ml); textile industrial factory wastewater (6.0 ± 4.5 ng/ml); wood and paper industrial factory wastewater (100 ± 3.0 ng/ml)	0.02 ng/ml
Wang et al.	2012c	HPLC	2,4-Dinitrophenylhydrazine	Beer samples (172–385 ng/ml)	0.6 ng/ml
Deng et al.	2012	Electrochemiluminescence	Tris(2,2-bipyridyl) ruthenium(II), hexamethylenetetramine	Indoor air of a new building (18.2–40.6 g/m^3); outdoor air of Seven Star Park and Guilin Sanjin Pharmaceutical Company was from 7.6 to 21.3 g/m^3	0.15 μg/m^3
Zali et al.	2013	GC-MS	–	Diphtheria-tetanus antigen (2.9×10^{-3}), DT vaccine (4.3×10^{-5}), enterotoxemia (0.133%) (w/v%), black disease (0.130%), hemorrhagic septicemia (0.050%)	979 ng/L

Alizadeh et al.	2013	Chemiresistor sensor	Chemically exfoliated graphene flakes, blended with poly(methyl methacrylate)	Formaldehyde concentration was linear between 0.05 and 5.0 ppm	10 ppb
Tian et al.	2013	Pd-functionalized SnO_2 sensor	Pd-functionalized mesoporous SnO_2 fibers	HCHO gas in air (200 ppm)	50 ppb
Yeh et al.	2013	GC-MS	2,4-Dinitrophenylhydrazine	Shredded squid 1 (22.3 mg/kg)	2.0 mg/kg
Kleinnijenhuis et al.	2013	HPLC-MS	2,4-Dinitrophenylhydrazine	Average formaldehyde concentration in rat blood was 2.3 mg/L	0.015%
Jiang et al.	2013	Chromatography and mass spectrometry	Purpald, alcohol oxidase	The range of formaldehyde(0.02–0.8 mM)	5 μM
Monakhova et al.	2013	^{1}H NMR spectroscopy	–	Hair straightening products (0.42–5.83%)	0.14%
Maneli et al.	2013	HPLC	2,4-Dinitrophenylhydrazine	"Brazilian keratin-type" hair straightening products (0.96%–1.4%)	6 ppb
Sassine et al.	2014	Colorimetry	2,4-Dinitrophenylhydrazine	Formaldehyde in air sample (6.4 ± 0.7, 14.6 ± 1.7 ppbv)	4 pptv for 100 cm
Yu et al.	2014a	Colorimetry	Single-crystal-to-single-crystal fashion (Cu(I)-MOF)	The range of formaldehyde (0.0016–1.6 ppm)	0.016 ppm
Li et al.	2014	DOAS (differential optical absorption spectroscopy)	–	HCHO concentrations in Shanghai ambient air at a research station in Fudan University, April 2010 to April 2011	1.0 ppb
Toda et al.	2014	Continuous flow fluorometry	10 mM H_2SO_4 0.01 M 2,4-Pentanedione (PD), 0.25 M acetic acid, and 2 M ammonium acetate	In forest air, particulate HCHO (0.18 μg/m^3); gas HCHO (3.9 μg/m^3)	0.048 μg/m^3 (HCHO(g)) and 0.0033 μg/m^3 (HCHO(p))
Zhang et al.	2014	Spectroscopy	MBTH, $NH_4Fe(SO_4)_2$, H_2O, Au/SiO_2 colloids	Fresh squid and shrimp samples (0.13–0.21 mg/kg)	0.17 mg/L
Tang et al.	2014	Spectrophotometry	Eosin Y, potassium bromate, phosphoric acid	Shrimp (8 ± 0.5 μg/g), yuba (25 ± 0.3 μg/g), shiitake (19 ± 0.7 μg/g)	0.00988 μg/mL
Chen et al.	2014	Gas sensor	Zeolitic imidazolate framework (ZIF)	Range of formaldehyde concentration (5–500 ppm)	5 ppm

(continued)

Table 14.1 (continued)

Authors	Year	Methods	Reagents	Samples (value)	Limit of detection
Ueta et al.	2015	HPLC	2,4-Dinitrophenylhydrazine	Real indoor air samples (14.7 ng/L; 15.7 ng/L; 17.4 ng/L)	2.5 ng/L
Jeong et al.	2015	Gas chromatography-mass spectrometry	O-(2,3,4,5,6-Pentafluoro-benzyl)-hydroxylamine hydrochloride	Herb wine (74.70 ± 5.12 ng/g); apricot liqueur (56.81 ± 5.47 ng/g); beer (45.47 ± 1.93 ng/g); red wine (40.90 ± 2.69 ng/g); white wine (18.38 ± 1.34 ng/g); sparkling wine (19.50 ± 0.17 ng/g); whisky (782.10 ± 6.10 ng/g)	0.01 mg/L
Yang et al.	2015	Spectrometry	G-Quadruplex DNAzyme and ABTS, H_2O_2	Indoor air (0.12 mg/m3)	0.01 mM
Fang et al.	2015	Gas-sensing performance	Ag-modified In_2O_3/ZnO nanobundles	Range of formaldehyde concentration (0.1–1.6 ppm)	100 ppb
Monkawa et al.	2015	Spectrometry	WST-8, FALDH, NAD+, EDTA	Water (0.005 to 0.5 ppm); gas (5–80 ppb)	15 ppb (water)
					1.5 ppb (gas)
Wong et al.	2015	Visual detection	Resin-linked sterically bulky amines, fluorescent alkynes	Formaldehyde concentration (0–1000 ppm)	10 ppm
Yanaga et al.	2015	Fluorometric approach	10 μmol/L ITS,10 mmol/L H2SO4	Formaldehyde concentration (0–1200 ppbv)	0.3 ppbv
			0.01 mol/L 2,4-Pentanedione, 0.25 μmol/L, acetic acid, 2 μmol/L ammonium acetate		
Gallia et al.	2015	HPLC	Acetonitrile HPLC gradient-grade and hydrochloric acid, 2,4-dinitrophenylhydrazine	Determined concentrations ranging from 0.0036% to 0.184%	0.41 g/mL
Tang et al.	2015	Spectrophotometry	Malachite green, potassium bromate, sulfuric acid	Tap water (0.2 ± 0.02 μg/g), Coca-Cola (0.5 ± 0.07 μg/g), shiitake (18.5 ± 0.05 μg/g)	1 ng/mL

Brewer et al	2015	Fluorescent probe	2 Aza Cope based trigger of FAP-1	Determined concentrations range in HEK293T cell	5 μM
Xu et al.	2016	Naphthalene-based fluorescent probe (AENO)	6-(1-Aminobut-3-en-1-yl) naphthalen-2-ol	Concentration of FA in HeLa cells (209 μM)	0.57 μM
Tang et al.	2016b	Fluorescent probe	1,8-Naphthalimide, hydrazine	Lysosomes in the living HeLa cells	5.02×10^{-6} M
Tang et al.	2016a	Fluorometry	Two-photon fluorescent FA probe (Na-FA)	Living HeLa cells, liver slides	7.1×10^{-7} M
Khataee et al.	2016	Flow injection chemiluminescence (CL) method	$KMnO_4$, CdS QDs	Underground water (9.86 ± 1.2 μg/L) Drinking water (9.52 ± 1.96 μg/L)	0.0003 μg/L and 1.2 μg/L
Yang et al.	2016	QCM resonator	Graphene oxide	Determined concentrations (3.5 ppm, 2.6 ppm, 1.7 ppm, 0.9 ppm, and 0.5 ppm)	0.06 ppm
Wahed et al.	2016	HPLC	2,4-Dinitrophenylhydrazine	Cereals (10.74 mg/kg); fish (26.2 mg/kg); leafy vegetables (5 mg/kg); non-leafy vegetables (2.5 mg/kg); fruits (3.08 mg/kg); milk (3.0 mg/kg)	0.39 mg/L
Backe W J	2017	Solid-phase extraction (SPE) and liquid chromatography/tandem mass spectrometry (LC/MS/MS)	Using electrospray ionization (ESI)	Pharmaceuticals, steroids, pesticides, and personal care products	1.0 ng/L in water
Barkhordari A et al.	2017	Needle trap device (NTD)	Packed with nanoporous silica aerogel	Aqueous and urine samples	0.01–0.03 μg/L(in aqueous sample), 2.8–7.3%(in urine)
Gu D C et al.	2017	HPLC-UV	2,4-Dinitrophenylhydrazine	Fresh squid contained less than 20 mg/kg; free-formaldehyde concentration in shredded squid was 48.5 mg/kg	15 mg/kg

To determine this smallest aldehyde molecule, different methods have been developed for different purposes. Sawicki and coworkers estimated formaldehyde with colorimetric procedure using 3-methyl-2-benzothiazolone hydrazone (Eugene Sawicki et al. 1961). After that, different methods and protocols were established such as ion chromatography (Lorrain et al. 1981), polarography (Septon and Ku 1982), radiometric analysis with dimedone ^{14}C (Szarvas et al. 1986), spectrophotometry with pararosaniline (Munoz et al. 1989), high-performance liquid chromatography (HPLC) (Houlgate et al. 1989), adsorption voltammetry (Chan and Xie 1997), gas chromatography (GC) (Ebeler et al. 1997), flow injection analysis (Bechmann 1998), preconcentration-chemical ionization mass spectrometry (Kato et al. 2001), fluorescence (Motyka and Mikuska 2004), interdigitated microelectrode array (Tomcik et al. 2005), gas sensor (Kawamura et al. 2005), polydimethylsiloxane microfluidic chip (Weng et al. 2009), $^{13}C/^{1}H$ NMR spectroscopy (Emeis et al. 2010; Monakhova et al. 2013), electro-spinning/netting nano-fiber/nets (Wang et al. 2012b), Pd-functionalized SnO_2 sensor (Tian et al. 2013), DOAS (differential optical absorption spectroscopy) (Li et al. 2014), fluorescent probe (Brewer and Chang 2015; Xu et al. 2016), and flow injection chemiluminescence (Khataee et al. 2016). In the recent decade, more methods have been invented and used than ever before. Different techniques were applied in the analysis of formaldehyde. Backe (2017) derivatized formaldehyde into 3,5-diacetyl-1,4-dihydrolutidine (DDL) using an acetylacetone solution. The method includes a derivatization procedure that is newly adapted to HPLC-MS/MS; therefore, structures were proposed for the product ions of the derivative DDL.

Among so many methods, the HPLC is one of the most approaches employed with different chemical reagents. Adducts between FA and other reagents can be selectively separated by HPLC and then quantified by an electrochemical approach. A fluorometric method to determine formaldehyde in candle smoke was reported by using HPLC. These studies focused on the analysis of formaldehyde levels in the air, water, and food. Consequently, the World Health Organization (WHO) suggests the 30-min exposure limit of 0.08 ppm (0.107 mg/m^3) (2001). Environmental Protection Agency of the USA suggested that the ADI of formaldehyde is 0.2 mg/kg body weight (EPA 1999).

The determination of concentration of FA and MDA in 20 rats showed a variation, respectively, of 10.0–900.0 ng/mL for FA and 18.0–800.0 ng/mL for MDA. Formaldehyde is a well-known environmental toxic hazard. It is also a product of oxidative deamination of methylamine catalyzed by semicarbazide-sensitive amine oxidase (SSAO). Increased SSAO-mediated deamination has been implicated in some pathophysiological conditions, such as diabetic complications. The measurement of formaldehyde in the enzymatic reactions and in vivo production using conventional methods was not straightforward due to limitations of selectivity and sensitivity. A novel high-performance liquid chromatography (HPLC)/electrochemical procedure for the measurement of formaldehyde has been developed. The measurement is based on the formation of adducts between formaldehyde and dopamine. These adducts can be selectively purified and concentrated using a batch method of alumina absorption, separated by HPLC, and electrochemically quantified. The method is highly selective and substantially more sensitive, i.e., detection of picomole levels of formaldehyde, than the conventional methods. The procedure not only facilitates the assessment of SSAO activity in vitro but also is useful for assessing formaldehyde in tissues and biological fluid.

3 2,4-Dinitrophenylhydrazine, One of the Most Commonly Used Reagents

In the determination of exogenous and endogenous formaldehyde, different reagents and techniques have been developed. Among the most used reagents, 2,4-dinitrophenylhydrazine (DNPH) is most used with different analytic instruments and techniques including spectrophotometry (Munoz et al. 1989), GC (Ebeler et al. 1997), GC-MS (Zhang et al. 2009), ion chromatography (Lorrain et al. 1981), HPLC (Houlgate et al. 1989), and HPLC-mass spectrometry (Iqbal and Novalin 2009). In 1989, Houlgate and colleagues used DNPH to analyze the mainstream cigarette smoke (5 ± 0.07 μg per cigarette) with HPLC. Among the different methods, HPLC with ultraviolet detector method is commonly used to monitor aqueous formaldehyde. In this method, sample collected impregnates with DNPH. Formaldehyde can be converted to the 2,4-dinitrophenylhydrazone derivative. Then the derivatives were separated by column and detected. The method was approved by the American Society for Testing Materials (ASTM) as a standard test method for determination of formaldehyde.

4 Determination of Brain Formaldehyde

The levels of brain formaldehyde depend upon the analytic methods, that is, different methods produce different values of concentrations of brain formaldehyde. This is reasonable because different methods have different sensitivities to formaldehyde, with different performance schemes. Maboudou and colleagues determined the rat brain formaldehyde with a gas chromatography-mass spectrometry (GC-MS) method. They homogenized rat brains with deionized water and derivatized the extract with DNPH to obtain formaldehyde hydrazine derivatives. The determination of concentration of formaldehyde in 20 rats showed a variation, respectively, of 0.33–30.00 μM for formaldehyde in the homogenate of the brain (Maboudou et al. 2002). Yu and his collaborators used a high-performance liquid chromatography/electrochemical procedure for measuring formaldehyde. The formaldehyde level of mouse brain tissues was estimated to be 11 μmol/kg tissues (Yu et al. 2003). Tong and his colleagues employed HPLC. As described by Su and his colleagues, brain tissues of porcine were homogenized, and the supernatant fractions (weight of tissue: ultrapure water = 1: 4) were analyzed by DNPH coupled with HPLC after centrifugation (3000 rpm, 4 °C, 10 min). The formaldehyde standard curve was made with analytic purified formaldehyde. The concentrations of formaldehyde in parietal lobe, frontal lobe, temporal lobe, occipital lobe, hippocampus, cerebellum, and brain stem of porcine brain (n = 5) were measured. The porcine brain formaldehyde was 75.5 ~ 83.4 μmol/kg fresh brain tissues measured by this improved method. Recoveries of spiked formaldehyde at low and high level were 95.96% ~ 102.04% with relative standard deviations less than 10% (Su et al. 2011). In addition, the FA level in solid tissues, for instance, teeth, was measured by HPLC-OPLC-MS. Results showed an increase in FA level in carietic teeth in comparison to healthy teeth (Rozylo et al. 2000).

5 Analysis of Formaldehyde in Urine and Body Fluid

Szarvas and coworkers estimated the formaldehyde concentration in human urine and blood with dimedone ^{14}C, a reagent for radiometric analysis (Szarvas et al. 1986). The formaldehyde in blood and urine varied between 0.01–0.02 mM and 0.08–0.13 mM, respectively. Gas chromatography (GC) has been used to compare the concentrations of formaldehyde between the expired air of tumor-bearing mice (1.43–2.98 μM) and control mice (0.77–1.01 μM) (Ebeler et al. 1997). The concentration of urine formaldehyde from healthy individuals ranges from 1.89 to 4.81 μM, as determined by Takigawa and colleagues using gas chromatography (Takigawa et al. 2007). Using selected-ion flow-tube mass spectrometry (SIFT-MS), the mean concentrations of urine formaldehyde in patients with prostate cancer and bladder cancer were shown to be 0.83 μM and 2.83 μM, respectively, compared with 0.37 μM in healthy controls (Spanel et al. 1999). Preconcentration of the body fluid samples rendered formaldehyde more detectable by SIFT-MS. Formaldehyde concentrations in HeLa-S3 cervical cancer, K562 leukemia, and MCF-7 breast cancer cell lines range from 1.5 to 4.0 μM (Kato et al. 2001b). The use of HPLC combined with an electrochemical procedure is an established method for the measurement of formaldehyde in vivo (Su et al. 2011). Using this method, the concentration of rabbit urine formaldehyde is about 18 nM and that of mouse liver, kidney, and brain tissues are 63, 51, and 11 nmol/g tissue, respectively (Yu et al. 2003). Another method to determine formaldehyde in blood samples by HPLC utilizes a fluorescence assay. Derivatization of human plasma with ampicillin leads to a fluorescent adduct which can be measured with a fluorescence spectrophotometer. The average level of human blood formaldehyde as determined by Fluo-HPLC is approximately 40 μM (Luo et al. 2001). Thin-layer chromatography is used to estimate changes in the levels of formaldehyde in methylation and demethylation (Kalász 2003). Yu and his workers reported that the rabbit urine samples were spiked with 0.25 nM formaldehyde and subjected to extraction with alumina and HPLC assessment (Yu et al. 2003). Among those methods to analyze urine formaldehyde, Su and his collaborators preferred the method of HPLC coupled with DNPH in the clinical determination of uric samples because it is sensitive and repeatable described as follows (Su et al. 2011).

6 An Example for Analysis of Urine Formaldehyde Using HPLC Coupled with DNPH Absorbance

Yu and her collaborators analyzed urine formaldehyde using HPLC coupled with DNPH absorbance and performed neuropsychological tests for aged people from the communities of Beijing city in a double-blinded manner (Yu et al. 2014b). Also, they measured concentrations of urine formaldehyde for AD patients compared with the age-matched participants. The average concentration of urine formaldehyde in AD patients is significantly higher than that of the controls (Li et al. 2016). Qiang and her colleagues also employed the method in the determination of formaldehyde levels in mouse brain, kidney, liver tissues, and the medium of cells (Qiang et al. 2014). The microbiota-gut-brain axis is regarded as the human "second brain," which is involved in neurodegenerative diseases such as Alzheimer's disease, vascular dementia, and Parkinson's disease (Hu et al. 2016). Liu and his collaborators found a marked increase of formaldehyde in cecum digestion contents of APP/PS1 transgenic mouse compared with C57BL/6j wild-type mouse (Liu et al. 2017).

Ten milliliters of urine was collected from each participant and immediately placed on ice or stored at −70 °C. To prevent the samples from repeated thaw-frozen cycles, the urine should be analyzed freshly or stored in aliquots. The urine sample (1 mL, thawed at 4 °C) was pipetted into a 1.5 ml Eppendorf tube and centrifuged (12,000 rpm for 10 min at 4 °C). A 0.4 mL aliquot of the supernatant was mixed with 2,4-dinitrophenylhydrazine (DNPH, final concentration 0.1 g/L in acetonitrile) and 0.1 mL trichloroacetic acid. Samples were vortexed vigorously for 30 s and then centrifuged (12,000 rpm for 10 min at 4 °C). The supernatant was added to a 2 mL glass vial, heated in a 60 °C water bath for 30 min, and then analyzed by HPLC. The mobile phase was 65% acetonitrile water solution. The flow rate was 0.8 mL/min and column temperature 35 °C. The formaldehyde-DNPH derivative was eluted from the HPLC column at a retention time of 6–7 min at the maximum wavelength of 355 nm as previously described (Su et al. 2011).

Urine formaldehyde was also determined with HPLC coupled with fluorescence (ampicillin) detection as described (Luo et al. 2001). Using this method, the concentrations of urine formaldehyde of age-matched control, mild cognitive impairment (MCI), moderate cognitive impairment, and server cognitive impairment were determined as 0.083 ± 0.001 mM, 0.196 ± 0.69 mM, 0.293 ± 0.050 mM, and 0.313 ± 0.046 mM, respectively (Tong et al. 2009). These values are significantly higher than those determined by HPLC coupled with DNPH, which generates the urine formaldehyde of AD patients as 13.70 ± 5.17 ($n = 62$) and that of age-matched participants as 9.61 ± 2.90 ($n = 69$). The HPLC with fluorescent ampicillin is more sensitive, but fluorescence is susceptible to interference, such as changes in temperature, acidity of solution, and concentrations of samples.

It should be noticed that formaldehyde has the capacity to react with proteins and produce modified derivatives. Thus, the accurate determination of blood formaldehyde is difficult since blood is rich in proteins. Formaldehyde is prone to react with serum proteins. On the other hand, determination of human urine formaldehyde by

HPLC coupled with DNPH is precise, convincible, and reproducible because urine contains much less proteins. Therefore, in clinical trials or epidemiological investigations, we would like to suggest the concentration values of urine formaldehyde should be more convincible.

7 Development of Methods with Nano-techniques

Nowadays, significant attention has been brought to the development of sensitive, specific, cheap, and reliable sensors for real-time monitoring formaldehyde. Several relatively low-cost and convenient methods without derivation or chromatography were developed, because the operation of derivation, elution, and extraction needs lots of time or money. Among various techniques, quartz crystal microbalance (QCM) (Yang and He 2016) could measure nanogram-scale changes in mass on the quartz crystal surface based on the Sauerbrey equation by recording its frequency shifts. QCM with oxygen oxide (GO) sensors showed high sensitivity, good reversibility and repeatability, and good selectivity to formaldehyde. GO-functionalized QCM resonators respond to formaldehyde in 60 s and are stable in the sensing characteristics at least up to 100 days.

Recently, a rapid and simple formaldehyde determination technique with a microfluidic chip was developed (Weng et al. 2009). This technique is based on the reaction between formaldehyde and acetylacetone in the presence of ammonium acetate to form yellow 3,5-diacetyl-1,4-dihydrolutidine which has a differential absorbability of around 410 nm violet light. Microfluidic device was designed to decrease the sample volume and to make multiple sample detection. Compared with the conventional formaldehyde detection system, the microfluidic chip method can realize rapid detection within 1 min, use a small sample volume about 1-2 μl, and reduce the cost significantly. As described by Tang and colleagues, molecular imprinting technology is a versatile and promising technology for practical applications in many areas, particularly chemical sensors (Tang et al. 2017). They presented a titanium dioxide nanotube array (TiO_2-NTA) for increasing its surface-to-volume ratio in a chemical sensor for detecting formaldehyde, a toxic common indoor pollutant gas.

8 Fluorescent Probe to Detect Formaldehyde in Living Cells

Methods for monitoring formaldehyde in living biological specimens are important to investigate the elevations of formaldehyde implicated in a variety of disease pathologies, such as neurodegenerative diseases and age-related cognitive impairment. Traditional methods for biological FA detection rely on sample destruction and/or extensive processing, resulting in a loss of spatiotemporal information. Therefore, specific probes for formaldehyde in cellular activity, especially kinetic study of formaldehyde in living cells, are urgently required.

Several methods of high selective detection of endogenous formaldehyde in living cell were exploited. Formaldehyde reacts with 2-aza-Cope in order to transform a homoallylic amine into an aldehyde coupled with a fluorogenic turn-on response (Brewer and Chang 2015). Formaldehyde probe-1 (FAP-1) is capable of detecting exogenous and endogenous FA in living cells. Using confocal microscopy, the images were taken 30 min after addition of FAP-1. The in vitro detection limit for FA was 5 μM. Another method detected the level of FA in HeLa cells by naphthalene-based fluorescent probe (AENO) (Xu et al. 2016). This probe AENO can quantitatively detect FA with a detection limit of 0.57 μM with the essentially pH-insensitive from 5.5 to 10.5. The fluorescence after incubation with AENO was linearly correlated with FA ranging from 0 to 1000 μM, and the endogenous FA in HeLa cells was calculated to be 209 μM. As reported by Tang and colleagues, a new organelle-targeted fluorescent formaldehyde probe (Na-FA-Lyso) was developed to detect the endogenous FA in the lysosomes in living cells (Tang et al. 2016a, b). The new probe Na-FA-Lyso was suppressed in its photoinduced electron transfer (PET) pathway and turned on after reacting with FA. The detection limit of the probe was calculated to be 5.02 μM. Yue and her colleagues detected formaldehyde in brain tissues by using Fluoral P (Yue et al. 2017. n. d.). Those powerful molecular probes improve the investigation of physiological and pathological roles of endogenous FA on the cellular and subcellular level.

To help fill up this technological gap, Brewer and Chang have designed, synthesized, and biologically evaluated a fluorescent probe for live-cell FA imaging that relies on a formaldehyde-induced aza-Cope rearrangement. A method exploited an FA-induced 2-aza-Cope reaction to transform a homoallylic amine into an aldehyde coupled with a fluorogenic turn-on response (Brewer and Chang 2015). FAP-1 is capable of detecting exogenous and endogenous FA in living cells. Using confocal microscopy, the images were taken 30 min after addition of FAP-1. The in vitro detection limit for FA was found to be 5 μM (Gromping and Cammann 1993). This fluorescent probe is expected to be developed into a convinced, convenient, and high effective method in determination of both exogenous and endogenous formaldehyde.

Abnormally high level of formaldehyde induced the cellular oxidative stress (Evans et al. 2016), resulting in abnormal modification and accumulation of protein, neuron death, and cognitive impairment (Jian and Zhu 2016; He 2016). To investigate the relation of oxidative stress and excess formaldehyde, which leads to neuron death and cognitive dysfunction, the technique for monitoring the dynamic distribution of formaldehyde in living cell has been performed by Chen and coworkers (Chen et al. 2017). In the cellular experiment, they added H_2O_2 to bEed.3 cells and found that the yield of formaldehyde is localized in the lysosome detected by the formaldehyde molecular probe (Fig. 14.1). They added formaldehyde to the cells and also observed that the signal of formaldehyde markedly increases inside lysosome and then decreases with time. Furthermore, they isolated lysosomes from rat brain and demonstrated the location of formaldehyde in the lysosome (Chen et al. 2017).

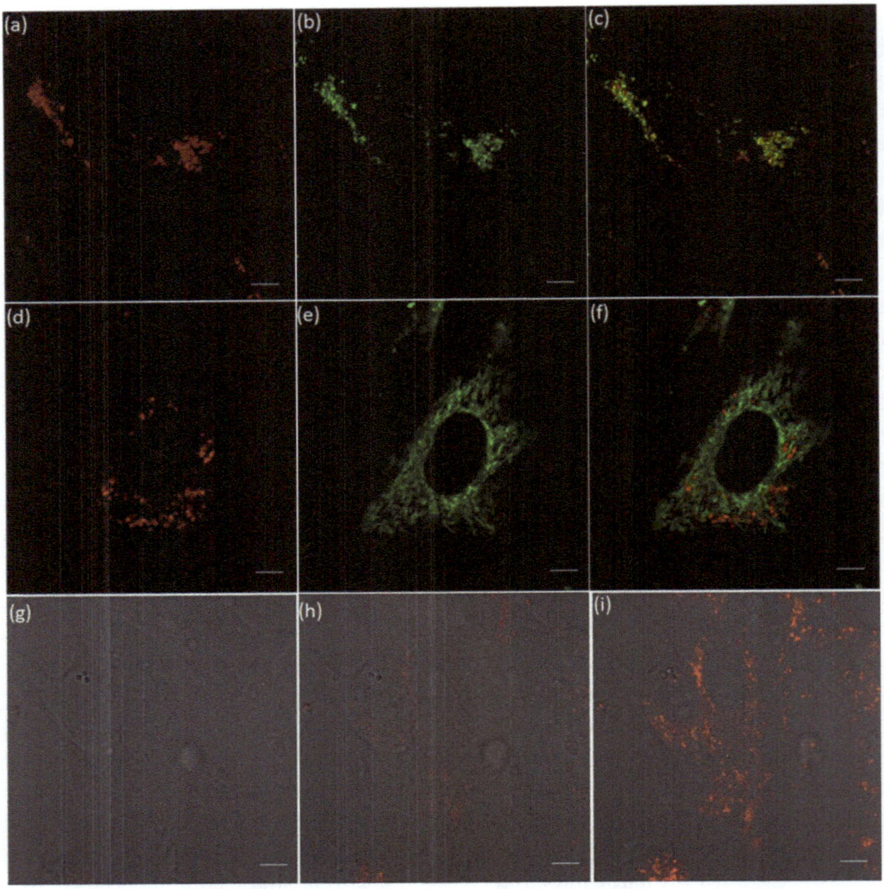

Fig. 14.1 Fluorescence image of the endogenous formaldehyde in the bEed.3 cells. Fluorescence image of FAP-1 (*red*, panel **a**); fluorescence image of Lyso Tracker Green DND-26 (*green*, panel **b**); merged image of panel **a** and panel **b** (panel **c**); fluorescence image of FAP-1 (*red*, panel **d**); fluorescence image of Mito Tracker Green FM (*green*, panel **e**); merged image of panel **d** and panel **e** (panel **f**); bright-field image of the cells as control (panel g); bright-field image of the cells treated with PAP-1 (panel **h**); bright-field image of the cells treated with PAP-1 and 0.1 mM formaldehyde (panel **i**). *Scale bar*: 10 μm (Chen et al. 2017) (The usage of this figure was authorized by the authors and chief editor of Prog. Biochem. Biophys)

9 Prospective

In recent years, formaldehyde is not only studied as an important exogenous carcinogenic substance but also one of the endogenous reactive carbonyl species produced by enzymatic oxidation of the amino acids, lipids, carbohydrates, as well as demethylation. Methods for monitoring formaldehyde in living biological specimens are important to investigate the elevations of FA that implicates a variety of disease pathologies, such as cognitive impairment, neurodegenerative diseases, heart disorders, cancers, diabetes, and chronic hepatitis. Barkhordari and colleagues designed a needle trap device (NTD) packed with nanoporous silica aerogel as a sorbent and used as a new technique for sampling and analysis of formaldehyde and acrolein in urine. The ranges of detection limit, quantification limit, and relative standard deviation (RSD) were 0.01–0.03 μg L^{-1}, 0.03–0.1 μg L^{-1}, and 2.8–7.3%, respectively. They believed that the technique can be applied as an effective and reliable method for sampling and analysis of aldehyde compounds from different biological matrices such as urine and exhalation (Barkhordari et al. 2017). Ultraperformance liquid chromatography tandem mass spectrometry was used to determine formaldehyde hemoglobin adducts in humans as a biomarker for formaldehyde exposure (Yang et al. 2017). A miniaturized sensor has been developed for detection of formaldehyde fumes (Zilberstein et al. 2017). The signal response has been assessed over low (20–120 ppb) as well as higher (1–15 ppm range) levels. A reagentless amperometric formaldehyde-selective chemosensor was established by Demkiv and collaborators based on platinized gold electrodes (Demkiv et al. 2017). As the technique advances in recent years, current efforts are underway to develop more straightforward detection methods with greater sensitivity, lower cost, and improved portability. The operational principles and sensing performance of sensor have been invented, and the convenient and high-performance formaldehyde-determining applications probably appear in the near future.

Acknowledgment This project was supported by grants from the Beijing Municipal Science and Technology Project (Z161100000217141; Z161100000216137), the National Key Research and Development Program of China (2016YFC1306300), the National Basic Research Program of China (973 Program) (2012CB911004), the Natural Scientific Foundation of China (NSFC 31301880, NSFC 31270868), Foundation of Chinese Academy of Sciences (CAS-20140909), the Queensland-Chinese Academy of Sciences Biotechnology Fund (GJHZ201302), Program for Liaoning Excellent Talents in University (LJQ2015057), and Dalian High Level Talent Innovation Support Plan (No. 2015R067).

Competing Financial Interests The authors declare no competing financial interests.

References

Abbasia S, Esfandyarpoura M, Taher MA, Daneshfar A (2007) Catalytic-kinetic determination of trace amount of formaldehyde by the spectrophotometric method with a bromate-Janus green system. Spectrochimica Acta Part A-Mol Biomol Spectrosc 67(3–4):578–581

Alizadeh T, Soltani LH (2013) Graphene/poly (methyl methacrylate) chemiresistor sensor for formaldehyde odor sensing. J Hazard Mater 248:401–406

Arvand M, Bozorgzadeh E, Shariati S, Zanjanchi MA (2012) Ionic liquid-based dispersive liquid-liquid microextraction for the determination of formaldehyde in wastewaters and detergents. Environ Monit Assess 184(12):7597–7605

Backe WJ (2017) A novel mass spectrometric method for formaldehyde in Children's personal-care products and water via derivatization with Acetylacetone. Rapid Commun Mass Spectrom 31(12):1047–1056

Barkhordari A, Azari MR, Zendehdel R, Heidari M (2017) Analysis of formaldehyde and acrolein in the aqueous samples using a novel needle trap device containing nanoporous silica aerogel sorbent. Environ Monit Assess 189(4):171

Bechmann IE (1998) Comparison of the formaldehyde content found in boiled and raw mince of frozen saithe using different analytical methods. Food Sci Technol -Lebensmittel-Wissenschaft & Technologie 31(5):449–453

Bertoni G, Ciuchini C, Di Palo V, Possanzini M (2005) Development of a passive sampler for long-term measurements of formaldehyde and total oxidants in air. Chromatographia 61(7–8):385–389

Bianchi F, Careri M, Musci M, Mangia A (2007) Fish and food safety: determination of formaldehyde in 12 fish species by SPME extraction and GC-MS analysis. Food Chem 100(3):1049–1053

Bolta SV, Zupancic-Kralj L, Marsel J (1998) Gas chromatographic determination of formaldehyde in air using solid-phase microextraction sampling. Chromatographia 48(1–2):95–100

Brewer TF, Chang CJ (2015) An Aza-cope reactivity-based fluorescent probe for imaging formaldehyde in living cells. J Am Chem Soc 137(34):10886–10889

Burini G, Coli R (2004) Determination of formaldehyde in spirits by high-performance liquid chromatography with diode-array detection after derivatization. Anal Chim Acta 511(1):155–158

Chan WH, Xie TY (1997) Determination of sub-ppbv levels of formaldehyde in ambient air using Girard's reagent T-coated glass fiber filters and adsorption voltammetry. Anal Chim Acta 349(1–3):349–357

Chan WH, Shuang SM, Choi MMF (2001) Determination of airborne formaldehyde by active sampling on 3-methyl-2-benzothiazolinone hydrazone hydrochloride-coated glass fibre filters. Analyst 126(5):720–723

Chen L, Jin HY, Wang LG, Sun L, Xu HY, Ding L, Yu AM, Zhang HQ (2008) Dynamic ultrasound-assisted extraction coupled on-line with solid support derivatization and high-performance liquid chromatography for the determination of formaldehyde in textiles. J Chromatogr A 1192(1):89–94

Chen EX, Yang H, Zhang J (2014) Zeolitic Imidazolate framework as formaldehyde gas sensor. Inorg Chem 53(11):5411–5413

Chen XX, Su T, He YG, He RQ (2017) Effect of oxidative stress on accommodation and transportation of formaldehyde by lysosome. Prog Biochem Biophys 44(6):486–494

Cui XJ, Fang GZ, Jiang LQ, Wang SZ (2007) Kinetic spectrophotometric method for rapid determination of trace formaldehyde in foods. Anal Chim Acta 590(2):253–259

Demkiv O, Smutok O, Gonchar M, Nisnevitch MM (2017) A Reagentless Amperometric formaldehyde-selective chemosensor based on platinized gold electrodes. Materials 10(5):503

Deng BY, Liu Y, Yin HH, Ning X, Lu H, Ye L, Xu QX (2012) Determination of ultra-trace formaldehyde in air using ammonium sulfate as derivatization reagent and capillary electrophoresis coupled with on-line electrochemiluminescence detection. Talanta 91:128–133

Ebeler SE, Clifford AJ, Shibamoto T (1997) Quantitative analysis by gas chromatography of volatile carbonyl compounds in expired air from mice and human. J Chromatogr B Biomed Sci Appl 702(1-2):211–215

Emeis D, de Groot AC, Brinkmann J (2010) Determination of formaldehyde in formaldehyde-releaser patch test preparations. Contact Dermatitis 63(2):57–62

EPA (1999) Integrated risk information system (IRIS) on formaldehyde. National Center for Environmental Assessment, Office of Research and Development, Washington, DC

Evans AM, Fameli N, Ogunbayo OA, Duan J, Navarro-Dorado J (2016) From contraction to gene expression: nanojunctions of the sarco/endoplasmic reticulum deliver site- and function-specific calcium signals. Sci China Life Sci 59(8):749

Fang F, Bai L, Song DS, Yang HP, Sun XM, Sun HY, Zhu J (2015) Ag-modified In2O3/ZnO Nanobundles with high formaldehyde gas-sensing performance. Sensors 15(8):20086–20096

Flyvholm MA, Menne T (1992) Allergic contact-dermatitis from formaldehyde – a case-study focusing on sources of formaldehyde exposure. Contact Dermatitis 27(1):27–36

Gallia CL, Bettinb F, Metrab P, Fidentea P, Dominicisb ED , Marinovicha M (2015) Novel analytical method to measure formaldehyde release from heated hair straightening cosmetic products: impact on risk assessment. Regul Toxicol Pharmacol 72(3):562–568
Gromping AHJ, Cammann K (1993) Development and comparison of different Fluorometric Hplc-methods with standard methods for the determination of formaldehyde in the atmosphere. Chromatographia 35(3–4):142–148
Gu DC, Zou MJ, Guo XX, Yu P, Lin ZW, Hu T, Wu YF, Liu Y, Gan JH, Sun SQ, Wang XC, Xu CH (2017) A rapid analytical and quantitative evaluation of formaldehyde in squid based on tri-step IR and partial least squares (PLS). Food Chem 15(229):458–463
He R (2016) Abnormal lysosome, formaldehyde Dysmetabolism and age-related cognitive impairment. Prog Biochem Biophys 43(12):1197
Hill AA, Lipert RJ, Fritz JS, Porter MD (2009) A rapid, simple method for determining formaldehyde in drinking water using colorimetric-solid phase extraction. Talanta 77(4):1405–1408
Houlgate PR, Dhingra KS, Nash SJ, Evans WH (1989) Determination of formaldehyde and acetaldehyde in mainstream cigarette-smoke by high-performance liquid-chromatography. Analyst 114(3):355–360
Hu X, Wang T, Jin F (2016) Alzheimer's disease and gut microbiota. Sci China Life Sci 59(10):1006
IARC (2004) Monographs on the evaluation of carcinogenic risks to humans. International Agency for Research on Cancer, Lyon
Iqbal MZ, Novalin S (2009) Analysis of formose sugar and formaldehyde by high-performance liquid chromatography. J Chromatogr A 1216(26):5116–5121
Ivanyi T, Heyden YV, Visky D, Hoogmartens LJ (2002) Minimal number of chromatographic test parameters for the characterisation of reversed-phase liquid chromatographic stationary phases. J Chromatogr A 954(1–2):99–114
Jeong HS, Chung H, Song SH, Kim CI, Lee JG, Kim YS (2015) Validation and determination of the contents of acetaldehyde and formaldehyde in foods. Toxicological Research 31(3):273–278
Jian X, Zhu MX (2016) Regulation of lysosomal ion homeostasis by channels and transporters. Sci China Life Sci 59(8):777–791
Jiang WZ, Adamec J, Weeks DP (2013) A small-scale, inexpensive method for detecting formaldehyde or methanol in biochemical reactions containing interfering substances. Anal Biochem 442(2):146–148
Kalasz H (2003) Biological role of formaldehyde, and cycles related to methylation, demethylation, and formaldehyde production. Mini Rev Med Chem 3(3):175–192
Karlberg AT, Skare L, Lindberg I, Nyhammar E (1998) A method for quantification of formaldehyde in the presence of formaldehyde donors in skin-care products. Contact Dermat 38(1):20–28
Kato S, Burke PJ, Koch TH, Bierbaum VM (2001) Formaldehyde in human cancer cells: detection by preconcentration-chemical ionization mass spectrometry. Anal Chem 73(13):2992–2997
Kawamura K, Kerman K, Fujihara M, Nagatani N, Hashiba T, Tamiya E (2005) Development of a novel hand-held formaldehyde gas sensor for the rapid detection of sick building syndrome. Sensors and Actuators B-Chemical 105(2):495–501
Khataee A, Lotfi R, Hasanzadeh A, Iranifam M (2016) A simple and sensitive flow injection method based on the catalytic activity of CdS quantum dots in an acidic permanganate chemiluminescence system for determination of formaldehyde in water and wastewater. Photochem Photobiol Sci 15(4):496–505
Kleinnijenhuis AJ, Staal YCM, Duistermaat E, Engel R, Woutersen RA (2013) The determination of exogenous formaldehyde in blood of rats during and after inhalation exposure. Food Chem Toxicol 52:105–112
Komazaki Y, Hiratsuka M, Narita Y, Tanaka S, Fujita T (1999) The development of an automated continuous measurement system for the monitoring of HCHO and CH3CHO in the atmosphere by using an annular diffusion scrubber coupled to HPLC. Fresen J Anal Chem 363(7):686–695
Largiuni O, Becagli S, Innocenti M, Stortini AM, Traversi R, Udisti R (2005) Formaldehyde determination in seawater. Preliminary application to coastal samples at Terra Nova Bay (Antarctica). J Environ Monit 7(12):1299–1304

Lefebvre MA, Meuling WJA, Engel R, Nohynek GJ (2012) Consumer inhalation exposure to formaldehyde from the use of personal care products/cosmetics. Regul Toxicol Pharmacol 63(1):171–176

Li B, Liu M, Zhang ZJ, Xu CL (2003) Flow-injection chemiluminescence determination of formaldehyde with a bromate-rhodamine 6G system. Anal Sci 19(12):1643–1646

Li JZ, Dasgupta PK, Luke W (2005) Measurement of gaseous and aqueous trace formaldehyde-revisiting the pentanedione reaction and field applications. Anal Chim Acta 531(1):51–68

Li X, Wang SS, Zhou R, Zhou B (2014) Urban atmospheric formaldehyde concentrations measured by a differential optical absorption spectroscopy method. Environmental Science-Processes & Impacts 16(2):291–297

Li T, Su T, He YG, Lu JH, Mo WC, Wei Y, He RQ (2016) Brain formaldehyde is related to water intake behavior. Aging Dis 7(5):561–584

Liu JF, Peng JF, Chi YG, Jiang GB (2005) Determination of formaldehyde in shiitake mushroom by ionic liquid-based liquid-phase microextraction coupled with liquid chromatography. Talanta 65(3):705–709

Liu KL, He YG, Yu LX, He RQ (2017) Elevated formaldehyde in the cecum of APP/PS1 mouse. Microbiol China. DOI:10.13344/j.microbiol.china.176008

Lorrain JM, Fortune CR, Dellinger B (1981) Sampling and ion chromatographic determination of formaldehyde and acetaldehyde. Anal Chem 53(8):1302–1305

Luo WH, Li H, Zhang Y, Wang CY (2001) Determination of formaldehyde in blood plasma by high-performance liquid chromatography with fluorescence detection. J Chromatogr B 753(2):253–257

Maboudou P, Mathieu D, Bachelet H, Wiart JF, Lhermitte M (2002) Detection of oxidative stress. Interest of GC-MS for malondialdehyde and formaldehyde monitoring. Biomed Chromatogr 16(3):199–202

Maneli MH, Mkentane K, Khumalo NP (2013) Lipid distribution and influence on hair structure. Int J Cosmet Sci 35(5):523–523

Marceaux JC, Dilks LS, Hixson S (2008) Neuropsychological effects of formaldehyde use. J Psychoactive Drugs 40(2):207–210

Mariscal A, Varo MC, Gomez E, Crehuet JF (2005) A fluorescence bioassay to detect residual formaldehyde from clinical materials sterilized with low-temperature steam and formaldehyde. Biologicals 33(3):191–196

Martos PA, Pawliszyn J (1998) Sampling and determination of formaldehyde using solid-phase microextraction with on-fiber derivatization. Anal Chem 70(11):2311–2320

Monakhova YB, Kuballa T, Mildau G, Kratz E, Wilhelm AK, Tschiersch C, Lachenmeier DW (2013) Formaldehyde in hair straightening products: rapid 1H NMR determination and risk assessment. Int J Cosmet Sci 35(2):201–206

Monkawa A, Gessei T, Takimoto Y, Sanari N (2015) Highly sensitive and rapid gas biosensor for formaldehyde based on an enzymatic cycling system. Sensors and Actuators B-Chemical 210:241–247

Motyka K, Mikuska P (2004) Continuous fluorescence determination of formaldehyde in air. Anal Chim Acta 518(1–2):51–57

Motyka K, Mikuska P, Vecera Z (2006) Continuous chemiluminescence determination of formaldehyde in air based on Trautz-Schorigin reaction. Anal Chim Acta 562(2):236–244

Motyka K, Onjia A, Mikuska P, Vecera Z (2007) Flow-injection chemiluminescence determination of formaldehyde in water. Talanta 71(2):900–905

Munoz MP, Rueda FJMD, Diez LMP (1989) Determination of formaldehyde in air by flow-injection using Pararosaniline and spectrophotometric detection. Analyst 114(11):1469–1471

Nengsih S, Umar AA, Salleh MM, Oyama M (2012) Detection of formaldehyde in water: a shape-effect on the Plasmonic sensing properties of the gold nanoparticles. Sensors 12(8):10309–10325

Perna RB, Bordini EJ, Deinzer-Lifrak M (2001) A case of claimed persistent neuropsychological sequelae of chronic formaldehyde exposure - clinical, psychometric, and functional findings. Arch Clin Neuropsychol 16(1):33–44

Possanzini M, DiPalo V (1997) Determination of formaldehyde and acetaldehyde in air by HPLC with fluorescence detection. Chromatographia 46(5–6):235–240

Qiang M, Xiao R, Su T, Wu BB, Tong ZQ, Liu Y, He RQ (2014) A novel mechanism for endogenous formaldehyde elevation in SAMP8 mouse. J Alzheimers Dis 40(4):1039–1053
Quesenberry MS, Lee YC (1996) A rapid formaldehyde assay using purpald reagent: application under periodation conditions. Anal Biochem 234(1):50–55
Rivero RT, Topiwala V (2004) Quantitative determination of formaldehyde in cosmetics using a combined solid-phase microextraction-isotope dilution mass spectrometry method. J Chromatogr A 1029(1–2):217–222
Rozylo TK, Siembida R, Memeth ZI, Tyihak E (2000) HPLC-OPLC-MS investigation of change of formaldehyde and its generators in human teeth of different physiological stage. Biomed Chromatogr 14(3):173–179
Sardi E, Tyihak E (1994) Simple determination of formaldehyde in Dimedone adduct form in biological samples by high-performance liquid-chromatography. Biomed Chromatogr 8(6):313–314
Sassine M, Varrault BP, Perraudin E, Chiappini L, Doussin JF, George C (2014) A new device for formaldehyde and total aldehydes real-time monitoring. Environ Sci Pollut Res Int 21(2):1258–1269
Sawicki E, Hauser TR, Stanley TW, Elbert W (1961) The 3-methyl-2-benzothiazolone Hydrazone test. Anal Chem 33(1):93–96
Septon JC, Ku JC (1982) Workplace air sampling and Polarographic-determination of formaldehyde. Am Ind Hyg Assoc J 43(11):845–852
Shrivastaw KP, Singh S (1995) A new method for spectrophotometric determination of formaldehyde in biologicals. Biologicals 23(1):47–53
Soman A, Qiu Y, Li QC (2008) HPLC-UV method development and validation for the determination of low level formaldehyde in a drug substance. J Chromatogr Sci 46(6):461–465
Spanel P, Smith D, Holland TA, Singary WA, Elder JB (1999) Analysis of formaldehyde in the headspace of urine from bladder and prostate cancer patients using selected ion flow tube mass spectrometry. Rapid Commun Mass Spectrom 13(14):1354–1359
Sritharathikhun P, Oshima M, Motomizu S (2005) On-line collection/concentration of trace amounts of formaldehyde in air with chromatomembrane cell and its sensitive determination by flow injection technique coupled with spectrophotometric and fluorometric detection. Talanta 67(5):1014–1022
Su T, Wei Y, He RQ (2011) Assay of brain endogenous formaldehyde with 2,4-Dinitrophenylhydrazine through UV-HPLC. Prog Biochem Biophys 38(12):1171–1177
Szarvas T, Szatloczky E, Volford J, Trezl L, Tyihak E, Rusznak I (1986) Determination of endogenous formaldehyde level in human-blood and urine by Dimedone-C-14 radiometric method. Journal of Radioanalytical and Nuclear Chemistry-Letters 106(6):357–367
Takigawa A, Yokoyama M, Tachibana S, Nagahara H, Ozawa K (2007) Anisotropic evaporation of forsterite in protoplanetary discs and its effect on infrared spectra. Meteorit Planet Sci 42:147–147
Tang XJ, Bai Y, Duong A, Smith MT, Li LY, Zhang LP (2009) Formaldehyde in China: production, consumption, exposure levels, and health effects. Environ Int 35(8):1210–1224
Tang YF, Chen H, Weng C, Tang XH, Zhang ML, Yang QQ, Hu T, Cai CQ (2014) The catalytic kinetic method for the determination of trace formaldehyde (FA) base on a bromate-eosin Y system. Spectrochimica Acta Part a-Molecular and Biomolecular Spectroscopy 125:126–130
Tang YF, Chen H, Weng C, Tang XH, Zhang ML, Hu T (2015) Determination of trace amount of formaldehyde base on a bromate-malachite green system. Spectrochimica Acta Part a-Molecular and Biomolecular Spectroscopy 135:506–510
Tang YH, Kong XQ, Xu A, Dong BL, Lin WY (2016a) Development of a two-photon fluorescent probe for imaging of endogenous formaldehyde in living tissues. Angewandte Chemie-International Edition 55(10):3356–3359
Tang YH, Kong XQ, Liu ZR, Xu A, Liu WY (2016b) Lysosome-targeted turn-on fluorescent probe for endogenous formaldehyde in living cells. Anal Chem 88(19):9359–9363
Tang X, Raskin JP, Lahem D, Krumpmann A, Decroly A, Debliquy M (2017) A formaldehyde sensor based on molecularly-imprinted polymer on a TiO nanotube Array. Sensors 17(4):675

Teshima N, Fernandez MSK, Ueda M, Nakai H, Sakai T (2011) Flow injection spectrophotometric determination of formaldehyde based on its condensation with hydroxylamine and subsequent redox reaction with iron(III)-ferrozine complex. Talanta 84(5):1205–1208

Tian SQ, Ding XH, Zeng DW, Wu JJ, Zhang SP, Xie CS (2013) A low temperature gas sensor based on Pd-functionalized mesoporous SnO2 fibers for detecting trace formaldehyde. RSC Adv 3(29):11823–11831

Toda K, Yunoki S, Yanaga A, Takeuchi M, Ohira SI, Dasgupta PK (2014) Formaldehyde content of atmospheric aerosol. Environ Sci Technol 48(12):6636–6643

Tomcik P, Jencusova P, Krajcikova M, Bustin D, Brescher R (2005) The detection of formaldehyde in textiles using interdigitated microelectrode array diffusion layer titration with electrogenerated hypobromite. Anal Bioanal Chem 383(5):864–868

Tong ZQ, Zhang JL, Luo WH, Wang WS, Li FX, Li H, Luo HJ, Lu J, Zhou JN, Wan Y, He RQ (2009) Urine formaldehyde level is inversely correlated to mini mental state examination scores in senile dementia. Neurobiol Aging 32(1):31–41

Trenholm RA, Rosario-Ortiz FL, Snyder SA (2008) Analysis of formaldehyde formation in wastewater using on-fiber derivatization-solid-phase microextraction-gas chromatography-mass spectrometry. J Chromatogr A 1210(1):25–29

Ueta I, Mochizuki S, Kawakubo S, Kuwabara T, Jinno K, Saito Y (2015) A novel miniaturized extraction capillary for determining gaseous formaldehyde by high-performance liquid chromatography. Anal Bioanal Chem 407(3):899–905

Wahed P, Razzaq MA, Dharmapuri S, Corrales M (2016) Determination of formaldehyde in food and feed by an in-house validated HPLC method. Food Chem 202:476–483

Wang S, Cui XJ, Fang GZ (2007) Rapid determination of formaldehyde and sulfur dioxide in food products and Chinese herbals. Food Chem 103(4):1487–1493

Wang YS, Tan X, Xue JH, Li GR, Shi LF, Yang HM, Liu L, Zhou B, Xiao XL (2011) Determination of trace formaldehyde in blood plasma by resonance fluorescence technology. Anal Chim Acta 690(2):234–239

Wang H, Ding J, Du XB, Sun X, Chen LG, Zeng QL, Xu Y, Zhang XP, Zhao Q, Ding L (2012a) Determination of formaldehyde in fruit juice based on magnetic strong cation-exchange resin modified with 2,4-dinitrophenylhydrazine. Food Chem 131(1):380–385

Wang XF, Cui FH, Lin JY, Ding B, Yu JY, Al-Deyab SS. (2012b) Functionalized nanoporous TiO2 fibers on quartz crystal microbalance platform for formaldehyde sensor. Sensors and Actuators B-Chemical 171:658–665

Wang T, Gao XL, Tong J, Chen LG (2012c) Determination of formaldehyde in beer based on cloud point extraction using 2,4-dinitrophenylhydrazine as derivative reagent. Food Chem 131(4):1577–1582

Wei YJ, Liu SH, Liu FJ, Liu J, Zhu Z, Li GL (2010) Direct measurement of formaldehyde and methanol emissions from gasohol engine via pulsed discharge helium ionization detector. Fuel 89(9):2179–2184

Weng X, Chon CH, Jiang H, Li DQ (2009) Rapid detection of formaldehyde concentration in food on a polydimethylsiloxane (PDMS) microfluidic chip. Food Chem 114(3):1079–1082

Wong KF, Deng JR, Wei XQ, Shao SP, Xiang DP, Wong MK (2015) Visual detection of formaldehyde by highly selective fluorophore labeling via gold(III) complex-mediated three-component coupling reaction. Org Biomol Chem 13(27):7408–7411

World Health Organization (WHO) (2001) Air quality guidelines, 2nd edn. WHO regional publications, European series NO 91. WHO Regional Office for Europe, Copenhagen, pp 97–91

Xu JC, Zhang Y, Zeng LT, Liu JB, Kinsella JM, Sheng RL (2016) A simple naphthalene-based fluorescent probe for high selective detection of formaldehyde in toffees and HeLa cells via aza-cope reaction. Talanta 160:645–652

Yanaga A, Hozumi N, IchiOhira S, Hasegawa A, Toda K (2015) Formaldehyde vapor produced from hexamethylenetetramine and pesticide: simultaneous monitoring of formaldehyde and ozone in chamber experiments by flow-based hybrid micro-gas analyzer. Talanta 148:649–654

Yang MQ, He JH (2016) Graphene oxide as quartz crystal microbalance sensing layers for detection of formaldehyde. Sensors and Actuators B-Chemical 228:486–490

Yang X, Wang Y, Liu W, Zhang Y, Zheng F, Wang S, Zhang D, Wang J (2015) A portable system for on-site quantification of formaldehyde in air based on G-quadruplex halves coupled with a smartphone reader. Biosens Bioelectron 75:48–54

Yang M, Ospina M, Tse C, Toth S, Caudill SP, Vesper HW (2017) Ultraperformance liquid chromatography tandem mass spectrometry method to determine formaldehyde hemoglobin adducts in humans as biomarker for formaldehyde exposure. Chem Res Toxicol 30(8):1592–1598

Yeh TS, Lin TC, Chen CC, Wen HM (2013) Analysis of free and bound formaldehyde in squid and squid products by gas chromatography-mass spectrometry. J Food Drug Anal 21(2):190–197

Yu PH, Cauglin C, Wempe K, Gubisne-Haberle D (2003) A novel sensitive high-performance liquid chromatography/electrochemical procedure for measuring formaldehyde produced from oxidative deamination of methylamine and in biological samples. Anal Biochem 318(2):285–290

Yu Y, Zhang XM, Ma JP, Liu QK, Wang P, Dong YB (2014a) Cu(I)-MOF: naked-eye colorimetric sensor for humidity and formaldehyde in single-crystal-to-single- crystal fashion. Chem Commun 50(12):1444–1446

Yu J, Su T, Zhou T, He YG, Lu J, Li J, He RQ (2014b) Uric formaldehyde levels are negatively correlated with cognitive abilities in healthy older adults. Neurosci Bull 30(2):172–184

Yue XP, Zhang YY, Xing W, Chen YT, Ge PC, Li TT, He RQ, Tong ZQ (2017) A sensitive and rapid method for detecting formaldehyde in brain tissues. Anal Cell Pathol. Accepted on 5 July, (ID 9043134)

Zali S, Jalali F, Es-haghi A, Shamsipur M (2013) Determination of free formaldehyde in vaccines and biological samples using solid-phase microextraction coupled to GC-MS. J Sep Sci 36(24):3883–3888

Zhang SP, Xie CS, Bai ZK, Hu ML, Li HY, Zeng DW (2009) Spoiling and formaldehyde-containing detections in octopus with an E-nose. Food Chem 113(4):1346–1350

Zhang ZM, Zhao C, Ma YJ, Li GK (2014) Rapid analysis of trace volatile formaldehyde in aquatic products by derivatization reaction-based surface enhanced Raman spectroscopy. Analyst 139(14):3614–3621

Zhao C, Li M, Jiao K (2006) Determination of formaldehyde by staircase voltammetry based on its electrocatalytic oxidation at a nickel electrode. J Anal Chem 61(12):1204–1208

Zilberstein G, Zilberstein R, Zilberstein S, Maor U, Baskin E, Zhang S, Righetti PG (2017) A miniaturized sensor for detection of formaldehyde fumes. Electrophoresis 38(17):2168–2174

Glossary

A

AChE	acetylcholinesterase
AD	Alzheimer's disease
ADH	alcohol dehydrogenase
ADH1	alcohol dehydrogenase 1
ADH3	alcohol dehydrogenase 3
ADH5	alcohol dehydrogenase 5
ADI	acceptable daily intake
AFM	atomic force microscope
AGEs	advanced glycation end products
ALA	α-linolenic acid
ALD	aldosterone
ALDH2	aldehyde dehydrogenase 2
ALKBH5	alkB homolog 5
ALS	amyotrophic lateral sclerosis
AMI	acute myocardial infarction
APP	amyloid precursor protein
ANG II	angiotensin II
ANS	8-anilino-1-naphthalenesulfonic acid
AO	acridine orange
AOF	amine oxidase (flavin-containing)
APP	amyloid precursor protein
ARB	Air Resources Board
ASTM	American Society for Testing Materials
ATP	adenosine triphosphate
AVP	arginine vasopressin
Aβ	amyloid β

R. He, *Formaldehyde and Cognition*, DOI 10.1007/978-94-024-1177-5

B

BBB	blood-brain barrier
BDNF	brain-derived neurotropic factor
BER	base excision repair
BHC	BRAF35 HDAC complex protein
BSA	bovine serum albumin

C

CaMKII	Ca^{2+}/calmodulin-dependent protein kinase II
CAN	China Nutrition Association
CAPE	caffeic acid phenethyl ester
CAT	catalase
CBS	cystathionine beta-synthase
CDK5	cyclin-dependent kinase 5
CIG	cornel iridoid glycoside
CKD	chronic kidney disease
CMCs	Chinese medicine clinics
CNS	central nervous system
CO_2	carbon dioxide
CORT	corticosterone
CpG islands	CpG dinucleotide
CR	caloric restriction
CSF	cerebrospinal fluid
5caC	5′-carboxylcytosine

D

DA	dopamine
DDL	3,5-diacetyl-1,4-dihydrolutidine
DHA	docosahexaenoic acid
DNMT	DNA methyltransferase
DNPH	2,4-dinitrophenylhydrazine
DOAS	differential optical absorption spectroscopy
DPCs	DNA–protein cross-links
DPX	DNA–protein cross-links
DR	dietary restriction
DSS	disuccinimidyl suberate
Dyrk1A	dual specificity tyrosine-phosphorylation-regulated kinase 1A

E

ECE	early children education
EFE-II	earthworm fibrinolytic enzyme-II
EFSA	European Food Safety Authority
EGFR	epidermal growth factor receptor
ELISA	enzyme-linked immunosorbent assay
EM	electron microscopy
EMSA	electrophoretic mobility shift assay
ENS	enteric nervous system
EPA	eicosapentaenoic acid
EPA	Environmental Protection Agency
EPM	elevated plus-maze test
ER	endoplasmic reticulum
ERK	extracellular regulated protein kinases
EXPOLIS	Air Pollution Exposure of Adult Urban Populations in Europe

F

F6P	fructofuranose-6-phosphate
FA	formaldehyde
FAD	formaldehyde dehydrogenase (homolog of mammalian GSNOR)
FAD	flavin adenine dinucleotide
FAE	formaldehyde-activating enzyme
FAP-1	formaldehyde probe-1
FBXL	F-box and leucine-rich repeat
FDH	formaldehyde dehydrogenase
fEPSP	field excitatory postsynaptic potential
fMRI	functional magnetic resonance imaging
FST	forced swimming tests
FTO	fat mass and obesity associated
5fC	5-formylcytosine

G

GABA	gamma-aminobutyric acid
GAPDH	D-glyceraldehyde-3-phosphate dehydrogenase
GC	gas chromatography
GC-MS	gas chromatography–mass spectrometer
GDH	glutamic dehydrogenase
GdnHCl	guanidine-HCl

Gfa	glutathione-dependent formaldehyde-activating enzyme/S-(hydroxymethyl)glutathione synthase
Gly	glycine
GRs	glucocorticoid receptors
GSCH2OH	S-(hydroxymethyl)glutathione
GSCHO	S-formylglutathione
GSH	glutathione
GSH-FDH	glutathione-dependent formaldehyde dehydrogenase
GSH-Px	glutathione peroxidase
GSHRx	GSH reductase
GSHST	GSH S-transferase
GSK-3β	glycogen synthase kinase 3β
GSNO	S-nitrosoglutathione
GSNOR	S-nitrosoglutathione reductase

H

H_2O_2	hydrogen peroxide
H_2S	hydrogen sulfide
H4MPT	tetrahydromethanopterin
HCHO	formaldehyde
HEXPOC	human exposure characterization of chemical substances
HFS	high-frequency stimulation
i.c.v.	intracerebroventricular injection
i.p.	intraperitoneal injection
hm6A	N6-hydroxymethyladenosine
HPA	hypothalamo–pituitary–adrenal gland
HPLC	high-performance liquid chromatography
HPS	3-hexulose-6-P synthase
Hu6P	hexulose-6-phosphate
5hmC	5-hydroxymethylcytosine
5-HT	5-hydroxltryptamine

I

IARC	International Agency for Research on Cancer
IRIS	Integrated Risk Information System
IURs	inhalation unit risk values

J

JARID jumonji, AT-rich interaction domain
JHDM jmjC domain-containing H3K9 demethylase
JMJD jumonji domain-containing proteins

K

KDMs lysine demethylases

L

LC/MS liquid chromatography and electrospray ionization mass spectrometry
LDH lactate dehydrogenase
lncRNAs long noncoding RNAs
LOAD late-onset AD
LSD1 lysine-specific demethylase 1
LTD long-term depression
LTM long-term memory
LTP long-term potentiation

M

MAO-A type A monoamine oxidase
MBD microtubule-binding domain
MCI mild cognitive impairment
MDA malondialdehyde
MDA methane dicarboxylic aldehyde
MDR medium-chain dehydrogenase/reductase
me1 monomethylated
me2 dimethylated
me3 trimethylated
5mC 5-methylcytosine
MFA metaformaldehyde
MMSE Mini-Mental State Examination
MoCA Montreal Cognitive Assessment
MPO myeloperoxidase

MRLs minimal risk levels
MS mass spectrometry
MT melatonin

N

N2a Neuro 2a
N5 10CH2=H4MPT, 5,10-methylene-tetrahydromethanopterin
N5 10CH2=THF, 5,10-methylenetetrahydropteroyl mono-L-glutamate
NAC N-acetyl cysteine
NBP DL-3-n-butylphthalide
NCLK neuronal cdc2-like protein kinase
NE norepinephrine
NER nucleotide excision repair
NFTs neurofibrillary tangles
NIOSH the National Institute for Occupational Safety and Health
NLMS National Longitudinal Mortality Study
NMDA N-methyl-D-aspartic acid
NMDAR N-methyl-D-aspartate receptor
NMR nuclear magnetic resonance
NOS nitric oxide synthase
NR1 N-methyl-D-aspartate, subtype 1
NR2B N-methyl-D-aspartate, subtype 2B

O

OEHHA Office of Environmental Health Hazard Assessment
OFT open-field tests

P

PD Parkinson's disease
PET photoinduced electron transfer
PFA paraformaldehyde
PHFs paired helical filaments
PHI 6-phospho-3-hexuloisomerase
PI3K phosphatidylinositol-3 kinase
PKG cGMP-dependent protein kinase
POCD postoperative cognitive dysfunction

PP2A phosphoseryl/phosphothreonyl protein phosphatase-2A
PP2A protein phosphatase-2A
ppb part per billion
ppm parts per million
PRD proline-rich domain
PRP Poria cum Radix Pini
PS1 presenilin-1
PSD poststroke dementia

Q

QCM quartz crystal microbalance

R

RAT Rhizoma Acori Tatarinowii
RBP radiation-induced brachial plexopathy
REL reference exposure level
Res resveratrol
RNA ribonucleic acid
ROS reactive oxygen species
RP Radix Polygalae
rRNA ribosomal RNA
Ru5P D-ribulose-5-phosphate
RuMP ribulose monophosphate

S

SAM S-adenosyl-L-methionine
SAM senescence-accelerated mouse
SAMP8 senescence-accelerated mouse-prone 8
SAMR 1 senescence-accelerated-resistant mice 1
SBS sick building syndrome
SD Sprague-Dawley
Ser serine
SHMT serine hydroxymethyltransferase
SNAP25 synaptosomal-associated protein 25
snRNA small nuclear RNA

SOD superoxide dismutase
SPs senile plaques
SSAO semicarbazide-sensitive amine oxidase
STM short-term memory

T

T2DM type 2 diabetes mellitus
TCM traditional Chinese medicine
TET ten-eleven translocated oncogene
TH tyrosine hydroxylase
THF tetrahydrofolate
THF tetrahydrofolic acid
THF/H4F tetrahydropteroyl mono-L-glutamate
ThS thioflavin S
ThT thioflavin T
TiO_2-NTA titanium dioxide nanotube array
TMAO trimethylamine oxide
TMAO-ase TMAO demethylaseTRPA1transient receptor potential cation channel, subfamily A, member 1
TSG 2,3,5,4′-tetrahydroxystilbene-2-O-β-D-glucoside
TUR toxics use reduction

U

UFFI urea formaldehyde foam insulation
UTX ubiquitously transcribed tetratricopeptide repeat

V

VD vascular dementia
VE vitamin E
VOCs volatile hydrocarbons
VSDRT variable spatial delayed response task

W

WHO World Health Organization

X

X X chromosome

Index

A
A375P, 81, 83, 215
A549 cell line, 233
Aβ_{1-40}, 37, 181, 182
A-β40, 230
Aberrant modification, 38
Aβ plaque, 113
Abrasive material, 23, 272
Abundant natural sources, 2
Acceptable daily intake (ADI), 272, 282
Acceptor, 29
Accumulation, 3, 10, 27, 31, 34, 36, 39, 49, 52, 74, 80, 82, 84, 85, 87, 89, 90, 103, 106, 110, 133, 135, 145, 148, 154, 168, 169, 174, 175, 177–180, 182, 200, 201, 222, 231, 234, 235, 246, 254, 255, 257, 262, 264, 287
Acetaldehyde, 111, 130, 131, 133, 169, 260
Acetylacetone, 275–277, 282, 286
Acetylacetone solution, 282
Acetylcholine (Ach), 72, 112, 199
Acori tatarinowii rhizoma, 224, 225
Acrolein, 31, 289
Active demethylation, 49
Acute myocardial infarction (AMI), 31
Adhesives, 4–6, 23, 155, 178, 272
Administration, 67, 68, 70, 71, 108, 111, 112, 136, 152, 153, 161, 173–175, 182, 194, 200, 210, 225, 226, 234, 246, 247, 258, 262–265
AD patients, 53, 55, 85, 103, 107–110, 136, 150–152, 160, 161, 174, 178, 179, 193, 198, 223, 224, 250, 252, 261, 265, 285
Adrenaline, 30, 130
Adrenocorticosteroid, 194
Adsorption voltammetry, 273, 282
Advanced glycation end products (AGEs), 169, 262
Aerospace syndrome, 156
Aetiology, 169, 200
Ageing, 200
Age-matched controls, 259, 285
Age-related cognitive impairments, 22, 48, 55, 65, 70, 74, 105, 106, 108–112, 130, 133, 136, 144, 148, 150, 152, 154, 157, 158, 161, 174, 177, 181, 182, 192, 194, 223–225, 264, 286
Aggregate formation, 37
Aggregation, 25, 30, 31, 67, 108, 121–137, 144, 150, 154, 168, 169, 171, 174, 175, 179, 181, 182, 247, 259
Aggression, 71, 192, 195
Aging, 39, 49, 52, 54–56, 64–67, 69, 70, 74, 80, 90, 103, 106, 110, 113, 114, 135, 136, 144–150, 152–154, 158, 161, 169, 174, 178, 182, 209–211, 214, 216, 222, 224, 225, 235, 236, 246, 252, 253
Aging process, 148, 209
Ag-modified In_2O_3/ZnO, 280
Air, 3–9, 12, 23, 80, 84, 87, 156, 157, 192, 196, 201–203, 222, 272–276, 278–280, 282, 284
Akkermansia, 108
AKT, 200, 227, 236
Alcohol, 3, 9, 10, 22, 24, 25, 34, 35, 111, 112, 153, 155, 160, 173, 199, 200, 211, 235, 254, 260

R. He, *Formaldehyde and Cognition*, DOI 10.1007/978-94-024-1177-5

Alcohol dehydrogenases (ADH), 3, 52, 211, 213, 235
Alcohol dehydrogenases 1 (ADH1), 22, 25, 260
Alcohol dehydrogenases 3 (ADH3), 24, 34, 35, 70, 86, 153, 154, 199, 254, 260, 264
Aldehyde dehydrogenase 2 (ALDH2), 34, 260
Aldehyde dehydrogenase 5 (ADH5), 82, 155
Aldehyde group, 27, 28
Aldehydes, 2, 24, 27, 28, 31, 34, 36, 73, 82, 100, 109, 112, 130–132, 136, 169, 199, 211, 255, 260, 261, 282, 287, 289
Algesthesia, 197–198
Alismatis rhizoma, 225
AlkB, 23, 50
AlkB homolog 5 (ALKBH5), 50
Allergic rhinitis, 199
Alois Alzheimer, 168
Alumina, 282, 284
Alzheimer's disease (AD), 25, 52, 67, 85, 193, 222, 231, 250, 252–254, 258, 259, 261, 262, 264–266
American Society for Testing Materials (ASTM), 283
ε-Amine group, 51
Amine oxidase, 22, 24, 51, 70, 86, 130, 153, 200, 236, 261
Amino acid, 12, 25, 31, 35, 36, 54, 82, 122, 123, 247, 288
α-Amino acids, 31
6-(1-Aminobut-3-en-1-yl) naphthalen-2-ol, 281
4-Amino hydrazine-5-mercapto-1,2,4-triazole (AHMT), 276
Ammonia, 30, 275, 278
Ammonium acetate, 279, 280, 286
Ampicillin, 107, 150, 274, 284, 285
β-Amyloid (Aβ), 108, 168, 181, 258
Amyloid β deposition, 258
β-Amyloid precursor protein (APP), 33, 53, 69, 109–111, 115, 152, 161, 181, 229, 251, 258, 259, 285
Amyotrophic lateral sclerosis (ALS), 198
Anabolism, 64, 71, 161, 265
Anaphylaxis, 192
An energy source in some yeast, 12
Anesthesia, 87, 149, 169, 255
Anesthesiologists, 148
Animal models, 65, 66, 69, 133, 169, 195, 198, 225, 230, 246, 252–264
Anthropogenic sources, 2, 4
Antibiotics, 134
Anti-apoptosis, 88, 227
Anti-inflammatory, 115, 227, 231, 236
Antioxidants, 34, 202, 227, 229, 231–234, 236, 238
Anti-oxidative, 115, 227
Antiseptics, 4
Antitumor drugs, 80
Anxiety, 68, 72, 107, 113, 114, 192, 193, 195, 196, 200–203, 255
Anxiety- and depression-like behaviours, 113
AP-1, 229
Apium graveolens, 226
APOE4, 181, 182
Apolipoprotein E, 181
Apoptosis, 67, 80–82, 85–92, 108, 133, 136, 171, 197, 215, 227, 229, 233, 234, 236, 252
APP/PS1 transgenic mice, 109–111, 115, 251, 259
APP/PS1-transgenic mouse, 152
APPV717I transgenic mice, 225, 230
Arginine, 36, 51, 71, 169
Artificial board, 4
Artificial carpets, 4
Artificial clothing, 4
Artificial the wall, 4
Ascitic fluid, 84
Asparagine, 25, 34, 36
Aspartame, 25, 34
Assemble, 124
Asthma, 10, 12, 234
Astrocytes, 108, 144, 236
Atherosclerosis, 36, 131, 261
Atrophy, 147, 157, 170–173
Attention, 11, 53, 67, 74, 90, 145, 158, 159, 192, 197, 198, 222, 235, 251, 258, 264, 265, 272, 286
Autism, 199
Autophagy, 89, 92
Axon, 168, 171, 177, 199, 230
Axon growth, 230
5-Azacytidine, 49

B

Bacillus subtilis, 101
Bacteria, 3, 12, 35, 100–108, 110, 112, 114, 115, 123, 134, 233, 251, 272
Balance, 50, 64, 66, 88, 154, 222, 231, 265
Bamboo shoot, 276
Base excision repair (BER), 49, 50
Bcl-2, 85, 87, 88, 227
BDNF–TrkB pathway, 88, 235

Beads-on-strings, 126, 177
Bean curd, 276
Beed.3 cells, 32, 287, 288
Behaviour, 192–200, 202
Berberidaceae, 227
Bifidobacterium dentium, 102, 105, 107
Bifidobacterium longum, 107
Biochemical compounds, 26–30, 246
Biological decomposition, 2
Biological metabolism, 12
Biological processes, 12, 33, 50, 169, 197
Biological rhythm, 193, 194
Biomacromolecules, 34, 36–38, 48, 135, 249
Biomarker, 27, 32, 69, 89, 123, 149, 150, 157, 158, 174, 200, 203
Bladder cancer, 84, 274, 284
Blood, 8, 11, 12, 25, 27, 28, 30, 52, 64, 86, 101, 113, 135, 144, 146, 153, 154, 156, 157, 174, 177, 201, 202, 210, 223–227, 236, 249, 255, 261, 262, 273, 274, 277, 279, 284, 285
Blood–brain barrier (BBB), 52, 101, 109, 153, 174, 177, 201, 255
Blood endothelial cell line (bEnd.3), 135, 249
Body mass index, 50
Body patterning, 154
Body weight, 9, 64, 66, 68, 84, 86, 156, 257, 272, 282
Brain, 10, 30–33, 35, 48, 66–74, 80, 84–87, 100, 122, 144, 168, 193–195, 200–202, 222, 246, 252–255, 257–259, 261, 263–265, 272, 274, 283–285, 287
  damage, 52, 87, 174
  lesion, 158, 170, 174
  stem, 252, 284
  tissues, 10, 210, 252, 253, 257, 274, 283, 284
Brain-derived neurotropic factor (BDNF), 88, 230, 235
Brain-patterning, 49
Breast cancer, 84
Breathing, 8, 23, 198
Building materials, 4–6, 144, 272
Bushen capsule, 223, 225, 226

C

C57 BL/6 mice, 33, 65, 68, 152, 252, 260, 262
C57 BL/6j wild-type mice, 68, 71, 263
C57 BL/6j wild-type mouse, 67, 109, 115, 161, 285
Ca^{2+} releasing, 92
Caenorhabditis elegans, 211, 215
Caffeic acid phenethyl ester (CAPE), 236, 237
Calmodulin-dependent protein kinase II (CaMKII), 175, 176, 182, 183, 229
Caloric restriction (CR), 214
Calorie, 214
Cancerigenic molecule, 34
Cancers, 10–12, 38, 51, 80–82, 84, 85, 87, 90, 92, 156, 169, 192, 198, 200, 202, 229, 249, 260, 284, 289
Candle smoke, 273, 282
Carbohydrate, 124, 288
Carbon dioxide (CO_2), 2, 10, 34, 38, 248, 249
Carbon-hydrogen bond, 29
Carbon–nitrogen bonds, 51
5′-Carboxylcytosine (5caC), 49
Carcinogen, 10, 11, 30, 80, 82, 90, 192, 272, 288
Carcinogenic effects, 2
Carcinogenic risk, 80
Carcinoma cells MRMT-1, 249
Carcinomas, 81–83, 90, 215, 249
Cardiovascular diseases, 247
Cardiovascular/neurodegenerative diseases, 27, 247
Carpet cleaners, 4
Caspase, 87, 88, 133
Caspase 3, 87, 88, 133, 227
Caspase 9, 88, 227
Cassia obtusifolia (CO), 231
Catalase, 23, 25, 234
Catalytic activity, 176, 213
Cecum, 109–111, 115, 152, 161, 251, 285
Cecum digestion contents, 109, 115, 285
Cell apoptosis, 67, 108, 136, 171, 229
Cell culture, 81, 86, 90, 133, 173, 178, 179, 215, 248–251
Cell cycle arrest, 87
Cell death, 91, 92, 131, 133, 134, 137, 150, 169, 171, 178, 182, 200, 227, 231, 252
Cell microenvironment, 144
Cell morphology, 82, 178
Cell proliferation, 91, 92, 111, 171, 215, 216
Cell senescence, 89, 92
Cellular and animal models, 246, 265
Cellular formaldehyde, 134–136, 144, 148
Cellular skeleton, 125, 171
Central dogma, 50
Central nervous system (CNS), 10, 12, 22, 52, 65, 87, 100, 109, 154, 157, 192, 199, 202, 222, 225, 234, 235, 258
Centrifugation, 283
Cerebellum, 234, 252, 284
Cerebrospinal fluid (CSF), 28

Cerebrovascular disorders, 200
cGMP-dependent protein kinase (PKG), 154
Characteristic pungent, 2
Characteristics, 2, 52, 53, 73, 123, 146, 147, 223, 225, 252, 262, 286
Chemical, 2, 3, 5, 8–10, 12, 25–30, 49, 64, 81, 100, 115, 122, 123, 126, 130, 148, 155, 170, 195, 198, 200–202, 210, 223, 227, 228, 230, 231, 235–238, 247, 248, 258, 274, 276, 282, 286
  reagents, 282
  synthesis, 12
Chemically exfoliated graphene flakes, 279
Chinese herbs, 223, 225, 227, 228, 230
Chinese medicine clinics (CMCs), 8
Chipboard, 6, 9
Chlamydophila pneumonia, 107, 108
Cholinergic neurons, 230
$^{13}C/^{1}H$ NMR spectroscopy, 282
Chromatin compaction, 51
Chromatin condensation, 85
Chromatin dynamics, 51
Chromatin structure, 49, 53
Chromosome, 11, 52, 84, 126, 211
  damage, 11
  stability, 48
Chronic accumulation, 39
Chronic dehydration, 66–68, 70, 71, 74, 112, 235, 264
Chronic inflammation, 157
Chronic latent neuroinflammation, 108
Circadian rhythm, 193, 194, 202, 233
Cistanches herba, 225
c-JNK pathway, 88
Clay, 4
Clinical, 10, 67, 69, 73, 110, 113, 122, 145, 148, 168, 178, 182, 192, 223, 225, 226, 235, 236, 265, 284, 286
Clinical trial, 113, 225, 235, 286
Cofactor, 29, 30, 35, 51, 102, 103
Cognition, 101, 103–108, 113, 114, 146, 148, 151, 158, 168, 173, 192, 195, 200, 222, 246, 252–255, 258, 260, 262, 264, 265
Cognitive ability, 48, 53, 66, 69, 70, 74, 112, 136, 144–161, 174, 182, 192, 226, 255, 259, 265
Cognitive change, 50, 80
Cognitive damage, 80
Cognitive dysfunction, 32, 51, 52, 66, 114, 148–150, 193, 199, 287
Cognitive impairment, 12, 22, 48, 64, 66–70, 73, 74, 87, 89, 101, 130, 144, 169, 192, 210, 223, 246, 252, 254, 262, 264, 272, 285–287, 289
Cognitive training, 148
Cohort study, 50
Colon carcinoma HT-29 cell, 81, 215
Common environment pollutant, 272
Concentration, 6–10, 12, 25, 56, 67–69, 71–74, 80, 82, 84–86, 90, 91, 108, 122, 144, 171, 192, 196, 198, 201, 202, 210–212, 214, 215, 227, 235, 236, 246, 249
Confect, 276
Confocal microscopy, 180, 249, 287
Consciousness, 158–160
Consciousness impairment, 158–160
Consecutive oxidation, 49
Construction materials, 6
Copulation, 215
Cornel iridoid glycoside (CIG), 230
Cosmetics, 4
CpG dinucleotide (CpG), 53
CpG islands, 48, 49
Cross-linking reaction, 12
Cross-linking reagent, 36
Cross-sectional survey, 157
Cultured cell, 49, 215
Curcumin, 228, 229
Curcuminoids, 229
Cyclin dependent kinase 5 (CDK5), 175, 176
Cynomolgus macaques, 86
CYP2E1, 260
Cystathionine beta-synthase (CBS), 234
Cysteine, 26, 36, 89, 169, 200, 231–233
Cysteine thiazolidination, 89
Cytochrome P-450 system, 22, 25, 29, 30
Cytoplasm, 26, 30, 32, 128, 129, 135, 171–173, 175, 183
Cytosine, 24, 33, 37, 48, 50, 52, 54
Cytoskeleton disassembly, 83
Cytotoxicity, 12, 25, 31, 67, 80, 82, 90, 108, 121–137, 144, 175, 229, 234, 252

D

Daf-16, 211
Daf-2, 211, 213
Daily life exposure, 5
Deamination, 30, 36, 49, 131, 181, 261
Death, 31, 32, 34, 38, 64, 91, 92, 131, 133, 134, 136, 150, 156, 168, 169, 171, 173, 178, 182, 193, 194, 198, 200, 202, 209, 210, 222, 227, 231, 258
Decoction, 225
Decoration materials, 6, 13
Defect, 55, 112, 144, 154, 199
Deferring senescence, 216

Deficiency, 71, 113, 153, 155, 156, 192, 223–225, 253, 260
Definition, 124, 158
Degradation, 22–24, 27, 38–40, 81, 82, 86, 127, 134, 135, 145, 153, 174, 213, 249
Deionized water, 283
Dementia, 52, 67, 69, 130, 136, 144, 145, 147, 148, 150, 153, 157, 158, 174, 175, 179, 181, 222, 224–227, 229, 230, 234–238, 257, 261
Demethylases, 23, 24, 33, 50, 51
Demethylating enzyme, 49
Demethylation, 12, 22, 33, 38, 48–56, 148, 210, 248, 284, 288
Demethylation of DNA, 12, 22, 49–51, 54–56, 210, 248
Demographic characteristics, 72, 146
Demographic features, 146
Denaturation of protein, 22
Deoxycytidine, 49, 54, 56
Deoxyribonucleic acid (DPCs), 38, 123, 229, 233, 237
Dephosphorylation, 123, 128, 171, 178, 182, 230
Deposition, 67, 68, 108, 109, 122, 130, 132–134, 150, 168–183, 202, 210, 246, 247, 252, 258, 259, 261, 262, 265
Depression-like behavior, 71, 113, 156, 255
Derivatives, 31, 34, 37, 100, 103, 282, 283, 285
Derivatization, 272, 282, 284
Derivatization procedure adapted to HPLC-MS/MS, 272, 282
Dermic route, 2
Detoxification, 34, 81, 101–103, 260
Development, 4, 31, 48, 49, 51, 53, 82, 100, 108, 115, 148, 154, 174, 181, 197, 222, 229, 235, 238, 259, 261, 262, 286
dFOXO, 211
Diabetes, 28, 30, 37, 50, 130, 131, 157, 200, 225, 234, 247, 261, 289
Diabetes mellitus, 247
3,5-Diacetyl-1,4-dihydrolutidine (DDL), 273, 282, 286
Diagnosis, 150, 161, 179, 201, 223, 225, 226
Dietary exposure, 4, 9
Dietary restriction (DR), 214, 215
Differential optical absorption spectroscopy, 279, 282
Digestion content, 109, 110, 115, 272, 285
Dimedone ^{14}C, 273, 284
Dimethylamine, 22, 26, 27
Dimethylaminoethanol, 22, 28, 29
Dimethylated, 51
Dimethylglycine, 22, 23, 28, 29
Dimethylglycine dehydrogenase, 23
2,4-Dinitrophenylhydrazine (DNPH), 27, 68, 109, 110, 150, 158, 196, 247, 248, 252, 253, 262, 273–275, 277–281, 283–286
2-Diphenylacetyl-l,3-indandione-l-hydrazone, 274
Diphtheria-tetanus-pertussis vaccine, 273
Diphtheria-tetanus vaccine, 273
Disassemble, 83, 170, 173
Dish-washing liquids, 4
Disinfectant, 2
Disturbances, 55, 80, 88, 113, 134, 168, 193, 195
DL-3-butylphthalide, 223, 226
DNA, 11, 12, 22, 48–56, 80, 82–87, 92, 123, 144, 168, 201, 210, 248
  adducts, 82, 83, 87, 92, 201
  damage, 80, 82, 83, 85, 87, 126, 150, 172, 177, 236
  demethylation, 49, 50, 54, 56, 84, 210
  double-strand break, 11, 87
  hypermethylation, 84
  methylation, 12, 48–50, 52–56, 148, 210, 262
  repair, 52, 82, 86, 87, 92
  replication, 49, 82, 127, 128
  strand breaks, 11, 87
  synthesis, 84
DNA methyltransferase (DNMT), 48, 49, 52–56, 261
DNA-protein cross-links (DPX), 38, 82, 84, 92
Docosahexaenoic acid (DHA), 232, 234
Donor, 25, 26, 54, 262
Dopamine (DA), 71, 72, 112–114, 157, 195, 197, 199, 200, 274
Dose-dependent, 83, 92, 229
Double-blinded experiment, 144, 150
Double-blinded manner, 285
Double strands, 87, 126, 127, 129, 132, 177
D-ribose, 28, 174, 247, 262
Drinking water, 9, 65, 68, 152, 173, 258, 259, 264, 281
Drosophila, 12, 154, 155, 209, 214, 257, 260
Drosophila melanogaster, 211, 257
Drug usage, 169, 202
Duodenum, 109, 111
Dyrk1A, 176
Dysmetabolism, 3, 28, 39, 48, 56, 64, 67–70, 100–115, 134–136, 144, 148–154, 158, 160, 169, 181, 201, 202, 223, 253, 261, 264

E

EARK1/2, 175
EcR, 213
Edema, 193, 198
Educational level, 146–148, 161
Effluents, 202
Eicosapentaenoic acid (EPA), 232, 234
Elderly participants, 146
Electric appliances, 5
Electrochemical approach, 282
Electrophilic carbon, 38
Electro-spinning/netting nano-fiber/nets, 278, 282
Elevated plus-maze tests (EPM), 71, 113, 194, 195
Elongator complex protein 3 and catalyzed 5mC demethylation, 49
Embalming fluid, 23, 272
Embryonic development, 48, 100
Embryonic stem cell, 48, 49
Emission rate, 6
Emission rate of formaldehyde, 6
Emotion, 68, 71, 101, 194–196, 200
Endogenous, 22–35, 37–39, 48–52, 54–56, 64, 69, 71–74, 80–82, 84, 86, 100, 103, 108, 109, 112, 114, 133–136, 144–155, 158, 160, 161, 171, 178, 179, 182, 193, 194, 196, 198–201, 203, 210, 222, 223, 231, 234, 235, 246, 253, 256, 259–262, 264, 265, 272, 283, 287, 288
Endogenous formaldehyde, 22–35, 37–39, 48–52, 54–56, 64, 69, 71–74, 80–82, 84, 86, 100, 103, 108, 109, 112, 114, 133–136, 144–155, 158, 160, 161, 171, 178, 179, 182, 193, 194, 196, 198–201, 203, 210, 223, 231, 234, 235, 246, 253, 259–262, 264, 265, 272, 283, 287, 288
Endogenous formaldehyde metabolism, 56, 103
Endoplasmic reticulum (ER), 81, 122, 135, 144, 234, 235
Endoplasmic reticulum (ER) stress, 85, 88, 92, 122, 234, 235
Enteric nervous system (ENS), 100
Enterica serovar Typhimurium, 33, 104–106, 108
Entorhinal cortex, 168
Environmental friendly materials, 7
Environmental protection agency (EPA), 22, 23, 210, 222, 272, 282
Enzyme, 22, 24, 26, 29, 30, 33–35, 48, 49, 52, 54, 71, 86, 101–106, 113, 115, 124, 130, 137, 153, 154, 157, 176, 195, 233, 234, 254, 260, 261
Enzyme-linked immunosorbent assay (ELISA), 72, 157
Epidemiological investigation, 69, 145, 156, 161, 182, 286
Epidermal growth factor receptor (EGFR), 83
Epigenetic changes, 82
Epigenetic drift, 210
Epigenetic modification, 48, 52, 53
Epigenetics, 48, 51–56, 82, 169, 210
Epilepsy, 199
Epimedii folium, 225
Epithelial cells, 8, 82, 88, 231
Epoxides, 199
Eppendorf tube, 285
Equilibrium, 36, 126, 144, 193, 272
Escherichia coli (*E.coli*), 104
Estrogen receptor, 229
Ethanol, 25, 29, 52, 111, 130, 169, 231, 235, 260, 261
Ethanol dehydrogenase, 52
Ethanolic fraction, 231
Ethyl 3-oxobutanoate, 275
Eukaryotes, 12, 50, 110, 134, 261
Examination, 11, 65, 66, 69, 104, 150, 161, 168, 172, 173, 178, 198, 200, 225
Excess, 8, 22, 32, 34, 36, 38, 39, 51, 54, 55, 74, 87, 88, 123, 136, 144, 154, 155, 173, 174, 196, 202, 233, 234, 246, 256, 257, 259, 261–263, 265, 287
Excitotoxicity, 85, 89, 92
Ex-GF mice, 113
Exhalation, 289
Exogenous formaldehyde, 34, 81, 84, 133, 144, 155–157, 200, 201, 255, 272, 283, 287, 288
Experimental model, 211
Explorative activity, 195
Exposure limit, 282
Exposure time, 8, 12, 86, 91
Extracellular regulated protein kinases (ERK), 83, 88, 92
Extra learning, 148
Extract, 28, 223, 225, 227, 230, 236, 251, 257, 277, 281, 283, 284, 286
Extragalactic, 3

F

Fabric softeners, 4
Factory, 4, 8, 12, 278
FAD-dependent amine oxidase family, 51
FANCD2, 82, 87
Fanconi anemia pathway, 87

Fat mass and obesity-associated gene (FTO), 50
Fatigue, 192, 195, 196, 265
Fe^{2+}-dependent AlkB family, 50
Fe^{2+}-dependent enzyme, 49
Fixative, 2, 122
Flavin adenine dinucleotide (FAD), 29, 51, 53, 154
Flavin-containing protein, 51
Flow injection analysis (FIA), 274, 282
Flow injection chemiluminescence, 275, 281
Fluorescence, 129, 130, 172, 177, 180, 249, 250, 275, 282, 284, 285, 287, 288
Fluorescent alkyne, 280
Fluorescent molecular probe, 272
Fluorescent probe, 32, 134, 135, 144, 281, 282, 286–288
Fluorescent techniques, 272, 282
Folate, 174
Folic acid, 262–264
Food ingredient, 169
Food preservative, 4, 272
Food restriction, 213
Forced swimming tests (FST), 71, 113, 194, 195
Formaldehyde (FA), 2–13, 23, 24, 27, 32, 33, 39, 40, 48–56, 64–74, 80–92, 100–115, 121–137, 144–161, 168–183, 192–203, 209–216, 222, 224, 226, 228, 232, 237, 246, 272–281
  abatement, 11, 222
  accumulation, 74, 80, 178, 180, 201, 231, 234, 235, 254, 257, 262, 264
  baseline, 80–82, 86, 92
  capturer, 81, 92, 229
  coming, 22
  dissimilation/oxidation enzymes, 106
  dysmetabolism, 39, 69, 100–115, 136, 152, 153, 223
  exposure, 2, 4–11, 80, 82, 83, 86–90, 113, 123, 144, 155, 156, 178, 192, 222, 229, 233, 234, 237, 255, 258, 272
  going, 22, 1
  inhalation, 4, 5, 22, 85, 86, 89, 157, 197, 210
  metabolism, 22, 39, 52, 54, 56, 68, 101–106, 110, 112, 114, 115, 135, 149, 153, 154, 181, 229, 254, 260, 262
  oxidation/dissimilation, 102, 104
  probe-1, 287
  stress, 114, 150, 176
Formaldehyde-based resins, 4
Formaldehyde dehydrogenase (FDH, a homolog of mammalian GSNOR), 3, 29, 35, 52, 54, 70, 103–105, 154, 155, 211, 213, 216, 260, 264
Formaldehyde fluorescence probe (FAP-1), 32, 249, 281, 287, 288
Formaldehyde-induced 2-aza-cope, 287
Formaldehyde-induced cytotoxicity, 229, 234
Formaldehyde-induced neuronal damage, 227, 228, 232
Formaldehyde-protein cross-linkage, 229
Formalin solution, 80, 255
Formate, 3, 34, 35, 258
Formic acid, 10, 26, 34, 38, 50, 174, 175, 235, 258, 260, 262
5-Formylcytosine (5fC), 49
10-Formyltetrahydrofolate, 34
FOXO3, 227
Free radicals, 177, 231, 234
Frontal lobe, 175, 181, 252, 284
Function, 12, 48, 66, 80, 90, 100, 122, 144, 168, 193, 196, 198–200, 213, 223, 230, 234, 237, 256, 265

G

Galaxy, 3
Gallic acid, 276
Gamma-aminobutyric acid (GABA), 107
GAP-43, 230
Gas-chromatography (GC), 273, 274, 276–280, 282–284
Gaseous formaldehyde exposure, 195, 255
Gaseous molecules, 12, 236
Gas sensor, 276, 279, 282
Gene expression, 48, 50–53, 153
Gene mutation, 38, 169
General indoor pollutant, 5, 286
Generation, 22, 23, 25, 29, 31, 36, 38, 39, 48, 56, 86, 181, 222, 225, 234, 256
Genetic damage, 11
Genetic information, 50, 122
Genetic mutations, 210
Geniposide, 181, 223, 227, 228
Genome, 49, 211
Genomic imprinting, 48
Genomic instability, 38, 87, 92
Genotoxic, 9, 82
Genotoxicitic markers, 9
Genotoxicity, 38, 80, 82, 229
Germ cells, 49
Germfree (GF) mice, 113

Gfa, 104–106
Ginseng radix et rhizoma rubra, 225
Ginsenoside Rg1, 88
Glass production, 4
Glia activation, 247, 262
Glucocorticoids, 194, 195, 200, 255
Glucose-regulated protein 78, 88
Glucosyltransferase, 50
Glutamine, 36
Glutaraldehyde, 130, 169
Glutathione (GSH), 29, 34–36, 102–104, 193, 227, 229, 231
Glutathione-dependent formaldehyde-activating enzyme, 29, 104
Glutathione peroxidase, 229
Glyceraldehyde, 28, 123
Glycerol, 22, 24, 29
Glycine, 26, 28, 29, 31, 102, 103
Glycogen synthase kinase, 67, 108, 175
Glycogen synthase kinase-3 beta (GSK-3β), 67, 108, 175, 227
Glycolaldehyde, 277
Golgi apparatus, 81, 135, 144
G-Quadruplex DNAzyme, 280
Gray matter, 157
Greenhouse gas emissions, 4
Grid cell, 168
GSH-Px, 34, 35, 229, 233
Gshrx, 34
GSH S-transferase (GSHST), 35
GSK-3β, 175, 183, 229
Gustatory receptor, 215
Gut bacteria, 35, 103–108, 114
Gut microbiota, 25, 33, 100–115
Gut microbiota-formaldehyde-brain cassette/axis, 105
Gut microorganisms, 115

H

H_2O_2, 25, 27, 29, 31, 236, 247, 249, 280, 287
Haematopoiesis, 11
Headaches, 192, 193, 196, 199, 265
Heart disorder, 289
Heat shock, 213, 214, 216, 237
HeLa cell, 81, 135, 215, 281, 287
HeLa-S3 cervical cancer, 274, 284
Helicobacter pylori, 107, 108
Heterotroph, 101
Hexulose-6-phosphate synthase (HPS), 101, 102, 105
High-confidence evidence, 149
Highly toxic, 2, 102
High-performance liquid chromatography (HPLC), 8, 27, 37, 68, 109, 110, 135, 150, 157, 158, 173, 196, 199, 210, 247–249, 252, 253, 257, 262, 272–275, 277–286
High-performance liquid chromatography/electrochemical procedure, 283, 284
Hippocampal formaldehyde, 54, 69, 152, 161, 256, 257
Hippocampus, 53, 55, 85, 86, 112, 133, 152, 153, 157, 173, 175, 176, 181, 182, 194, 195, 201, 225, 230, 233, 234, 252, 253, 256, 257, 259, 262, 265, 284
Histamine, 199
Histidine, 26, 36, 51
Histones, 22, 24, 33, 51, 53, 55, 127, 148, 262
Hoechst-33258, 133, 172
Homeodynamic pathways, 214
Homeostasis, 22, 39, 40, 56, 64, 66, 134, 262, 265
Homoallylic amine, 287
Homogenate, 26, 283
Homology, 211
Hormesis, 214
Hospitals, 148, 150, 174, 193
Host microbiome, 108
Household product, 4, 144, 272
HSP70, 237
Human activities, 35, 48, 100, 157
Human brain, 222, 236
Human carcinogen, 10, 11, 80
Human endothelial HUV-EC-C cell, 81, 90, 215
Human exposure characterization of chemical substances (HEXPOC), 9
Hydrazine, 273, 275, 276, 281
Hydrogen atoms, 2
Hydrogen peroxide, 30, 32, 33, 36, 38, 51, 251, 276
Hydrogen sulfide (H_2S), 86, 88, 234, 235, 256
Hydrogenation of CO ice, 3
Hydroxyl radicals, 25
5-Hydroxltryptamine (5-HT), 71, 72, 112–114, 195, 264
Hydroxymethyl, 22, 24, 35, 37
5-Hydroxymethylcytosine (5hmC), 49, 210
2-Hydroxyproprianaldehyde, 31
Hyperphosphorylation, 122, 150, 154, 168, 169, 171, 173–175, 177, 179, 182, 227, 230, 235, 253, 258, 259
Hypodiploid DNA, 86
Hypohalous acids, 31
Hypothalamic-pituitary-adrenal gland (HPA), 194

Hypothalamus, 157, 195
Hypoxia-ischemia, 154, 169

I
Icariin, 227, 228
IGF-1, 211
Imbalance, 54, 64, 66, 70, 103, 110, 115, 145, 152, 153, 155, 169, 181, 253, 254, 264
Imethylarginine dimethylaminohydrolase 1, 88
Imine, 36
Immunocytofluorescence, 172
Immunological system, 2
In vitro, 27, 31, 37, 38, 49, 55, 56, 89, 108, 124–127, 129–131, 133, 175, 182, 194, 198, 247, 249, 258, 272, 283, 287
In vivo, 12, 22–25, 27–36, 39, 40, 54–56, 89, 123, 127–129, 133, 134, 136, 148, 169, 173, 175, 181, 194, 247, 257, 264, 272, 282, 284
Incineration, 2, 23
Indigestion, 193, 196
Indoor exposure, 5, 10, 11
Indoor pollutants, 4, 5, 286
Industry, 272
Inflammation, 87, 100, 107, 131, 157, 169, 229, 231, 261
Inflammatory, 31, 33, 80, 87, 100, 197, 198, 200, 231, 249
Inflammatory pain, 87, 198
Inflammatory reaction, 197
Inhalation formaldehyde, 4–9, 34, 85, 89, 156, 157, 197
Inhalational, 2, 4, 5, 8, 25, 80, 84, 89, 156, 197, 201, 210
Insomnia, 201
Insulation insecticides, 272
Insulin, 169, 211, 215
Interdigitated microelectrode array, 275, 282
International Agency for Research on Cancer (IARC), 11, 80, 192, 272
Intervention, 11, 111, 112, 136, 153, 222–238
Intestinal flora, 22, 33, 38
Intestinal microbe flora, 169
Intestinal microbiota, 33–36, 69, 100, 107, 109, 110, 115, 136, 251, 272
Intestine, 30, 37, 109, 110, 215
Intracerebroventricular injection, 181, 255, 256
Intraperitoneal administration, 256
Intraperitoneal injection, 28, 70, 84, 86, 108, 153, 182, 194, 231, 234, 247, 255–257, 262
Ion chromatography, 273, 282, 283
Ion flow-tube mass spectrometry, 274, 284
Iron(III) chloride-sulfamic acid solution, 274
Irritating odour, 2
Irritation, 10, 82, 196, 197

J
Japanese encephalitis vaccine, 273
JmjC domain-containing H3K9 demethylase
JmjC domain-containing proteins, 23
JTT matrix-based model, 104
Jumonji domain-containing proteins, 23, 24

K
KDMs, 33
Keratinocytes, 83, 86, 200
α-Ketoglutarate, 50
Key developmental function, 115
Kidney–marrow deficiency, 223, 224
Kinase, 67, 124, 130, 137, 154, 169, 175–177, 227, 229, 230
K562 leukemia, 274, 284
Knockout, 54, 82, 88, 216, 260, 261
K48 polyubiquitination, 89

L
Laboratories, 5, 8, 13, 48, 64, 73, 82, 100, 130, 135, 169, 197, 213, 246, 252, 262
Lachnospiraceae, 108
Lactobacillus fermentum NS9, 102, 105
Lactobacillus brevis, 102, 105, 107
Lactobacillus helveticus, 107
Lactobacillus helveticus NS8, 114
Largest chipboard manufacturing, 9
Larvae, 213
Laser printers, 5
Late-onset, 181
LC/MS, 36, 281
LDH leakage, 90
L-DOPA, the precursor of dopamine, 113
Learning, 48, 53–55, 71, 72, 107, 112, 148, 153–156, 158–160, 194, 195, 201, 213, 224, 225, 227, 230, 231, 233, 247, 253, 256, 257, 260–264, 272
Learning and memory, 52, 53, 55, 56, 107, 112, 144, 153–156, 159, 213, 224, 227, 230, 231, 253, 254, 256, 260, 262, 263
Leptin pathway, 89
Lesions, 12, 30, 155, 174, 196
Leukaemia, 10–12, 192

Leukemia, 83
LiCl, 175
Life span, 209, 214, 215, 253, 257
Ligusticum chuanxiong hort, 225
Lime, 202
Limited water intake, 246
α-Linolenic acid, 234
Lipid, 22, 26–28, 34, 38, 86, 247, 288
Lipid oxidation, 27, 28, 247, 248
Lipid peroxidation, 27, 34, 86
Liquid chromatography (HPLC), 36, 37, 275, 281–283
Liver, 10, 25, 28–30, 84, 86, 153, 202, 227, 258, 260, 273, 274, 281, 284, 285
Living habits, 7
Local communities, 150, 161
Locomotor, 195
Locomotors behavior, 156
Longevity, 210, 211, 213, 215, 257
Longitudinal studies, 147
Long non-coding RNAs (lncRNAs), 50
Long-term depression (LTD), 176
Long-term potentiation (LTP), 112, 150, 176, 194, 256, 257, 259
Lotion, 272, 278
Lysine, 28, 33, 37, 51, 53, 169
Lysine-specific demethylase 1 (LSD-1), 23, 51, 84
Lysine-specific demethylase 1A, 51
Lysosome, 12, 32, 39, 40, 81, 134–136, 144, 169, 202, 249, 281, 287
Lyso Tracker Green DND-26, 249, 288

M
Macrophages, 31, 87, 125
Magnetic resonance imaging (MRI), 158
Mainstream cigarette smoke, 273, 283
Malachite green, 280
Malaise, 193
Malignant melanoma SK-MEL-28 cells, 81, 215
Malondialdehyde (MDA), 27, 28, 86, 181, 226, 229, 233, 247, 248
MAPK pathway, 88
Mass spectrometry Mc deamination dependent BER, 49
MCF-7 breast cancer, 229, 274, 284
5MC oxidative demethylation, 49
Medicines, 4, 150, 223–231, 238
Medium-chain dehydrogenase/reductase (MDR), 211
Meiosis, 50
Melamine resin, 4, 276
Melatonin, 156, 192–194, 232, 233
Melon seed, 276
Melting temperature, 126, 177
Memory, 66, 67, 71, 74, 86, 87, 107, 144, 173, 192–194, 201, 213, 222, 224, 225, 227, 230, 231, 233–236, 246, 272
  decline, 108, 113, 154, 155, 192, 194, 252, 256, 257
  formation, 48, 52–54, 56, 175, 256, 257
  retrieval ability, 256
Mesoporous SnO_2 fibers, 279
Metabolism, 11, 12, 22–40, 52, 64, 68, 71, 81, 101–106, 110–115, 135, 145, 148, 149, 152–155, 161, 181, 193, 194, 202, 223, 229, 231, 247, 253, 260, 262, 265
Metabolite, 111, 113, 154, 157, 169, 171, 174, 195, 210, 235
Metaformaldehyde (MFA), 100
Metal ions, 169
Methanol, 2, 10, 22, 25, 26, 28, 33, 34, 38, 52, 67, 101, 105, 108, 144, 148, 152, 153, 161, 169, 173–175, 181, 182, 196, 210, 235, 246, 258–260, 265, 273
Methanol exposure, 174, 235
Methanol metabolism, 52, 174
Methanotrophic bacteria, 101, 102
Methionine, 26, 49
Methods, 2, 31, 37, 104, 105, 109, 111, 123, 126, 177, 201, 202, 213, 216, 235, 238, 265, 272–289
Methyl acetoacetate, 278
Methylamine, 22, 30, 31, 36, 51, 105, 181, 200, 236, 261
Methylase, 51, 261, 262
Methylated amine, 51
Methylation, 25, 33, 56, 148, 176, 261, 284
3-Methyl-2-benzothiazolinone hydrazone hydrochloride (MBTH), 274
3-Methyl-2-benzothiazolone hydrazone, 273, 282
5-Methylcytosine (5mC), 48, 210
Methyl dehydrogenation, 49
1-Methyl-1-(2,4-dinitrophenyl)hydrazine, 275
Methyl donor, 25, 49, 52
5,10-Methylene tetrahydrofolate, 26
5,10-Methylene tetrahydrofolic acid, 49
5,10-Methylene-thf, 26
Methyl group, 26, 33, 48–50, 53, 54, 84
Methylobacterium extorquens AM1, 102, 103
Methylol adducts, 36, 38
Methylotroph, 101, 103
Methylotrophic organisms, 101, 102

Mice, 28, 30, 31, 35, 50, 54, 55, 65, 82, 107, 136, 153, 173, 193–196, 210, 211, 213, 225, 246, 252, 253, 257, 260, 261, 263, 284
Microbiota-gut-brain, 33, 106, 109, 114, 285
Microbiota-gut-brain axis, 33, 106, 109, 112, 114, 285
Microcirculation, 154, 169
Microenvironment, 84
Microfluidic chip, 277, 282, 286
Microglia, 108
Micronucleus frequency, 8
MicroRNA (miRNA), 86, 197
Microsomes, 29
Microtubule, 122–125, 127–129, 131, 132, 136, 144, 168, 170–173, 177–179, 183, 230, 235, 258
Mild cognitive impairment (MCI), 150, 157, 160, 225, 226
Mini-mental state examination (MMSE) scores, 73, 108, 151, 157, 161
Mini-mental status examination (MMSE), 265
Mitochondria, 28, 29, 34, 80, 81, 85–88, 92, 135, 223, 225, 230, 234, 236
Mitochondrial dysfunction, 80, 85, 88, 92
Mito Tracker Green FM, 249, 288
MoCA scores, 146, 147
Moderate cognitive impairment, 223, 285
Molecular, 2, 32, 50, 83, 105, 106, 156, 172, 182, 209, 211, 227, 232, 246–266, 272, 286, 287
  biology, 50
  mechanisms, 197, 211, 227, 257
Monkeys, 67, 108, 125, 136, 152, 161, 174, 175, 181, 196, 201, 258, 259
Monoamine oxidase (MAO-A), 30, 236
Monoamine oxidase inhibitor, 236
Monomethylated, 51
Montreal cognitive assessment (MoCA), 69, 146, 265
Morning urine, 149, 157
Morphogenesis, 123, 197
Morphology, 86, 90, 128, 134, 171, 178–180, 197, 209, 227, 230
Morris water maze, 153, 154, 236, 256, 257, 260, 262, 263
Motor neuron, 198, 215
Mouse, 30, 67, 108, 125, 152, 173, 229, 247, 252, 253, 257, 258, 260–262, 264, 265, 283
Mouse model, 33, 68, 71, 108, 115, 254, 260, 261
Mouse primary hippocampal cultures, 231
MPO-H_2O_2-chloride system, 31
Mth, 213
Multi-targets, 223
Muscle, 22, 26, 30, 86, 123, 198, 215, 234, 261
Mushroom, 154, 275–277
Mutagenesis, 80, 82
Mutants, 125, 129, 132, 210, 211, 213, 216, 251
Myelin oxidation, 27, 247
Myeloperoxidase (MPO), 23, 31

N

N2a cell, 31, 82, 88, 108, 135, 154, 171–173, 175–179, 229, 248, 250, 263
N-acetyl cysteine (NAC), 231, 232
NAD+, 280
N6-adenosine-methyltransferase, 50
NADP oxidoreductase, 34
Nanogram-scale, 286
Nanoporous silica aerogel, 272, 281, 289
Nanoporous silica aerogel sorbent, 272, 289
Nano-techniques, 286
NAPDH, 29
Naphthalene-based fluorescent probe (AENO), 281, 287
1,8-Naphthalimide, 281
Nasal airway, 82
Nasal cancer, 11, 82
Nasal mucosa, 199
Nasal respiratory tissue, 84
Nasogastric tube administration, 86
Nasopharyngeal cancer, 10, 11, 80
Nasopharynx, 192
National Longitudinal Mortality Study (NLMS), 198
Natural food, 272
Naturally, 3, 12, 100, 272
Nausea, 193, 196
Navigation, 168
NCBI protein, 104, 105, 115
Necrosis, 80, 90–92
Needle trap device (NTD), 281, 289
Nerve cell, 48, 201, 222
Nerve growth factor, 85, 230
Neural differentiation, 82
Neural maturation, 82
Neural plasticity, 48
Neural stem cell, 82
Neural system, 213, 215, 216, 262
Neural transmitters, 112
Neurite, 154, 173, 175, 178–180

Neuro-2a (N2a) cell line, 125, 248, 251
Neuroblastoma, 67, 81, 82, 88, 90, 125, 135, 169, 181, 227, 229, 230, 248
Neuroblastoma cell, 67, 125, 181, 227, 229
Neurodegeneration, 33, 114, 115, 130, 133, 176, 235, 252, 265
Neurodegenerative disease, 48, 52, 54, 101, 103, 106–108, 122, 134, 137, 145, 154, 168, 171, 199, 200, 231, 247, 254, 285, 286, 289
Neurofibrillary tangles (NFT), 108, 129, 168–171, 174, 175
Neurofilaments, 199
Neurogenesis, 82, 197
Neuroglia, 201
Neurological disorders, 12, 56, 131, 261
Neurological system, 2
Neuron, 31, 71, 80, 101, 122, 144, 168, 193, 194, 199, 202, 210, 215, 223, 225, 227, 228, 230, 232, 252, 256, 258
Neuronal cell, 80, 85, 90, 92, 144, 223, 227, 230, 264
Neuronal cell death, 227, 287
Neuronal disorders, 222
Neuron survival, 230
Neurophysiology, 201
Neuroprotective, 223–227, 231, 235
Neuroprotective effect, 85, 223, 224, 227, 229–231
Neuropsychiatry, 192, 222, 265
Neuropsychological assessment, 150, 201
Neuropsychological test, 149, 225, 265, 285
Neurotoxicity, 88, 90, 122, 156, 199, 229, 233, 234, 236
Neurotransmitter, 71, 100, 107, 112, 113, 197, 200, 202, 216
Neutrophils, 31
NF-κB, 229
N6-formyladenosine (f6A), 50
N6-hydroxymethyladenosine (hm6A), 50
Nitric oxide (NO), 66, 155, 213, 227, 234, 260
N6-methyladenosine, 50
N-methylated histone, 51
N-methyl-D-aspartate receptor (NMDAR), 176, 177, 257
Nociceptive stimulation, 87
Nonenzymatic process, 35, 51
Nonhuman primate, 174
Non-invasive biomarker, 150, 157
Nonmethylated k, 51
Non-regulated sources, 4, 5
Norepinephrine (NE), 71, 112–114, 200
Nosebleeds, 199
Noticeable phenotypes, 50
NR1, 176
NR2B, 176
NS capsule, 112
NS lactobacillus, 111, 112
Nuclear fragmentation, 88
Nuclear magnetic resonance (NMR), 37, 277, 282
Nucleophilic site, 38
Nucleotide excision repair (NER), 49, 87
Nucleus, 88, 126, 128, 172, 173, 175, 177, 183
Nylon 6 NFN membranes, 278

O

Obesity, 24, 50
Occipital lobe, 252, 284
Occupational contact, 80, 86
Occupational exposure, 5, 8, 11
Occurrence, 4, 66, 182, 195, 198
Octanol, 213
O-demethylation, 248
Odor, 26, 100, 157, 211, 215
Offices exposure, 5, 7
Olfaction, 196, 197, 199
Olfactory bulb, 157, 197
Olfactory memory, 173, 258
Olfactory receptor, 215
Oligomers, 37, 134, 179, 181
Omega-3 fatty acids, 234
Oncogenesis, tumor, 49
One-carbon circle, 26, 35
One-carbon compounds, 101
One-carbon cycle, 148, 262
One-carbon pool, 48, 51
Oocyte growth, 86
Open-field tests (OFT), 71, 113, 194, 195
OR83b, 215
Orally, 2
Organism, 12, 28, 33, 36, 101, 102, 144, 209–211, 214–216, 233, 246, 265, 272
Outer space, 3
Ovalbumin-immunization, 85
Oxidation of lipid, 22, 246, 247
Oxidative demethylation, 33, 50
Oxidative reaction, 2
Oxidative stress, 12, 22, 27, 30–33, 38, 80, 85, 87–89, 92, 135, 136, 145, 148, 149, 157, 158, 169, 193, 214, 227, 229, 231, 233, 234, 236, 237, 247, 248, 250, 251, 256, 261, 287
Oxidoreductases, 28, 34
2-Oxoglutarate (2OG), 33, 49

O-(2,3,4,5,6-pentafluoro-benzyl)-hydroxylamine, 276
O-(2,3,4,5,6-Pentafluoro-benzyl)-hydroxylamine hydrochloride, 280
Oxygen, 2, 3, 31–33, 51, 87, 126, 197, 213
Oxygen atom, 2
Oxygen oxide, 286

P

P25, 176
P35, 176
P38, 88, 200
P53, 82, 87, 227
Paired helical filament (PHF), 67, 122, 128, 129, 170, 171, 179, 183
Panax notoginseng, 227
Panic, 192
Paraformaldehyde (PFA), 100, 257
Paraoxonase-1, 88
Paraquat, 213
Parietal lobe, 175, 181, 252, 284
Parkinson's disease (PD), 285
Partial correlation analysis, 147
Particle board, 6, 8, 199
Particles minus 2.5 nanometre (pm2.5), 201
Passive demethylaiton, 49
Pathogenesis, 33, 108, 110, 112–114, 130, 181, 223, 230
Pathological, 30, 65–67, 69, 80, 89, 113, 122, 156, 168, 171, 174, 175, 177–179, 227, 229, 247, 250–252, 258, 259, 262, 287
Pathology, 11, 100, 114, 168, 174, 175, 181, 182, 200, 224, 259
Patho-mechanism, 10, 55, 137, 144, 151
Patient autopsy, 161
PC12 cell line, 88, 125, 234
Pd-functionalized SnO_2 sensor, 279, 282
Pectin, 25, 258
Pentafluorobenzyl-hydroxyl-amine hydrochloride, 276
2,4-Pentanedione, 275, 279, 280
Perioperation, 149
Peripheral blood lymphocytes, 8, 11, 86
Peroxidase, 31, 34, 227, 229, 233
Personalized medicine, 223
Phenelzine, 236, 237
Phenol-formaldehyde, 194
Phenol formaldehyde resin, 4
Phenotype, 50, 213, 216, 261
Phenyl hydrazine, 275
Phlegm-obstructing intelligence, 223, 224
Phosphate, 27, 131, 170, 247, 248
Phosphatidylinositol-3 kinase (PI3K), 230, 236
6-Phospho-3-hexuloisomerase (PHI), 101, 102, 105, 115
Phosphoric acid, 279
Phosphorylation, 31, 53, 67, 83, 88, 123, 124, 130, 137, 168–183, 227, 229, 246, 258
Photochemical reactions, 2, 3
Photocopiers, 5
Photoinduced electron transfer (PET), 287
Physical, 25, 74, 82, 100, 198, 201, 210, 223, 236, 238
Physiological condition, 34, 64, 123, 144
Plants combustion processes, 2
Plasma, 25, 30, 197, 274, 277, 284
Plywood adhesives, 272
Plywood workers, 11
Point mutations, 82
Polarography, 273, 282
Pollution, 4–7, 9, 12, 13, 196, 202, 203, 272
Polydimethylsiloxane microfluidic chip, 282
Polygoni multiflori radix praeparata, 225
Polygonum cuspidatum, 229
Polymer, 100, 130, 134, 172, 173, 183
Polymerization, 125, 129, 131, 133, 155, 172, 182, 183
Poly-methylmethacrylate, 279
Polyvinyl alcohol, 155
Porcine brain, 210, 252, 253, 284
Poria cum radix pini (PRP), 224
Post hoc Newman–Keuls comparisons, 146
Post-operation, 149
Post-operative cognitive dysfunction (POCD), 148–150, 161
Post-operative complications, 149
Post-operative infections, 149
Post-stroke dementia (PSD), 56, 157
Post-transcriptional regulation, 50
Potassium bromate, 275, 276, 279, 280
Potassium current, 85
Potassium ferricyanide (PF), 273
Power plants, 2, 23
Ppk23, 215
Preclinical, 122, 168, 182
Preconcentration, 274, 284
Preconcentration-chemical ionization mass spectrometry, 274
Precursor, 67, 108, 113, 144, 174, 181, 197, 231, 258
Prefrontal cortex, 86, 233
Presenilin 1 (PS1), 53, 230
Preservative, 2, 4, 8, 272
Prevalence, 10, 145

Primary hippocampal neuron, 82, 85, 180
Primary neuron, 175, 180
Primary open-angle glaucoma (POAG), 171, 198, 199
Primitive atmosphere, 2, 3
Probiotics, 102, 103, 105–107, 111, 112, 115
Procarcinogens, 29
Production of forest, 2
Production of jungle fire, 2
Programmed cell death, 89, 92
Progression, 31, 49, 64, 67, 84, 112, 122, 136, 137, 144, 157, 160, 177, 179, 193, 247, 258
Prokaryotes, 211, 261
Prokaryotic pathway, 102
Promoter, 48, 53
Propagating fecundity, 211
Prospective cohort study, 50, 198
Prostate cancer, 84, 274, 284
Protection equipment, 13
Protein, 2, 22, 48, 50, 53, 54, 67, 82, 87–89, 101, 122, 144, 169, 197, 213, 227, 246, 247, 253, 254, 258, 260, 261, 265, 285
  aggregations, 25, 30, 31, 67, 89, 122, 131, 133–136, 144, 247
  complexes, 51
  crossing linkage, 2, 181
  denaturation, 22, 26, 248
  linkage, 2, 181
  misfolding, 133
Protein phosphatase 2A (PP2A), 176, 177, 182, 183, 230
Proteotoxicity, 80, 85, 89, 92
Protofibrils, 37, 181
Proton, 37, 51
Psychosomatic disorders, 195
Pterin cofactors, 103
Puerariae lobatae radix, 225
Pungent, 196
Purines, 28, 37

Q
Qi deficiency and blood stasis, 223
Quartz crystal microbalance (QCM), 281, 286

R
Rabbit, 86, 88, 123, 284
Radiometric analysis with dimedone ^{14}C, 273, 282
Radix polygalae (RP), 224
Rat, 29, 32, 52, 83, 84, 86, 87, 89, 133, 154, 161, 173, 176, 195, 196, 233, 234, 236, 249, 255, 256, 265, 274, 283
Reactive aldehydes, 31, 200
Reactive oxygen, 31, 32, 87, 126, 213
Reactive oxygen species (ROS), 31–33, 87, 88, 126, 213, 234, 249, 250
Real-time monitoring, 286
Recoveries, 73, 230, 257, 284
Reducing property, 12
Reducing sugar decomposition, 28, 248
Reference exposure level (REL), 6
Relative standard deviations (RSD), 284, 289
Renal dialysis, 236
Renovated homes, 6, 193
Reproductive system, 2, 12, 86
Residential, 5–7, 12, 146
Residential renovation, 6
Resin, 6, 8, 192
Respiratory, 12
Respiratory complications, 149
Respiratory irritation, 10, 193, 197
Respiratory tract, 12, 80, 192, 222
Resveratrol, 74, 81, 82, 111, 177, 202, 213, 228–230
Reversible RNA modifications, 50
Rhesus macaque, 174, 235
Rhesus monkeys, 182, 246, 258
Rhinitis, 192, 199
Rhizoma acori tatarinowii (RAT), 224, 225
Rhodamine B, 276
Ribosomal RNA (rRNA), 50, 126, 129
Ribulose monophosphate (RuMP) pathway, 101, 102, 105
Risk factor, 36, 39, 54, 66, 67, 70, 114, 130, 134, 149, 157, 181, 199, 202, 223, 237, 252
RNA, 12, 22, 24, 33, 34, 36–38, 48–52, 54, 56, 126, 128, 129, 136, 137, 148, 197, 210, 247, 248, 262
  and histone, 22, 51, 52, 54, 148, 248, 262
  molecule, 197
  transcription, 128, 129
RNAi, 213
Rodents, 39, 152–154, 174, 177, 182, 197, 200, 214, 246, 252–255, 264, 265

S
S-adenosy-L-methionine (SAM), 25, 26, 48, 51
Salmonella enterica subsp., 33, 104–106, 108
Sarcosine (N-methylglycine), 22, 28, 29
Sarcosine dehydrogenase, 28, 29
Sauerbrey equation, 286
Scaffolding theory, 148
Schiff-base, 36
Sea cucumber, 276

Selective serotonin reuptake inhibitors (SSRIs), 113
Selenium, 233
Semicarbazide-sensitive amine oxidase (SSAO), 22, 30, 31, 36, 51, 70, 86, 130, 131, 153, 157, 158, 181, 200, 236, 254, 261, 264
Senescence-accelerated mouse (SAM), 253, 254
Senescence-accelerated mouse-prone 8 strain (SAMP8), 30, 35, 69, 153, 154, 253–255
Senescence-resistant strain 1 (SAMR1), 30, 35, 69, 153, 253, 254
Senile plaque (SPs), 67, 108, 129, 136, 152, 153, 168, 181, 258, 259, 261
Senility, 154, 169
Sensitization, 196
Ser396, 172–176
Serine hydroxymethyltransferase (SHMT), 22, 23, 26, 102, 103, 105, 106
Serine pathway, 102, 103, 106
Serotonin (5-HT), 112–114, 200
Server cognitive impairment, 285
S-formyl GSH, 35
β-Sheet, 37, 181
Shenwu capsule, 223, 225
Shoe-care agents, 4
Shrank, 172
Shrimp, 276, 277, 279
SH-SY5Y cell, 81, 87, 88, 90, 125, 131, 133, 169, 177, 227
Shuttle box, 71, 72, 112, 153, 195, 264
Shuttle box assay, 71, 153
S-(hydroxymethyl)glutathione synthase, 102, 104, 106
Sick building syndrome (SBS), 10, 192
Signaling pathways, 82, 83, 86, 87, 210
Single-crystal-to-single-crystal fashion (Cu(I)-MOF), 279
Sir2, 215
SIRT-1, 88, 234
Skin-care products, 272
Skin contact, 2
SK-N-SH cells, 88, 230
Sleep disorders, 192
Sleeve-fish, 276
Smallest aldehyde, 2, 255, 282
Small intestine, 109, 110
Small molecule compound, 52
Small nuclear RNA (snRNA), 50
Smart soup, 223–225
Smoking, 11, 155
S-nitrosoglutathione (GSNO), 213
S-nitrosoglutathione reductase (GSNOR), 154, 213, 260
Social relationship, 215
SOD1, 215
Soil, 3, 272
Somatic cell, 49
Somnifacient, 193
Spatial learning and memory impairment, 52, 153, 256
Spatial memory, 54–56, 87, 156, 193, 256, 262
Spatial recognition, 173, 258
Spectrophotometer, 273, 276, 278, 279, 284
Spectrophotometry with pararosaniline, 282
Spinal cord, 87, 197
Spleen-yang impaired by damp, 223
Squamous cell carcinoma, 82
Stabilization, 171, 178
Standard curve, 252, 253, 283
Standard test method, 283
Starvation, 169, 213, 214, 216
Staurosporin, 85
Stem cell differentiation, 82, 111
Sterically bulky amine, 280
Stimulation, 83, 87, 89, 177, 256
Stress resistance, 209, 213–216, 257
Substrates, 22, 29–31, 50, 51, 103, 124
Sugar decomposition, 28, 246, 248
Sulfuric acid, 276, 280
Supernatant fraction, 283
Superoxide dismutase (SOD), 226, 227, 229, 233, 234, 236
Surface-to-volume ratio, 286
Survival time, 214
Synapse, 52, 53, 175, 176, 223, 230, 252
Synaptophysin, 230
Synaptosomal-associated protein 25 (SNAP25), 157, 197
Synaptosome, 71, 195
Syndromes, 10, 156, 223
Synthese, 174

T

Tail intravenous injection, 257
Tail vein injection, 84
Tau, 30, 67, 89, 108, 109, 122, 144, 168, 202, 210, 227, 229, 230, 235, 246, 252–254, 257–259, 262, 265
Tau hyperphosphorylation, 31, 67, 68, 108, 109, 122, 130, 136, 152, 153, 168–179, 182, 183, 202, 210, 229, 230, 247, 252, 254, 257, 259, 262, 265
Tau hyperphosphorylation and aggregation, 31, 130–133, 150, 153, 154, 169, 171

Tauopathy, 130, 133, 137, 172
Teeth, 274, 284
Temporal lobe, 175, 181, 225, 226, 252, 284
Ten-eleven translocated oncogene, 49
TET1, 49
TET2, 49
TET3, 49
Tetanus toxoid vaccine, 273
Tetrahydrobiopterin, 197
Tetrahydrofolate (THF), 26, 34
Tetrahydrofolic acid (THF), 34, 35, 49, 104
Tetrahydroxystilbene glucoside, 230
TH-DA-NE cassette, 113
Theophylline, 236, 237
Thinking, 158
Thin-layer chromatography, 284
Thioflavin S (ThS), 133, 169, 172, 183
Thiols, 35
Thr181, 172–176, 229
Thymine DNA glycosylase, 37
Time- and dose-dependent, 92
Titanium dioxide nanotube array (TiO_2-NTA), 272, 286
TMAO, 23, 26
Toluene, 156, 193
Toxicity, 2, 35, 64, 80, 90, 92, 133, 134, 144, 171, 173, 181, 200, 222, 224, 226, 228, 232, 237, 261
Toxicosis, 10, 133, 196, 200–202, 260
Toxoplasma gondii, 107, 108
Traditional Chinese medicine (TCM), 181, 223–225, 238
Transcription, 33, 48–53, 86, 126, 128, 129, 176, 210
  complex, 49
  factors, 52, 210
Transgenic model, 33, 52, 69, 84, 109, 111, 115, 153, 225, 230, 246, 251, 258–262, 285
Transgenic rodents and *Drosophila*, 246
Transient receptor potential cation channel, subfamily A, member 1 (TRPA1), 197, 198
Trichloroacetic acid, 285
Trigger, 26, 32, 37, 38, 67, 85, 89, 91, 92, 108, 122, 149, 168–177, 182, 281
Trimethylamine, 22–24, 26, 27, 251
Trimethylamine oxide (TMAO), 22, 26
Trimethylamine oxide demethylase (TMAOase), 23
Trimethylated, 51
Tripe, 276
Triphlorethol-A, 236, 237
Tris(2,2-bipyridyl)ruthenium(II), hexamethylenetetramine, 278
TrkA, 230
tRNA, 50, 129
Tryptophan (Trp), 26, 36, 113, 169
Tumor, 80, 82, 84, 85, 90, 169, 249
  cell, 80, 84
  suppression, 82
Tumorigenesis, 3
Type 2 diabetes mellitus (T2DM), 247
Tyrosine, 36, 176, 195, 197, 200, 255
Tyrosine hydroxylase (TH), 71, 113, 195, 197, 200

U

Ubiquitin, 171
Ultraviolet detector, 283
Ultraviolet radiation, 213
United States, 9, 272
Unpleasant odour, 10
Urea-formaldehyde resin, 4, 6
Urine, 28, 69, 72, 73, 81, 84, 110, 144–151, 157, 158, 161, 178, 196, 201, 261, 264, 284–286, 289
Urine formaldehyde, 69, 72, 73, 84, 145–151, 157, 158, 161, 196, 284–286

V

VAP-1, 30
Variable spatial delayed response task (VSDRT), 174, 259
Vascular dementia (VD), 69, 130, 153, 226, 285
Vehicle exhaust, 4, 23
Ventilation, 5–7, 10, 13, 202
Ventilation facilities, 13
Vermicelli, 276
Vertigo, 196
Victims, 10, 196, 201
Vitamin E, 232, 233

W

Water, 2, 39, 64, 110, 134, 152, 192, 194, 196, 202, 210, 230, 246, 272
Water deprivation, 68, 70, 73, 264

Water intake, 65, 66, 68, 70–74, 112, 194, 235, 236, 246, 264
Water maze test, 153, 154, 236, 256, 257
Western blotting, 131, 134, 153, 172, 173, 176, 182, 257
Winter indoor formaldehyde, 7
Wood-based furnitures, 4
Wood-based materials, 4, 6
Working memory, 67, 152, 161, 174, 233, 258, 265
World Health Organization (WHO), 2, 5, 8, 11, 196, 222, 282
Worm-like conformation, 123, 131, 133
Worms, 211, 213, 214

X
X-linked mental retardation
Xylene, 193

Y
Yeast, 214, 215
Y-maze, 71, 173, 258
Yu fire disturbing, 223

Z
Zeolitic imidazolate framework (ZIF), 279

Kolibri 360 GmbH
Ferdinand-Jühlke-Straße 7
99095 Erfurt, Deutschland
produktsicherheit@kolibri360.de